T0189904

INNOVATIONS IN MICRO IRRIGATION TECHNOLOGY

Research Advances in Sustainable Micro Irrigation

Volume 10

INNOVATIONS IN MICRO IRRIGATION TECHNOLOGY

Edited by
Megh R. Goyal, PhD, PE, Senior Editor-in-Chief
Vishal K. Chavan, MTech, Co-editor
Vinod K. Tripathi, PhD, Co-editor

APPLE ACADEMIC PRESS

Apple Academic Press Inc. | Apple Academic Press Inc.
3333 Mistwell Crescent | 9 Spinnaker Way
Oakville, ON L6L 0A2 | Waretown, NJ 08758
Canada | USA

©2016 by Apple Academic Press, Inc.

First issued in paperback 2021

Exclusive worldwide distribution by CRC Press, a member of Taylor & Francis Group
No claim to original U.S. Government works

ISBN 13: 978-1-77463-564-3 (pbk)
ISBN 13: 978-1-77188-150-0 (hbk)

Library and Archives Canada Cataloguing in Publication

Innovations in micro irrigation technology/edited by Megh R. Goyal, PhD, PE, senior editor-in-chief; Vishal K. Chavan, MTech, co-editor, Vinod K. Tripathi, PhD, co-editor.

(Research advances in sustainable micro irrigation ; volume 10)
Includes bibliographical references and index.
Issued in print and electronic formats.
ISBN 978-1-77188-150-0 (bound).--ISBN 978-1-77188-291-0 (pdf)
1. Microirrigation--Technological innovations. I. Goyal, Megh Raj, Editor II. Chavan, Vishal K., editor III. Tripathi, Vinod K., editor IV. Series: Research advances in sustainable micro irrigation; v. 10

| S619.T74166 2016 | 631.5'87 | C2016-900856-8 | C2016-900857-6 |

Library of Congress Cataloging-in-Publication Data

Names: Goyal, Megh Raj, editor. | Chavan, Vishal K., editor. | Tripathi, Vinod K., editor.
Title: Innovations in micro irrigation technology/ Megh R. Goyal, PhD, PE, senior editor-in-chief; Vishal K. Chavan, MTech, co-editor, Vinod K. Tripathi, PhD, co-editor.
Description: Toronto : Apple Academic Press, 2016. | Series: Research advances in sustainable micro irrigation; volume 10 | Includes bibliographical references and index.
Identifiers: LCCN 2016005630 (print) | LCCN 2016006095 (ebook) | ISBN 9781771881500 (hardcover : alk. paper) | ISBN 9781771882910 ()
Subjects: LCSH: Microirrigation. | Irrigation--Technological innovations.
Classification: LCC S619.T74 1534 2016 (print) | LCC S619.T74 (ebook) | DDC 631.5/87--dc23
LC record available at http://lccn.loc.gov/2016005630

Apple Academic Press also publishes its books in a variety of electronic formats. Some content that appears in print may not be available in electronic format. For information about Apple Academic Press products, visit our website at **www.appleacademicpress.com** and the CRC Press website at **www.crcpress.com**

CONTENTS

LIST OF CONTRIBUTORS

Hussein Mohammed Al-Ghobari
Department of Agricultural Engineering, College of Food and Agriculture Sciences, King Saud University, Riyadh 11451, Kingdom of Saudi Arabia

Khodran H. Alzahrani
Department of Agricultural Extension and Rural Sociology, Faculty of Food and Agricultural Sciences, King Saud University, Saudi Arabia. E-mail: khodran@ksu.edu.sa

Mirza B. Baig
Department of Agricultural Extension and Rural Sociology, Faculty of Food and Agricultural Sciences, King Saud University, Saudi Arabia

B. Ahmed Bakry
Researcher, Agronomy Department, Agriculture and Biology Research Division, National Research Center, El-Behoose St., Dokki, Cairo, Egypt. Mobile: +20126834233, E-mail: bakry_ahmed2004@yahoo.com; website: www.nrc.sci.eg

Vishal Keshavao Chavan
Assistant Professor and Senior Research Fellow in SWE, Agriculture University, Akola, Maharashtra, Website: www.pdkv.ac.in; E-mail: vchavan2@gmail.com

Suresh Kumar Devarajulu
Professor, Department of Agricultural Economics, Tamil Nadu Agricultural University (TNAU), Coimbatore – 641003, Tamil Nadu, India. E-mail: rithusuresh@yahoo.com

A. Ramadan Eid
Water Relations and Field Irrigation Department (Agricultural and Biological Division), National Research Center, Cairo, Giza, Egypt

Mohamed Said Abdalla El Marazky
Department of Agricultural Engineering, College of Food and Agriculture Sciences, King Saud University, Riyadh 11451, Kingdom of Saudi Arabia. Permanent address: Agriculture Engineering Research Institute, Agriculture Research Center, P.O. Box 256, Cairo, Dokki, Giza, Egypt. E-mail: melmarazky@ksu.edu.sa; elmarazky58@gmail.com

Said Mohammad Fawzi
Department of Agricultural Engineering, College of Food and Agriculture Sciences, King Saud University, Riyadh 11451, Kingdom of Saudi Arabia

Megh R. Goyal
Retired Professor in Agricultural and Biomedical Engineering from General Engineering Department, University of Puerto Rico – Mayaguez Campus; and Senior Technical Editor-in-Chief in Agriculture Sciences and Biomedical Engineering, Apple Academic Press Inc., PO Box 86, Rincon – PR – 00677 – USA. E-mail: goyalmegh@gmail.com

Ahmed Abdel-Kareem Hashem Abdel-Naby
Agricultural Engineering Department, Faculty of Agriculture at Suez Canal University, Egypt; PhD research Scholar, Agricultural and Biological Engineering, Purdue University, West Lafayette, IN 47907–1146, USA. E-mail: ahashem@purdue.edu

Marvin E. Jensen
Retired Research Leader at USDA – ARS. 1207 Spring Wood Drive, Fort Collins, Colorado 80525, USA, E-mail: mjensen419@aol.com

K. R. Kakumanu
International Water Management Institute, South Asia Regional Office, ICRISAT Campus, Patancheru, Hyderabad 502324, India. Phone: 040-3071-3741, E-mail: k.krishnareddy@cgiar.org

Ravinder Paul Singh Malik
Tamil Nadu Agricultural University, Coimbatore, India

Kadiri Mohan
International Water Management Institute, South Asia Regional Office, ICRISAT Campus, Patancheru, Hyderabad 502324, India. Phone: 040-3071-3749, E-mail: k.mohan@cgiar.org

Siddig E. Muneer
Professor of Rural Sociology and Rural development, Department of Agricultural Extension and Rural Sociology, Faculty of Food and Agricultural Sciences, King Saud University, P.O. Box 2460, Riyadh 11451. Tel: 00966501661327, E-mail: siddigmuneer@gmail.com

Kuppannan Palanisami
International Water Management Institute, South Asia Regional Office, ICRISAT Campus, Patancheru, Hyderabad 502324, India. Phone: 040-3071-3732, E-mail: k.palanisami@cgiar.org; palanisami.iwmi@gmail.com

Sabreen Kh. A. Pibars
Water Relations and Field Irrigation Department (Agricultural and Biological Division), National Research Center, Cairo, Giza, Egypt. E-mail: saberennrc@yahoo.com

S. Raman
Former Head of Water Management Scheme, Navsari, Agricultural University, Gujarat. E-mail: raman261@rediffmail.com

Rama Rao Ranganathan
International Water Management Institute, New Delhi, India

R. K. Sivanappan
Former Professor and Dean, College of Agricultural Engineering and Technology, Tamil Nadu Agricultural University (TNAU), Coimbatore. Mailing address: Consultant, 14, Bharathi Park, 4th Cross Road, Coimbatore–641043, India. E-mail: sivanappanrk@hotmail.com

Alla S. Taha
Department of Agricultural Extension and Rural Sociology, Faculty of Food and Agricultural Sciences, King Saud University, Saudi Arabia

M. H. Taha
Agronomy Department, Faculty of Agriculture, Cairo University, Giza, Egypt

Vinod Kumar Tripathi
Assistant Professor, Center for Water Engineering and Management, Central University of Jharkhand, Ratu-Loharghat Road, Brambe, Ranchi-Jharkhand – 835205, India, Mobile: +918987661439; E-mail: tripathiwtcer@gmail.com

LIST OF ABBREVIATIONS

ASABE	American Society of Agricultural and Biological Engineers
CU	coefficient of uniformity
DIS	drip irrigation system
DOY	day of the year
EPAN	pan evaporation
ET	evapotranspiration
ETc	crop evapotranspiration
FAO	Food and Agricultural Organization, Rome
FC	field capacity
FUE	fertilizers use efficiency
gpm	gallons per minute
ISAE	Indian Society of Agricultural Engineers
kc	crop coefficient
kg	kilograms
Kp	pan coefficient
lps	liters per second
lph	liter per hour
msl	mean sea level
PE	polyethylene
PET	potential evapotranspiration
pH	acidity/alkalinity measurement scale
PM	Penman-Monteith
ppm	one part per million
psi	pounds per square inch
PVC	poly vinyl chloride
PWP	permanent wilting point
RA	extraterrestrial radiation
RH	relative humidity
RMAX	maximum relative humidity
RMIN	minimum relative humidity

RMSE	root mean squared error
RS	solar radiation
SAR	sodium absorption rate
SDI	subsurface drip irrigation
SW	saline water
SWB	soil water balance
TE	transpiration efficiency
TEW	total evaporable water
TMAX	maximum temperature
TMIN	minimum temperature
TR	temperature range
TSS	total soluble solids
TUE	transpiration use efficiency
USDA	US Department of Agriculture
USDA-SCS	US Department of Agriculture-Soil Conservation Service
WSEE	weighed standard error of estimate
WUE	water use efficiency

LIST OF SYMBOLS

A	cross sectional flow area (L^2)
AW	available water ($\ominus_w$, %)
Cp	specific heat capacity of air, in J/(g·°C)
CV	coefficient of variation
D	accumulative intake rate (mm/min)
d	depth of effective root zone
D	depth of irrigation water in mm
Δ	slope of the vapor pressure curve ($kPa°C^{-1}$)
e	vapor pressure, in kPa
e_a	actual vapor pressure (kPa)
E	evapotranspiration rate, in g/(m2·s)
Ecp	cumulative class A pan evaporation for two consecutive days (mm)
eff	irrigation system efficiency
E_i	irrigation efficiency of drip system
E_p	pan evaporation as measured by class-A pan evaporimeter (mm/day)
Es	saturation vapor pressure, in kPa
E_{pan}	class A pan evaporation
ER	cumulative effective rainfall for corresponding two days (mm)
e_s	saturation vapor pressure (kPa)
$e_s - e_a$	vapor pressure deficit (kPa)
ET	evapotranspiration rate, in mm/year
ETa	reference ET, in the same water evaporation units as Ra
ETc	crop-evapotranspiration (mm/day)
ET_o	the reference evapotranspiration obtained using the Penman-Monteith method, (mm/day)
ET_{pan}	the pan evaporation-derived evapotranspiration
EU	emission uniformity
F	flow rate of the system (GPM)

F.C.	field capacity (v/v,%)
G	soil heat flux at land surface, in W/m^2
H	the plant canopy height in meter
h	the soil water pressure head (L)
I	the infiltration rate at time t (mm/min)
IR	injection rate, GPH
IRR	irrigation
K	the unsaturated hydraulic conductivity (LT^{-1})
K_c	crop coefficient
Kc	crop-coefficient for bearing 'Kinnow' plant
Kp	pan factor
K_p	pan coefficient
n	number of emitters
P	percentage of chlorine in the solution*
Pa	atmospheric pressure, in Pa
P.W.P.	permanent wilting point (Θ_w%)
Q	flow rate in gallons per minute
q	the mean emitter discharges of each lateral (lh^{-1})
R	rainfall
r_a	aerodynamic resistance (s m^{-1})
Ra	extraterrestrial radiation, in the same water evaporation units as ETa
R_e	effective rainfall depth (mm)
R_i	individual rain gauge reading in mm
R_n	net radiation at the crop surface (MJ m^{-2}day^{-1})
Rs	incoming solar radiation on land surface, in the same water evaporation units as ETa
RO	surface runoff
r_s	the bulk surface resistance (s m^{-1})
S	the sink term accounting for root water uptake (T^{-1})
Se	the effective saturation
S_p	plant-to-plant spacing (m)
S_r	row-to-row spacing (m)
SU	statistical uniformity (%)
S_ψ	water stress integral (MPa day)
t	the time that water is on the surface of the soil (min)

T	time in hours
V	volume of water required (liter/day/plant)
V_{id}	irrigation volume applied in each irrigation (liter tree^{-1})
V_{pc}	the plant canopy volume (m^3)
W	the canopy width
W_p	fractional wetted area
z	the vertical coordinate positive downwards (L)

Greek Symbols

α	related to the inverse of a characteristic pore radius (L^{-1})
γ	the psychrometric constant (kPa°C^{-1})
θ	volumetric soil water content (L^{3}L^{-3})
$\theta(h)$	the soil water retention (L^{3}L^{-3}),
θr	the residual water content (L^{3}L^{-3})
θ_s	the saturated water content (L^{3}L^{-3})
θ_{vol}	a volumetric moisture content (cm^3/cm^3)
λ	latent heat of vaporization (MJ kg^{-1})
λE	latent heat flux, in W/mo
ρa	mean air density at constant pressure (kg m^{-3})

PREFACE

Due to increased agricultural production, irrigated land has increased in the arid and sub-humid zones around the world. Agriculture has started to compete for water use with industries, municipalities and other sectors. This increasing demand along with increments in water and energy costs have made it necessary to develop new technologies for the adequate management of water. The intelligent use of water for crops requires understanding of evapotranspiration processes and use of efficient irrigation methods.

Every day, news on water scarcity appears throughout the world, indicating that government agencies at central/state/local level, research and educational institutions, industry, sellers and others are aware of the urgent need to adopt micro irrigation technology that can have an irrigation efficiency up to 90% compared to 30–40% for the conventional gravity irrigation systems. I stress the urgent need to implement micro irrigation systems in water scarcity regions.

Among all irrigation systems, micro irrigation has the highest irrigation efficiency and is most efficient. Micro irrigation is sustainable and is one of the best management practices. Micro irrigation systems are often used for farms and large gardens, but are equally effective in the home garden or even for houseplants or lawns. The water crisis is getting worse throughout the world, including India, Middle East and Puerto Rico where I live. We can therefore conclude that the problem of water scarcity is rampant globally, creating the urgent need for water conservation. The use of micro irrigation systems is expected to result in water savings, increased crop yields in terms of volume and quality. The other important benefits of using micro irrigation systems include expansion in the area under irrigation, water conservation, optimum use of fertilizers and chemicals through water, and decreased labor costs, among others. Recently, it has been proven that recycled wastewater can be used in drip irrigation with an adequate filtration system. The worldwide population is increasing at a rapid rate and it is imperative that food supply keeps pace with this increasing population.

The mission of this compendium is to serve as a reference manual for graduate and under graduate students of agricultural, biological and civil engineering; horticulture, soil science, crop science and agronomy. I hope that it will be a valuable reference for professionals that work with micro-irrigation/wastewater, and water management; for professional training institutes, technical agricultural centers, irrigation centers, Agricultural Extension Service, and other agencies that work with micro irrigation programs.

After my first textbook on drip/trickle or micro irrigation management by Apple Academic Press Inc., and response from International readers, I was motivated to bring out for the world community this ten-volume informative series on "*Research Advances in Sustainable Micro Irrigation*." This book series will complement other books on micro irrigation that are currently available on the market, and my intention is not to replace any one of these. This book series is unique because it is simple with world-wide applicability to irrigation management in agriculture. This series is a must for those interested in irrigation planning and management, namely, researchers, scientists, educators and students.

The contribution by all cooperating authors to this book series has been most valuable in the compilation of this volume. Their names are mentioned in each chapter and in the list of contributors of each volume. This book would not have been written without the valuable cooperation of these investigators, many of whom are renowned scientists who have worked in the field of micro irrigation throughout their professional careers. I am glad to introduce Dr. Vinod Kumar Tripathi and Vishal K. Chavan as Co-editors for this volume. Dr. Tripathi is an Assistant Professor and Distinguished Research Scientist in Wastewater Use in Micro Irrigation at Center for Water Engineering and Management, Central University of Jharkhand Ranchi, Jharkhand, India. Engineer Chavan is an Assistant Professor-cum-Senior Research Fellow in Micro Irrigation and Dryland Agriculture at Dr. Punjabrao Deshmukh Krishi Vidyapeeth Akola, Maharashtra, India. Both have contributed to AAP book series on micro irrigation. Without their support and extraordinary work, readers will not have this quality publication.

I will like to thank editorial staff, Sandy Jones Sickels, Vice President, and Ashish Kumar, Publisher and President at Apple Academic Press, Inc., (http://appleacademicpress.com/contact.html) for making every effort to publish the book when the diminishing water resources is a major issue worldwide. Special thanks are due to the AAP production staff for their work on the manuscript and for the quality production of this book.

We request that the reader offer us your constructive suggestions that may help to improve the next edition.

I express my deep admiration to my family for understanding and collaboration during the preparation of this ten volume book series. Throughout my professional career, I have been able to apply my expertise in "micro irrigation technology" to come up with new ideas/developments, etc. in order alleviate problems of water scarcity and salinity. My salute to those who are involved in micro irrigation technology for their interest, devotion, and vocation. As an educator, there is a piece of advice to one and all in the world: *"Permit that our Almighty God, our Creator and excellent Teacher, irrigate the life with His Grace of rain trickle by trickle, because our life must continue trickling on..."*

—*Megh R. Goyal, PhD, PE, Senior Editor-in-Chief*

December 30, 2015

WARNING/DISCLAIMER

User Must Read It Carefully

This book volume 10 on "*Innovations in Micro Irrigation Technology*" presents technological advances in irrigation for economical crop production. The editor, the contributing authors, the publisher, and the printer have made every effort to make this book as complete and as accurate as possible. However, there still may be grammatical errors or mistakes in the content or typography. Therefore, the contents in this book should be considered as a general guide and not a complete solution to address any specific situation in irrigation. For example, one size of irrigation pump does not fit all sizes of agricultural land and to all crops.

The editors, the contributing authors, the publisher and the printer shall have neither liability nor responsibility to any person, any organization, or entity with respect to any loss or damage caused, or alleged to have caused, directly or indirectly, by information or advice contained in this book. Therefore, the purchaser/reader must assume full responsibility for the use of the book or the information therein.

The mention of commercial brands and trade names are only for technical purposes. It does not mean that a particular product is endorsed over another product or equipment not mentioned. The author, cooperating authors, educational institutions, and the publisher Apple Academic Press, Inc., do not have any preference for a particular product.

All weblinks that are mentioned in this book were active on December 31, 2015. The editors, the contributing authors, the publisher and the printing company shall have neither liability nor responsibility, if any of the weblinks are inactive at the time of reading of this book.

ABOUT THE SENIOR EDITOR-IN-CHIEF

Megh R. Goyal, PhD, PE, is a retired professor in agricultural and biomedical engineering from the General Engineering Department in the College of Engineering at University of Puerto Rico–Mayaguez Campus; and senior acquisitions editor and senior technical editor-in-chief in agriculture and biomedical engineering for Apple Academic Press Inc.

He received his BSc in engineering in 1971 from Punjab Agricultural University, Ludhiana, India; his MSc in 1977 and PhD in 1979 from the Ohio State University, Columbus; and his master of divinity in 2001 from Puerto Rico Evangelical Seminary, Hato Rey, Puerto Rico, USA.

He spent one-year sabbatical leave in 2002–2003 at the Biomedical Engineering Department at Florida International University in Miami, Florida, USA. Since 1971, he has worked as soil conservation inspector (1971); research assistant at Haryana Agricultural University (1972–1975) and Ohio State University (1975–1979); research agricultural engineer/professor at the Department of Agricultural Engineering of UPRM (1979–1997); and professor in agricultural and biomedical engineering in the General Engineering Department of UPRM (1997–2012).

He was first agricultural engineer to receive the professional license in Agricultural Engineering in 1986 from College of Engineers and Surveyors of Puerto Rico. On September 16, 2005, he was proclaimed as "Father of Irrigation Engineering in Puerto Rico for the twentieth century" by the ASABE, Puerto Rico Section, for his pioneer work on micro irrigation, evapotranspiration, agroclimatology, and soil and water engineering. During his professional career of 45 years, he has received awards such as Scientist of the Year, Blue Ribbon Extension Award, Research Paper Award, Nolan Mitchell Young Extension Worker Award, Agricultural

Engineer of the Year, Citations by Mayors of Juana Diaz and Ponce, Membership Grand Prize for ASAE Campaign, Felix Castro Rodriguez Academic Excellence, Rashtrya Ratan Award and Bharat Excellence Award and Gold Medal, Domingo Marrero Navarro Prize, Adopted Son of Moca, Irrigation Protagonist of UPRM, and Man of Drip Irrigation by Mayor of Municipalities of Mayaguez/Caguas/Ponce and Senate/Secretary of Agriculture of ELA, Puerto Rico.

He has authored more than 200 journal articles and textbooks, including *Elements of Agroclimatology* (in Spanish) by UNISARC, Colombia, and two *Bibliographies on Drip Irrigation*. Apple Academic Press Inc. (AAP) has published several of his books, including *Management of Drip/Trickle or Micro Irrigation, Evapotranspiration: Principles and Applications for Water Management, Sustainable Micro Irrigation Design Systems for Agricultural Crops: Practices and Theory, among others*. Readers may contact him at goyalmegh@gmail.com.

ABOUT THE CO-EDITORS

Vishal Keshavrao Chavan, MTech, is currently working as a Senior Research Fellow in the Office of the Chief Scientist under Dr. Mahendra B. Nagdeve of the AICRP for Dryland Agriculture at Dr. Punjabrao Deshmukh Krishi Vidyapeeth, Akola, the premier agricultural university in Maharashtra, India. His work included the evaluation of the basic mechanism of clogging in drip irrigation system. His area of interest is primarily micro irrigation and soil and water engineering.

He obtained his BTech in agricultural engineering in 2009 from Dr. Punjabrao Deshmukh Krishi Vidyapeeth, Akola, in Maharashtra, India; and his MTech from the University of Agricultural Sciences, in Raichur, Karnataka, India, in soil and water engineering. An expert on clogging mechanisms, he has published four popular articles and three research papers in national journals, and he attends international conferences.

Vinod Kumar Tripathi, PhD, is working as assistant professor in the Center for Water Engineering and Management at the School of Natural Resources Management of Central University of Jharkhand, Brambe, in Ranchi, Jharkhand, India. He obtained his BTech in agricultural engineering from Allahabad University, India; his MTech from the G. B. Pant University of Agriculture and Technology, in Pantnagar, India, in irrigation and drainage engineering; and his PhD in 2011, from the Indian Agricultural Research Institute, New Delhi, India.

He has developed a methodology to improve the quality of produce by utilizing municipal wastewater under drip irrigation during PhD research.

His areas of interest are geo-informatics, hydraulics in micro-irrigation, and development of suitable water management technologies for higher crop and water productivity. He has taught hydraulics, design of hydraulic structures, water and wastewater engineering, geo-informatics to postgraduate students at Central University of Jharkhand, Ranchi, India.

He is an international expert and has evaluated funded plans for rural water supply and sanitation; and his PhD. in 2011, from Indian Agricultural Research Institute, New Delhi, India. He has published more than 32 peer-reviewed research publications and bulletins, and has attended several national and international conferences.

OTHER BOOKS ON MICRO IRRIGATION TECHNOLOGY BY APPLE ACADEMIC PRESS, INC.

Management of Drip/Trickle or Micro Irrigation
Megh R. Goyal, PhD, PE, Senior Editor-in-Chief

Sustainable Micro Irrigation Design Systems for Agricultural Crops: Practices and Theory
Megh R. Goyal, PhD, PE, Senior Editor-in-Chief

Management of Clogging in Micro Irrigation: Theory and Practices
Megh R. Goyal, PhD, PE, Senior Editor-in-Chief

Book Series: Research Advances in Sustainable Micro Irrigation
Senior Editor-in-Chief: Megh R. Goyal, PhD, PE

CHAPTER 1

MICRO IRRIGATION TECHNOLOGY IN INDIA: A SUCCESS STORY

R. K. SIVANAPPAN

CONTENTS

1.1   INTRODUCTION

In India, the water demand is increasing in all sectors due to escalated population. Agriculture draws about 80% of the total fresh water at present and the remaining are used for drinking, industries etc. But the allocation of water for agriculture will have to be reduced from 80% to 70% in the next 10–15 years to allot more water for other sectors [16]. The only way to overcome the problem is to increase the productivity per unit quantity of water and per unit area and unit time. This is possible, since the scientists have developed the agriculture technology and water management practices to increase the production by using less water. Micro irrigation technology can save our planet from water scarcity in agriculture.

This success story of drip irrigation technology is narrated by the author, who has been working on drip irrigation since 1969.

1.2   HISTORICAL DEVELOPMENTS IN MICRO IRRIGATION [3, 4]

Development of micro or drip irrigation can be traced back to experiments in Germany in 1860s. Farmers laid clay pipes with open joints about 0.8 meters below the soil surface in an effort to combine irrigation and drainage. The first work in drip irrigation in the United States was a study by House in Colorado in 1913. An important break-through was made in Germany in 1920, when perforated pipe drip irrigation was introduced.

In the early 1930, the peach growers in the State of Victoria – Australia, developed irrigation system with 5 cm galvanized iron pipes laid along the line of trees. Water was supplied through a triangular hole, cut on a pipe with a chisel at each tree. During the early 1940, Symcha Blass, an Israeli Engineer, observed that a large tree near a leaking faucet exhibited a more vigorous growth than the other trees in the area, which were not reached by the water from the faucet. This led him to the concept of an irrigation system that would apply water in small amount, literally drop by drop. The technique, as developed by Blass, was subsequently refined by him and various manufactures. Around 1948 in the United Kingdom, greenhouse operators began to try a similar method with some modifications. When Blass conceived the idea of drip irrigation in the 1940s, the materials needed to build a low-pressure system at a reasonable cost were not available. Only with the rapid development of the plastics industry after World War II, appropriate materials for making chemically resistant, flexible pipes of small diameter were produced economically.

Initially, the system was installed underground (called subsurface drip irrigation, SDI). Later because of the primitive filtration techniques of the time and frequent clogging, the system was moved above ground (called surface drip irrigation). This development made it easy to check the tubes for clogging and maintained the chief advantage of the system – the direct application of water to the root zone of the plant.

One of the refinements made by Blass in his original system was a coiled emitter. It consisted of a spiral tube in a hard casing. The tube served to reduce the flow pressure by lengthening the flow path of the water, thereby making it possible to discharge the water at low pressure while dripping.

In the early 1960s, experiments in Israel reported spectacular success when they applied the Blass system in the desert areas of the Negev and Arava. The conditions for agriculture in the desert areas were distinctly adverse not only with saline water, but also with high temperatures, low relative humidity and sandy soils. For, example, a field trial in the Arava – Israel produced a yield of 65.0 metric tons of winter tomatoes per hectare under drip irrigation, compared with 39.0 metric tons with sprinkler irrigation.

In 1969, drip irrigation pipes were began to be sold outside Israel as a commercial business. Drip irrigation units in diverse forms are installed widely in USA, Australia, Israel, Mexico and to a lesser extent in Canada, Cyprus, France, Iran, New Zealand, United Kingdom, Greece and in India. In India, the area under drip irrigation is only 50 Ha in 1975 to more than 1.53 MHa in 2009. It was estimated that about 0.44 million-ha area was under drip irrigation world-wide in 1980 and was increased to about 60,89,530 ha in 2006 and 103,10,441 ha in 2012 (http://www.icid.org/sprin_micro_11.pdf) [6].

However, it is rather peculiar that drip irrigation should appear so suddenly on an international scale and to such an extent that it has warranted eight international congresses on drip irrigation, during 1971 – today. ICID volunteered to organize the event commencing from 5th International Micro Irrigation Congress held at South Africa in 2000 with an objective of creating awareness among its members about latest developments in micro irrigation technology to enhance crop production. Since 2012, the event is renamed as International Micro Irrigation Symposium. Following congresses have been held since 1971 (http://www.icid.org/conf_microirri_past.html) [5].

Date	Place and Country	Details
15–23 October 2011	Tehran, Iran	8th International Micro Irrigation Congress
		Theme: Innovation in Technology and Management of Micro-Irrigation for Crop Production Enhancement
13–15 September 2006	Kuala Lumpur, Malaysia	7th International Micro Irrigation Congress
		Theme: Advances in Micro Irrigation for Optimum Crop Production and Resource Conservation
22–27 October 2000	Cape Town, South Africa	*6th International Micro Irrigation Congress, 22–27 October 2000, Cape Town, South Africa
		Theme: Micro-Irrigation Technology for Developing Agriculture
2–6 April 1995	Orlando, Florida, USA	5th International Micro Irrigation Congress, 2–6 April 1995, Orlando, Florida, USA
		Theme: Microirrigation for a Changing World: Conserving Resources/Preserving the Environment
23–28 October 1988	Albury – Wodonga, Australia	4th International Micro Irrigation Congress
18–21 November 1985	Fresno, California, USA	3rd International Micro Irrigation Congress
7–14 July 1974	San Diego, CA, USA	2nd International Micro Irrigation Congress
6–13 September 1971	Tel Aviv, Israel	1st International Micro Irrigation Congress

1.2.1 MICRO/DRIP IRRIGATION RESEARCH IN INDIA

Though the drip irrigation system is fast becoming popular in other parts of the world, it was in the experimental stage during 1970's in India and large-scale adoption took place only after many drip irrigation companies were established in Maharashtra and Tamil Nadu from 1985 onwards. Research studies have been conducted at the following institutions.

1. Tamil Nadu Agricultural University (TNAU), Coimbatore
2. Water Technology Center, Indian Agricultural Research Institute (IARI), New Delhi
3. University of Udaipur, College of Agriculture, Jobner, Rajasthan
4. University of Udaipur, College of Agricultural Engineering, Udaipur, Rajasthan
5. Haryana Agricultural University, Hissar
6. Central Arid Zone Research Institute (CAZRI), Jodhpur, Rajasthan
7. University of Agricultural Science, Dharwad, Karnataka
8. Mahatma Phule Krishi Vidyapeeth, Rahuri, Maharashtra
9. Bidhan Chandra Krishi Vishwa Vidyalaya, Kalyani, West Bengal
10. Center for Water Resources Development and Management, Calicut, Kerala
11. Jyothi Farm, Baroda, Gujarat
12. Indian Council of Agricultural Research Complex for North East Hill Region, Shillong
13. Commercial installations
    a. Jain irrigation – Maharashtra
    b. Nagarjuna Drip irrigation – Hyderabad
    c. LG brother Drip irrigation – Coimbatore

1.2.1.1 Research Trials at Coimbatore

At the College of Agricultural Engineering of Tamil Nadu Agricultural University, studies on drip irrigation were carried out from 1969 onwards by Professor R. K. Sivanappan and his colleagues [10–16]. A low cost drip irrigation system was designed and fabricated with locally available materials to determine the water use for various vegetable and fruit crops in collaboration with Horticultural College. This system operated at low pressures with main and lateral tubings. One-millimeter holes were provided at the desired intervals in the lateral tube. To avoid clogging and spray action, sockets were provided in the openings. The sockets can be adjusted to allow more or less water according to the crop requirements. During the 1970s, Extensive experiments were conducted to study the effects of this system on water use and yield for various crops namely: tomato, okra, radish, beetroot, eggplant, sweet potato, chilies, banana, cotton, and sugarcane. Demonstrations plots were also established in farmer's field for banana, cotton and tomato crops in and around Coimbatore. The results have indicated that the water required in the drip system was only 25–33% of that required under the surface irrigation and the yield was invariably superior in drip system when compared to the surface method (see Table 1.1).

TABLE 1.1   Water Used and Yield Obtained in Drip and Control Method, Coimbatore – India

Crop	Water use, cm		Yield, kg/ha		Rainfall, cm per season
	Control	Drip	Control	Drip	
Banana	166.4	40.0	19 kg/plant	16.5 kg/plant	61.0
Beetroot	86.7	17.7	571	887	-
Bhendi (okra)	53.5	8.6	10,000	11,310	24.1
Brinjal (Eggplant)	69.1	24.4	12,400	11,900	17.1
Chilly	109.7	41.7	4,233	6,086	20.7
Cotton	70.0	15.0	2,600	3,250	13.0
Papaya	228.0	73.3	13.40 kg/plant	23.48 kg/plant	81.6
Radish	46.4	10.8	1,045	1,186	-
Sugarcane	131.8	72.8	86,000	75,000	33.5
Sweet potato	63.1	25.2	4,244	5,888	12.1
Tomato	49.8	10.7	6,187	8,872	24.1

Source: Sivanappan, R. K. and O. Padmakumari, 1980. Drip Irrigation. Tamil Nadu Agricultural University, Coimbatore, [10, 15].

Experiments on cotton (Var. CBS-156 and MCU-9) have shown that the water saving with drip irrigation was 47% when compared to surface irrigation. The germination, moisture status and the crop response were superior in drip-irrigated plots. During 1979, experiments were conducted with one lateral supplying water for two rows of crops to reduce the cost of drip system (skip-drip trail) for tomato crop. The skip-drip trail method gave have yield of 12,000 kg per ha.

Experiments were also conducted to study the crop response and growth of papaya under drip irrigation. The germination was highest in drip plots. The crop responded well to drip irrigation such that flowering in drip irrigated plots started one month earlier than the control plots. The papaya fruit yield was 23.48 kg/plant in drip irrigation compared to 13.4 kg/plant in basin method of irrigation. In another experiment, the tapioca yield under drip irrigation was 20% higher compared to check basin method of irrigation.

The salient findings of the other Agricultural Universities, Central Institutions and Private drip irrigation companies are detailed in the following subsections.

1.2.1.2   IARI, New Delhi

Extensive field studies were conducted at the WTC, IARI farm to optimize the spacing of emitters, the water application rates, the duration of irrigation based on hydraulics of soil moisture front.

1.2.1.3   Udaipur, Rajasthan

The drip system was designed and field evaluated for potato. It was reported that besides water saving, the tuber yield was higher and the effect of frost was minimum and weed growth was the least.

1.2.1.4   Hissar, Haryana

Comparative studies of drip versus surface method were conducted in small plots with onion, sugarbeet, potato, radish and okra. It was found that the drip method produced higher yield and resulted in greater water use efficiency (WUE).

1.2.1.5   Other Research Stations

Drip irrigation experiments were conducted at Dharwad – Karnataka state for grapes and cabbage; Rahuri in Maharashtra for cotton; for bhendi, tomato, brinjal, sugarcane, at Kalyani in West Bengal; for potato, at Calicut; Kerala for coconut; banana crops, at the Center for Water Resources Development and Management and Jothi farm in Baroda and farm in Jalgaon and many commercial institutions like Jain Irrigation, etc. In all these institutions, the results have been very promising (see Tables 1.2 and 1.3) for water saving, yield increase, and benefit–cost ratio (BCR) for various crops under drip irrigation. Besides these research

TABLE 1.2   Water Use and Yield for Various Crops Under Drip and Conventional Irrigation Methods in India

Crop	Yield, Quintals (100kg)/Ha			Water use, cm		
	Conventional	Drip	Increase in yield, %	Conventional	Drip	Water saving, %
Banana	675.00	875.00	52	176.00	97.00	45
Beet root	45.71	48.87	7	88.71	17.73	79
Bitter gourd	154.34	214.71	39	24.50	11.55	53
Brinjal	280.00	320.00	14	90.00	42.00	53
Cabbage	195.80	200.00	2	66.00	26.67	60
Chillies	42.33	60.88	44	109.71	41.77	62
Cotton	23.30	29.50	27	89.53	42.00	53
Grapes	264.00	325.00	23	53.20	27.80	48
Lady finger	152.61	177.24	16	53.68	32.44	40
Mosambi ('000 No)	100.00	150.00	50	166.00	64.00	61
Papaya	13.40	23.48	75	228.00	73.30	68
Pomegranate ('000 No)	55.00	109.00	98	144.00	78.50	45
Radish	70.45	71.86	2	46.41	10.81	77
Ridge gourd	171.30	200.00	17	42.00	17.20	59
Sweet potato	42.44	58.88	39	63.14	25.50	60
Tomato	320.00	480.00	50	30.00	18.40	39
Watermelon	240.00	450.00	88	33.00	21.00	36

Source: National Committee on the Use of Plastics in Agriculture (NCPA), 1990. Status, potential and approach for adoption of drip and sprinkler irrigation system. Pune [8].

TABLE 1.3 Benefit–Cost Ratio (BCR) for Various Crops Under Drip Irrigation System, India

Crop	Spacing, m × m	Drip system cost, Rs./ha	Benefit–cost ratio	
			Excluding water saving	Including water saving
Acid lime - Citrus sp.	4.57 × 4.57 (15 ft × 15 ft)	9,000	1.76	6.01
Banana	1.52 × 1.52 (5 ft × 5 ft)	18,000	1.52	3.02
Coconut	7.62 × 7.62 (25 ft × 25 ft)	7,000	1.41	5.14
Grapes – 1	3.04 × 3.04 (10 ft × 10 ft)	12,000	13.35	32.32
Grapes – 2	2.44 × 2.44 (8 ft × 8 ft)	16,000	11.50	27.08
Mango	7.62 × 7.62 (25 ft × 25 ft)	7,000 for (25' × 25')	1.35	8.02
Orange	4.57 × 4.57 (15 ft × 15 ft)	9,000	2.60	11.05
Papaya	2.13 × 2.13 (6 ft × 6 ft)	8,000 for (6' × 6')	1.54	4.01
Pomegranate	3.04 × 3.04 (10 ft × 10 ft)	12,000	1.31	4.04
Sugarcane	Between lateral 1.83 (6ft)	20,000 for 1.83(6')	1.31	2.78
Vegetable	Between lateral 1.83 (6ft)	20,0000	1.35	3.09

advances in India, the area under drip irrigation has not increased at the expected level, even-though the potential area is estimated as 27 million-ha in India (see Table 1.4). The drip irrigation area has been low in the beginning and has increased in the recent years (see Table 1.5). At the same time, the area under drip irrigation was about 4,36,590 ha in 1981 and about 60,89,534 ha in 2006 in the world [9].

TABLE 1.4   Theoretical Potential Area for Drip Irrigation in India

Crop	Area, million-ha
Coconut and plantation crops	3.0
Cotton	1.7
Fruits	3.9
Spices & condiments	1.4
Sugarcane	4.3
Vegetables	3.6
Total	27

Source: Report of the task force on micro irrigation, January, 2004. G.O.I. Ministry of Agriculture, New Delhi.

TABLE 1.5   Growth of Drip Irrigation in India

Year	Drip irrigation area, ha
1985	1,000
1991	55,000
1999	254,000
2001	310,000
2003	500,000
2006	903,000
2009	153,1007

Source: Development of micro irrigation technology, micro irrigation WTC, IARI, New Delhi, 2009 [7].

1.2.2   MICRO IRRIGATION SYSTEMS

Micro irrigation is one of the latest and advanced methods of irrigation, which is becoming increasingly popular in areas having problems of water scarcity and poor quality of water. In drip irrigation, water drips into the soil at low rates (2 to 20 lph) from a system of small diameter pipes (12 or 16 mm) fixed with outlets called emitters or drippers (Figure 1.1). In the micro irrigation systems (micro sprinklers, micro sprayers, misters and foggers), water is applied close to plants so that only part of the soil in which the roots grow is wetted and it permits the irrigators to limit the

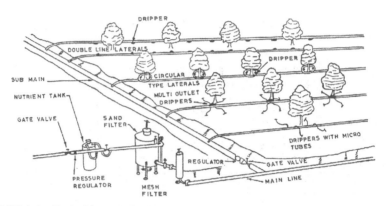

FIGURE 1.1   Typical layout of a drip irrigation system.

watering close to the consumptive use of the plants. In micro irrigation, water applications are more frequent (usually daily or alternate days) than with other methods and this provides a very favorable high moisture level always in the soil in which plants can flourish/grow well with more yield (Figure 1.2).

Micro irrigation can be adopted for undulating terrain. It is also suitable in most soils. In clay soils, water must be applied slowly (less discharge per hour) to avoid surface water ponding and run off. But in sandy soil,

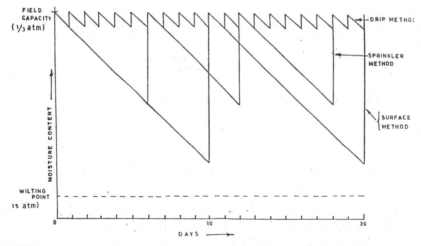

FIGURE 1.2   Moisture availability for crops in different irrigation methods.

higher application rate will be needed to ensure adequate lateral wetting of the soil. Drip irrigation is also suitable for water of poor quality (saline water), since water is given daily and salt is pushed to the periphery of the moist zone i.e. away from the root zone of the crop. Thus the method is best suited in areas of water scarcity, marginal water quality, the undulating topography (Hilly area), restricted soil depth; and where labor cost is high and the crop value is high.

1.2.2.1   Limitations of Micro Irrigation System

- Salinity hazard in the absence of leaching of salt built up.
- Sensitivity to clogging of system components.
- High cost of the system.
- Requirements of high skill in design, installation and operation.
- High wind velocity affects the pattern of water spread in micro sprinkler.

There are different types of micro irrigation systems such as surface drip, subsurface drip, micro sprinklers, bubbler system, spray system, pulse irrigation system and bi-wall system etc.

1.3   IMPACT OF MICRO IRRIGATION TECHNOLOGY IN TAMIL NADU

Indian agriculture continues to be the mainstay of economy and accounts for 25% of the nation's Gross Domestic Product (GDP), 15% of exports and 60% of the employment. Having achieved laudable success in agricultural production in the last 50 years (50 to 241 Million-tons), India has transformed herself from a food deficit to a food surplus country. Still there are many challenges, which Indian agriculture is facing in the fast changing scenarios. Relating to the natural resources and production base, water has emerged as the most crucial factor for sustaining the agricultural sector.

India accounts for 16% of the world's human population and nearly 30% of the cattle with only 2.4% of the land and 4% of the water resources. Even if the full irrigation potential is exploited, about 50% of the country's cultivated area will remain unirrigated, particularly with current level of irrigation efficiency. Tamil Nadu accounts for 6.5% of India's population with only 4% of land area and 3% of water resources. The availability of water per person per year is 2000 M^3 for India and 650 M^3 for Tamil Nadu. The share of water for agriculture would reduce further with increasing demand from other sectors. But the demand for water for agricultural purposes is estimated to increase to produce increasing quantities of food, horticultural produce and raw material for the industry. Efficient management of water is, therefore, key to future growth of Indian agriculture.

The requirement of water by different sectors by 2025 is estimated to be 109 MHM, but the share of water for agriculture is expected to get reduced from the present level of 84% to 69% by 2025. On the other hand, the demand for water for agricultural purposes is estimated to increase from 47 MHM in 1985 to 74 MHM in 2025. During the same period, the demand for non-agricultural use of water will multiply four fold from 7 MHM to 28 MHM.

Misplaced and inappropriate policies leading to indiscriminate use of water, lack of appropriate technologies, poor technology transfer mechanisms and inadequate and defective institutional support systems have led to serious agro-ecological and sustainability problems in irrigated areas. While the water table rise and water logging to an extent of 8.5 M-ha is the problem in canal command areas along with secondary salinization, receding water table at a rate of as high as one meter annually along with underground water pollution in many states are the daunting problems in tube-well irrigated areas. The water use efficiency (WUE) in canal command area is about 30–40%, is one of the lowest in the world, against 55% in China.

The vulnerability of Indian agriculture is bound to be severe, unless the present trend of water use and management is changed. The International Water Management Institute (IWMI) forecasts that by 2025, 33% of India's population will live under absolute water scarcity conditions. The per capita water availability, in terms of average utilizable water resources in the country, has dropped drastically from 6008 M^3 in 1947 to current 2000 M^3 and is expected to dwindle to 1450 M^3 by 2025. The International

Conference on Water and the Environment, Dublin and the 1992 United Nations Conference on Environment and Development, the 1992 Earth Summit in Rio De Janeiro, the 2000 Millennium Summit, and the 2002 Earth Summit Ten years Later, has drawn world's attention to this crisis.

Rain-fed lands are not only low in productivity and sustainability and are more prone to risks, as compared to those in irrigated areas, but are also the location for (proportionality) greater concentrations of poor and hungry persons. This can be obviated to some extent by expanding irrigated areas through improving water management and water use patterns. Presently, the problem facing the country is not only the development of water resources, but the management of the developed water resources in a sustainable manner. By adopting efficient water management practices, the bulk of India's agricultural lands could be rendered as irrigated. Micro irrigation is one such practice.

Micro irrigation has already been adopted by some countries for transforming their agriculture. India introduced this technology on a commercial scale in the eighth plan and during the past decade about 0.5 M-ha could be covered under drip irrigation, mostly for horticultural crops. However the coverage so far has been minuscule in the face of the fact that almost 27 M-ha could be covered through this improved system in another 10–15 years.

1.3.1 WATER RESOURCES OF INDIA AND TAMIL NADU

In Tamil Nadu, almost all the available water resources have been used. As the demand of water is increasing day by day, there is a need to go for advanced method of irrigation like drip or sprinkler method to save water and to increase productivity and production. Although India is blessed with good water resources, yet it is not distributed evenly and hence water scarcity exists in number of states and pockets. Further to increase production/productivity and to mitigate other problems, introduction of drip irrigation in large scale is very essential, but the progress is very slow. The Land and Water resources of India and Tamil Nadu and the availability of land and water for each person are given below.

Details	India	Tamil Nadu
Total geographical area	329 M-ha	13 M-ha
Land availability per person	0.15 Ha	0.10 Ha
Surface water available	195 M.H.M	2.42 M.H.M
(utilization)	40%	95%
Ground water available	43 M-HM	2.63 M-HM
Water availability per person/year	2000 M³	650 M³
Irrigated area (gross)	97 M.Ha	3.5/3.6 M.Ha
Drip irrigated area	1.53 M.Ha	0.135 M.Ha
Sprinkler irrigated area	0.65 to 0.7 M-ha	0.03 M-ha

1.3.2 MICRO IRRIGATION

Micro irrigation is suitable for all row crops and especially for wide spaced high value crops. The required quantity of water is supplied to each plant or each row daily at the root zone through a pipe network. The main advantages of micro irrigation as compared to gravity (surface) irrigation are:

- increased water use efficiency;
- higher yield;
- decreased tillage;
- high quality crop;
- saving of fertilizer up to 35%;
- less labor due to less weed growth.

Micro irrigation is an accepted method of irrigation by the farmers in many states including Tamil Nadu. They are convinced of the usefulness of the system but adoption is very slow due to the high initial investment cost. It has emerged an appropriate water saving and production augmenting technique.

1.3.3 EXTENT OF MICRO IRRIGATION IN INDIA AND TAMIL NADU

The research on drip irrigation was carried out in this country as early as 1969 by the author and his associates in TNAU and tried to develop the micro irrigation system by drilling 2 mm holes in the lateral and providing with socket and also with micro tubes attached to the laterals. They got very encouraging results, both from the point of view of water saving and in the increase in the productivity of many crops: fruits, vegetables, and commercial crops like cotton and sugarcane.

The Government of India (GOI) constituted a *National Committee in the use of plastics in agriculture* (NCPA) in 1981 [8]. This termed as the first milestone for the development of drip technology for different crops in the different agro-climatic conditions of the country. The first national seminar on micro irrigation was organized at TNAU, Coimbatore in 1982 and it gave an impetus for the adoption of micro irrigation by farmers as the GOI announced a subsidy of 35% for drip irrigation based on the recommendation of the seminar. In the subsequent years, drip irrigation companies were established in Tamil Nadu. Maharashtra, Karnataka and Andhra Pradesh have also played an important role in popularizing the micro irrigation. The NCPA through the 17 different plasticulture development centers played a very important role in technological development of the micro irrigation in India and also through organizing number of seminars at state and national levels [8]. The research institutions/universities have conducted training programs to the farmers and as well as to the field staff. The private sector has also played significant important role in capacity building of not only for the farmers, but also for all other stake-holders.

On financial side, NABARD has advised banks to finance full cost of the system, and adjust the subsidy as and when received. While sanctioning loans under Rural Infrastructure Development Fund (RIDF), NABARD stipulates a condition that the state government should form WUA and transfer the water distribution to these WUAs. Large number of WUA was formed in various states. These WUA played a leading role in promoting micro irrigation.

The efforts made by GOI by way of institutional development, strengthening of R and D efforts and financial assistance, had a significant efforts on the adoption of the micro irrigation by the farmers [2]. The efforts of the GOI, some state governments and the drip companies in promotion of MI have resulted in bringing about 0.5 M.Ha in 2003 and 1.531 M.Ha in 2009 under MI (see Table 1.6). Although the program for promoting

TABLE 1.6   Growth of Drip Irrigation in India

Year	Micro irrigation area, Ha
1985	1000
1991	55,000
1999	2,54,000
2001	3,10,000
2003	5,00,000
2009	15,31,007

drip irrigation was taken up throughout the country, yet it is seen that the maximum coverage has been in sates of Maharashtra, Andhra Pradesh, Karnataka, and Tamil Nadu accounting for nearly 80% of the coverage area under drip irrigation in the country. The efforts made by the government of Maharashtra and Jain Irrigation Co., Jalgoan (who were the pioneers in introducing the drip irrigation technology in the country) are note worthy. The government of Andhra Pradesh has also taken up steps for promoting MI in the state by launching the Andhra Pradesh Micro Irrigation Project (APMIP). The APMIP has also tapped Rural Infrastructure Development Fund (RIDF) of NABARD to implement the scheme covering an area of 0.25 M-ha in 2 years.

The program of MI has been constrained by several factors like non-competitive unit cost in the earlier years, increase in custom and exercise duty besides other taxes on the raw materials, the provision of inadequate allocation of subsidy, enormous delay in sanction and release of subsidy and inability of the state government to share the proportionate subsidy under the centrally sponsored scheme. Further growth of DI is unbalanced, limiting mainly to the Southern peninsular states of India.

Although the pioneer in research on micro irrigation is TNAU (1969 onwards), yet the adoption of micro irrigation in Tamil Nadu is not sufficiently high compared to Maharashtra state. The drip technology was taken up only after 1986 in Maharashtra, but the extent of drip irrigated area is much more than 50% of the total area in India, whereas the total drip irrigated area is only about 1.53 M.Ha in India. The total area under drip irrigation in the world is about 6 M-ha. It is proposed to bring about 0.5 to 0.6 M-ha under drip irrigation in the next 10 years in Tamil Nadu (i.e., 10% of the total irrigation at that time). The task force on micro irrigation has stated that the theoretical potential area of drip irrigation in India is about 27 M.Ha (Table 1.7).

1.3.4 IMPACT OF MICRO IRRIGATION

A review of the selected literature on drip technology is strongly suggestive of economic viability of this technology for many crops of India. What is no less striking is that this economic viability exists even without taking into account GOI and State governments subsidy available on drip irrigation under various development programs. The cost for installing the system (Rs./Ha) is estimated as 15,000 to 20,000 for widely spaced crops like coconut, mango, etc.; and 70,000 to 80,000 for closely spaced crops like cotton, sugarcane, vegetables, etc., at present. The cost depends upon the crop type, spacing, crop water needs, source of water supply, etc. The economics of micro irrigation has been calculated and payback period has been worked out after interviewing more than 50 farmers for different crops (Table 1.8). It is observed that the payback period is about one to two years for most of the crops and the benefit–cost ratio varies from 2 to 6.50. The impact in introducing the micro irrigation system is grouped under.

- benefit to the nation; and
- benefit to the farmers.

TABLE 1.7   Theoretical Potential Area for Drip Irrigation in India

Crop	Area in M-ha
Cotton	1.7
Vegetables	3.6
Spices and Condiments	1.4
Sugarcane	4.3
Fruits	3.9
Coconut and Plantation Crops	3.0
Total	27

TABLE 1.8 Cost-Benefit Pay Back Periods for Drip Irrigated Crops

Parameter	Drip irrigated crop									
	Sugarcane	Banana	Cotton	Papaya	Grapes	Pomegranate	Ber	Tomato	Strawberry	Rose
	2.75×2.75'×5.5'	5'×5' 3'×5'×6'	3'×5'×6'	6'×6' 8'×8'	10'×6'	14'×14'	15'×15'	45×45×165cm	9'×12'×9'	3'×2'×5'
Cost of the system/acre	19,000	−19,000	19,000	16,000	17,000	12,000	12,000	19,000	75,000	10,000 microtube
Water used	20,000 l/day/ac	15–20 L/day/P	8–10 LPD/P	15 LPD/P	12–20 LPD/P	50–60 60 LPD/P LPD/P 12,000–15,000 LPD/acre		20,000 to 24,000 LPD/acre	2 LPD/pl	10 LPD/pl
Yield in T/ac	80 t	30 t	1400 kg	750 kg Latex 60 T fruit	20 T	9 T	10 T	30 T	3 T	1000/day
Pay back period	1 year	1 year	1¹ᐟ² year 3 crops	1¹ᐟ² year 1 crop season	< 1 year	< 1 year	< 1 year	1 season 6 months	2 season 2 years	< 1 year
B:C ratio	3.44	3.08	1.77	4.09	3.64	7.03	6.51	1.91	2.34	2.71

TABLE 1.8 Continued

Parameter	Drip irrigated crop									
	Sugarcane	Banana	Cotton	Papaya	Grapes	Pomegranate	Ber	Tomato	Strawberry	Rose
	2.75×2.75× 5.5'	5'×5' 3'×5'×6'	3'×5'×6'	6'×6' 8'×8'	10'×6'	14'×14'	15'×15'	45×45× 165cm	9'×12'×9'	3'×2'×5'
Extra Income due to drip irrigation over conventional method in Rs	31,620	49,320	14,360	72,040	2,64,200	1,51,280	1,33,280	49,280	67,000	1,19,190

1.3.4.1   Benefit to the Nation

a. Saving in infrastructural cost on irrigation projects: With the adoption
    of micro irrigation, there was a saving of irrigation water required
    (40–70%) for the crops. Under the principle of "water saving is water
    created", there would be benefit to the nation in the form of saving in
    cost or creating irrigation infrastructure for increasing the irrigated
    area. In places where there is no water for extending irrigation facility,
    the same can be done by converting from surface irrigation to drip irri-
    gation. The task force on micro irrigation has worked out, for a cover-
    age of an additional area of 17 M-ha under irrigation by the end of 9th
    plan of GOI, the saving of water will be about 5.9 M-HM. Based on
    the saving in irrigation water and the total value of the savings which
    otherwise will have incurred for creating equal amount of irrigation
    potential, there will be saving of Rs. $450,000 \times 10^6$ per year.
b. Saving in subsidized electricity supplied to agriculture sector due to
    reduction in electricity consumption with micro irrigation. It is esti-
    mated about Rs. $37,670 \times 10^6$ per year.
c. Saving in subsidy for the fertilizer due to saving in fertilizer by adopt-
    ing drip/fertigation. Micro irrigation can save about 30–40% of the
    fertilizer used. This is about Rs. $55,000 \times 10^6$ per year.
d. Employment generation:
    •   due to micro irrigation industry;
    •   semi-skilled persons required for installation and maintenance;
    •   direct employment in agriculture since more area is brought under
        irrigation;
    •   indirect employment – post harvest, transportation, marketing, etc.

1.3.4.2   Benefits to the Farmers (due to increased yield and good quality of produce)

Micro irrigation led agriculture should be viewed as one of the eco technological approaches to attain sustained and enhanced agriculture production and productivity. Through micro irrigation, the green revolution could
be transformed into an evergreen revolution to ensure the sustainability,
profitability and equity. Since micro irrigation greatly enhances water,

fertilizer and energy use efficiency and promotes precision agriculture, the evergreen revolution could be achieved without the burden of environmental degradation.

1.4 KUPPAM MODEL—DRIP TECHNOLOGY

Drip irrigation technology was first introduced in Israel in 1960's and subsequently it was adopted in many countries including India. The area under drip irrigation has progressively increased in the world and in India during the 30 years (Table 1.9). The area under drip irrigation is only about 2% and 1.5% of the total irrigated area in the world and India, respectively.

In Israel, there is no surface irrigation at all and it was replaced by drip/mini-sprinkler and sprinkler irrigation systems. The Israeli scientists have developed appropriate drip irrigation and crop production technologies to use water efficiently and increase the production per unit area and per unit of water. Soil is sandy and made to fertile by application of organic manures. In fact, desert has been converted into a cultivated area. Further the main crop/cropping patterns are only horticultural crops, vegetables (especially in greenhouses) and flowers. Therefore even the small farmers, who own 2 to 3 acres, are living comfortably. The productivity in all crops is very high: vegetables and cotton are at least 4 to 6 times more than that of Indian productivity. Therefore, Indian farmers are interested to know and adopt/follow Israeli technology in irrigation and crop production in India.

In the last 10 years, many chief ministers of various States, bureaucrats, scientists, and farmers from India have visited Israel to study and

TABLE 1.9   Area Under Drip Irrigation in the World and in India (ha)

Year	World	India
1970	56,000	—
1988	10,55,000	1,000
1991	16,00,000	55,000
1999	28,00,000	2,54,000
2001	30,00,000	3,10,000
2009	6 M.Ha	15,31,007

learn about the water management and agronomical (crop management) practices to use the water efficiently and increase the production.

Though research in drip irrigation was started as early in 1969 in India at TNAU, Coimbatore, for various crops, yet it was mainly to economize water and increase the area of irrigation from the available scarce water resources. The result was that the water saving in drip irrigation was about 40–70% and it was observed that the productivity of all the crops was also increased by 20–100% apart from saving of labor, fertilizer and improving high quality of produce. However, the study was not integrated with drip and production technology. Researchers were able to bring/publish the findings of the drip technology; daily quantity of water for different crops; the cost of the system for various crops; cost-benefit ratio etc. As a result, many farmers adapted drip irrigation in states like Maharashtra, Tamil Nadu, Andhra Pradesh and Karnataka. The Government of India and some state Governments also encouraged the promoters of drip irrigation to mitigate water problem.

During this period, numbers of enterprising industrialists have ventured to start manufacturing drip irrigation equipments and its accessories. India's capacity in manufacturing drip equipments including all accessories are on par with the products made in any parts in the world. One or two companies are exporting drip equipments to USA, Europe, Africa, Australia, etc. This shows that India has worldwide reputation for drip technology. Indian agronomists also developed appropriate crop production technologies to enhance the crop yield. However, the only lacuna is that all the three branches have not been properly integrated to get optimum production and hence our farmers are not able to get the expected yield, on par with Israel and other countries using drip system of irrigation. This has resulted our Government and farmers approach Israel to obtain knowledge on both irrigation and crop production technologies. For this Government/private individuals are spending lot of money to get their advice/expertise.

1.4.1   MODEL FARMS IN INDIA WITH ISRAELI'S TECHNOLOGY

Three model projects were developed by Israelis in three States in India for vegetables and cotton, to prove their technology in larger areas: first by the private company in Tamil Nadu, the second by the Agricultural Universities in Maharashtra, and the third by the Andhra Pradesh Government, as listed below:

Place/State	Area, ha	Crops
1. Udumelpet, Tamil Nadu	300	Vegetable
2a. Punjabrao Deshmukh Krishi Vidyapeth, Akola	200	Cotton
2b. Maratwada Agricultural University, Parbhani Maharashtra State		
3. Kuppam, Andhra Pradesh	1st phase – 80	Vegetable
	2nd phase – 800	
	3rd phase – 4000	

1.4.1.1 Vegetable Cultivation in Ellayamuthur Farm, Udumelpet – Tamil Nadu

This project namely Terra Agro Technologies Ltd., was conceived in the year 1996–97 by PRICOL company at Coimbatore and was implemented with the technology support from M/s. Tandi Agriculture Ltd, Israel for scientific farming of vegetables and from m/s Hovev Agriculture Ltd, Israel for hot air dehydration of vegetables. Government Institutions like IIHR – Bangalore and TNAU – Coimbatore also extended their support whenever needed in implementing this project.

About 300 ha of land was developed (out of 400 Ha) to grow vegetable crops (tomato, cauliflower, beetroot, carrot, onion, cabbage, green pepper, etc.) using drip irrigation with fertigation and improved crop production technologies. This project was a 100% export oriented fully integrated agro project. The water for drip irrigation was pumped by paying price for the water to the government from the nearby Amaravathi river. The investment cost of the project was very high including the cost for water and infrastructural facilities as per the guidelines of Israel consultants. The yield obtained in this farm was 10–15% higher than the prevailing yield in this area (Tomato-35 tons/ha). The total cost of the project was 500 million Rs. Though the project was conceived very well, but it is running on loss due to marketing of the vegetables.

1.4.1.2   Cotton Cultivation in Punjabrao Deshmukh Krishi Vidyapeth Akola and Marathwada Agricultural University Parbhani in Maharashtra State

This is the 2nd project under Israeli technology contemplated to increase the WUE and cotton yield in Maharashtra State. The average yield of cotton in India is less than 300 kg/ha (Lints) whereas the world average is about 600 Kg/ha and in Israel, it is about 1500 kg/ha. The irrigated farmers take an yield of about 15–20 quintal (Kappas) in India and in Israel the yield is about 45–60 quintal/Ha (1 Quintal is equivalent to 100 kg). This raised the question "what is the Israel technology?" The main reason is that the Israeli farmers use drip irrigation and fertigation technology and using higher plant population to get outstanding yields. The use of controlled drip irrigation gives farmers an opportunity to induce early plant growth and higher yield. Taking this into considerations, the government of Maharashtra decided to establish pilot/model project for growing cotton, based on Israeli technology at PKV, Akola and at Marathwada Agricultural University at Parbhani.

The project was started in 1997–1998 at both universities. This was implemented by Agricultural Development Co. (International) Ltd. Israel in the University farm in an area of 200 ha at Akola. The Plastro supplied drip line with inline drippers at a spacing of 50 cm in the lateral line. The Israeli technologies adopted were:

1. deep plowing (land preparation);
2. increasing plant population – by reducing distance between plants;
3. planting the crop on raised beds;
4. early planting;
5. adopting drip irrigation and fertigation; and
6. high fertilizer application with more split dosages.

The drip layout was such that the same layout could be used for most of the crops grown in the area like cotton, sunflower, safflower, groundnut, maize, soya bean, chilly, pulses, etc. The cost of the system was about 20,000–22,000 Rs./acre without automation of the system. The distance between lateral line was 1.92 m (16 mm LLDPE pipe) and the distance between plants was 20–25 cm as advised by Israeli consultant and in practice it was more than 60 cm by the farmers in this area. The raised bed of 96 cm was compacted for uniform spread of water. It was anticipated that the yield will be about 25 quintal/ha in the first year, 35 quintal/ha in the 2nd year and 45 quintal/ha in the 3rd year according to Israeli experts. Somehow the project was not successful and it was dropped and the Israeli company left the campus. The reason for the failures are: the production technologies followed in Israel are not suited for the Indian conditions including adoption of narrow spacing, and fully mechanized with incorporation of the biomass/organic farming etc.

1.4.1.3 Vegetable Crops in Kuppam, Andhra Pradesh

The Andhra Pradesh government initiated the development of the dry area of Kuppam block in Chittor district by adopting Israeli Technology for better land and water use. The BHC agro (India) Pvt. Ltd. Hyderabad, an Israeli company established during 1996, was given the responsibility of using their technology to increase the living conditions of the small and marginal farmers in the drought prone district by the Andhra Pradesh Government. The strategies adopted in the project were:

- joining agricultural holdings for viability;
- improved technology inputs including use of machinery;
- efficient use of water by drip/sprinkler method;
- efficient use of fertilizer by fertigation;
- capacity building of the farmers through training;
- exploring international market; and
- cooperative spirit among farmers.

The following production technologies were introduced in the project:

- deep plowing – better soil preparation and to absorb rain water;
- use of drip and sprinkler irrigation;
- use of machinery;
- scientific usage of improved package of practices based on soil and water analysis;
- crop planning based on market requirements; and
- producing quality products.

The important technology is rational use of water resources. In India, due to fragmentation of holdings, every farmer is drilling his own well (even for one acre or less). By consolidating the fragmented holdings, there is no need to go for a well by each land holder, but provide a well and use the ground water optimally through drip irrigation. This will normally solve water problem and also minimize the need to drill more bore wells.

1.4.2 PHASE I: ON-FARM DEMONSTRATION FARM—RUN BY THE ISRAELI COMPANY

By consolidating the farmers' land and raising needed crops after development of land and getting water by drilling bore and irrigating through drip with fertigation, an area of 182 acres of land in Kuppam block benefitted about 162 farmers. Mostly vegetable crops (Gherkin, baby crop, beans, pepper, potato, cucumber, okra, tomato etc. of 3–4 months duration) were selected. The impact of this project in 3 years (1997–2000) were: Increase in crop productivity over traditional farming system by 4–70% in different crops with water saving from 33–50%. The net return varied from Rs. 10,000 to 35,000 for various crops/per season and a minimum of 3 crops/year were taken. Based on the success of the demo project, 2nd phase was taken up with the grant received from JICA, Japan and from the government of Andhra Pradesh.

1.4.3 PHASE II: IN FARMER'S FIELD—INSTALLATION OF THE SYSTEM WITH FREE OF COST

The area covered was about 1600 acres in 54 villages and benefiting 792 farms in Kuppam block. Crops cultivated were vegetables (beans, tomato, chilly, brinjal, potato, onion, etc.) started in 1999 and completed the entire area in 2002. In this, 4 or 5 farmers joined together and shared the water facilitating optimum utilization of water. The cost was about 40,000 Rs./ acre, which was at no cost to farmers. All the farmers were enthusiastic in the beginning and the project was successful. The farmers in the project were given agriculture field guidelines services on modern technology starting from land preparation to post harvest technology and marketing by BHC Agro (India) Pvt. Ltd. on daily basis. The government of Andhra Pradesh paid the consultancy charges to the company for the benefit of the farmers. The following were the impact of the project:

- acceptability of the technology by farmers;
- paved the way for further expansion to 10,000 acres in the Kuppam block (constituency) under phase III;
- cultivation of export oriented crops.

The author of this chapter interviewed the farmers and they were happy about the project. Some farmers were able to take 4 crops in a year. Though some farmers refused to take the free facilities in the early stages, they took it afterwards. Based on the experience, Andhra Pradesh government extended the technology in Phase III-SGSY special project to cover about 10,000 acres in the same area. In the phase III, the investment on infrastructure and irrigation equipments was 40,000 Rs./acre (with bore) and 37,000 Rs./acre without bore. Of which, 50% was a subsidy from DRDA – (75% GOI and 25% GOAP) and 50% bank loan (the farmers has to get and pay to the work). The projects were studied by NABARD/ICRISAT and also by the author independently. The observations were:

- the project is technically feasible, economically viable and socially acceptable;
- efficient use of available ground water. This will help in maintaining the water table in a sustained manner;
- improved crop production;
- efficient water management and fertilizer use;
- creating employment opportunities for marginal and small farmers;
- improved the socio economic condition of rural population;
- this is a model project, which can be replicated elsewhere in the State/in the country.

1.4.4   PHASE III: EXTENSION OF ISRAELI TECHNOLOGY ON 10,000 ACRES

Total number of beneficiaries was about 8,000 covering 10,000 acres. The total project cost was 465 milion-Rs. Farmers were selected below poverty line. The crops selected were vegetables and commercial crops having high WUE. The design was based on 65% area under drip and 35% area by sprinkler or leave this area (35% of area) as fallow for rotation.

Of the three projects, described in this section that were implemented using Israeli Technologies namely irrigation, crop production, and marketing, the first two projects started in an encouraging trend, but these did not sustain/succeed. The first project was completely collapsed and the second project has been continuing without any impact. This does not mean failure of technology. In both cases, the environments, the total involvement of the people and the management including marketing have been not raised up to the expectation. The third project Kuppam is running successfully. This is mainly due to the total government support, involvement of farmers and market provided by the Israeli company and therefore this model project is well known by all those interested in water management and agricultural development in the country.

In this connection, it is argued by the scientists, drip manufacturers and some of the farmers in the country whether the Israeli technology is not available in India and what is the need to give enormous consultation fee and to import Israeli materials when Indian companies are producing everything here on par with any standard.

1.4.5   INDIAN TECHNOLOGY IN DRIP IRRIGATION AND CROP PRODUCTION

The study conducted by using drip irrigation at Tamil Nadu agricultural University (TNAU), Coimbatore from 1969–1985 has clearly revealed that the water saving ranged from 40–70% and yield increases from 10% to 75%, for vegetables or orchard crops (see Table 1.2 in this chapter). Farm trials were also conducted in farmer's field in Tamil Nadu and Maharashtra States, prove its success. Subsequently many researchers from various institutions also confirmed these findings. In India, the research has not been envisaged in a holistic way involving and integrating disciplines like irrigation, crop production and marketing even though expertises in these disciplines are available. Since the Israeli technology is given by the consultancy companies, it is possible for the company to muster all the expertise and give service to the satisfaction of the clients, i.e., farmers as a businessman and farming as an industry. Therefore the same improvements/yield/benefits can be shown by the Indian technology itself wherever the technology needed can be fine-tuned to suit Indian conditions. There are drip companies in India, who can show the same results if not better than the Israeli irrigation company, provided they are given the suitable remuneration. According to author, there is no need to hire foreign companies for giving advice in drip irrigation and crop production technology by paying huge money as consultancy charge. The government should identify suitable company/individual experts in allotting the work so that the objectives can be achieved with cheap costs. There are already success stories in India by the enlightened and counterpoising farmers in the use of drip irrigation for horticultural crops. The size of holdings varies from 10 to 300 acres. The experiences gained already can be very well exploited and the technologies updated and used. The only problem is marketing. Organizations like APEDA, National Horticultural Board (NHB), and Ministry of Food Processing etc. can go a long way to help in marketing of agricultural produce through domestic markets and export.

There is great potential (up to 10 million-ha) to bring millions of ha under drip irrigation especially for vegetable, horticulture and commercial crops in the years to come in the country, where ground water is very limited, but large area is available for cultivation (see Table 1.10).

TABLE 1.10   Areas (million-ha) Sown and Irrigated (Suitable for Drip Irrigation) in India

Crop	Area, 2000		Expected area, 2020/25	
	Sown	**Irrigated**	**Sown**	**Irrigated**
Coconut/Arecanut	1.5	0.9	2.0	1.0
Cotton	11.5	7.5	12.0	8.8
Flowers	0.8	0.3	1.6	0.90
Fruits	4.0	1.2	4.2	2.2
Plantation crops	2.8	1.0	3.0	1.6
Sugarcane	5.5	5.5	6.0	6.0
Tobacco	0.6	0.4	0.6	0.6
Vegetables	8.3	4.2	10.0	7.4
Total	**35.0**	**21.0**	**39.4 = 40**	**28.5 = 29**

Note: Out of 29 million-ha, it is possible to bring 33% of this area (10 million-ha) under drip irrigation in 2020–2025.

Hence instead of engaging Israeli/other foreign company, it is recommended that Indian companies/experts can be engaged to achieve the goal with less cost and giving employment opportunity for local people in collaboration with APEDA and other marketing facilities available with the government for the produces.

1.5   DRIP TECHNOLOGY: 40 TONS OF BANANA PER ACRE PER YEAR

This section describes the drip technology that can be practiced by the farmers in a big way to produce the required banana yield with less water and increase their income. Banana is an important commercial fruit crop in India. The total area of banana cultivation, production and productivity in India and Tamil Nadu are given in Table 1.11.

The crop is grown only under irrigated conditions except in hilly areas, where the rainfall is spread over for about 8–10 months in a year. Extensive studies have been conducted in cultivation of banana including the use of tissue culture plants. The author has done detailed research on the use of drip irrigation for banana crop since 1970 and also collected data from the farmers in Tamil Nadu and Maharashtra states. The results have indicated that it was possible to save about 40–50% of water and increase the yield by about 30–40%. By use of tissue culture plants of high yielding varieties like 'Grand Naine' and adoption of improved crop growing technologies including drip irrigation and fertigation, it was possible to make a break through in water saving and optimum productivity of banana. Unfortunately, all the proven technologies have not been adopted by the farmers. Therefore, it is time now to go for this in a big way in order to save water and increase the profit to the farmers by increasing the yield of the crop.

TABLE 1.11   Area, Production and Productivity of Banana in India and Tamil Nadu

Location	Area Cultivated, ha	Total production, million-tons	Productivity, tons per	
			ha	acre
India	4, 82, 816	16.17	34.30	13.72
Tamil Nadu	84,542	3.54	41.92	16.77

Source: Horti Statistics (2001), Directorate of Horticulture and Plantation Crops, Chennai [1].

Nutrient management plays a major role in increasing the productivity of banana. Conventional fertilizer application has low uptake of nutrients (40–60%) and some portion is lost due to leaching or fixation. This can be overcome by fertigation, where water soluble nutrients are applied through drip at the root zone at frequent intervals. Fertigation also saves labor and time with uniform distribution of nutrients.

Organic manures are very much needed and they are applied 7 times once at basal and six times at monthly intervals. Inorganic fertilizers applied through fertigation include urea, single super phosphate, muriate of potash, ammonium sulfate and sulfate of potash at monthly intervals up to 7 months for general dosage and special dose at 5 days intervals from the day of planting to 315 days. In addition, micronutrients like magnesium sulfate, zinc sulfate, ferrous sulfate, boron and borocol can be applied depending on the need.

Inter-cultivation operations include hoeing and weeding up to 3 months. Other practices include propping to the bunches, covering the bunches with blue polythene sheet etc. Proper monitoring of lateral tubes and drippers is needed. The field has to be provided with wind-breakers in areas where wind is a problem.

Drip irrigation is an advanced method of micro irrigation. In this method water is applied, through a network of pipes, at the root zone of individual plant (Figure 1.1). The requirement of water is given daily to each plant though the drippers/emitters based on the evapotranspiration (ET) requirements of the crop. It varies from 5 to 20 liters per day depending upon the climate and the crop stage. Normally in surface irrigation, banana crop requires about 1800–2000 mm of water in 12 months. If drip irrigation is practiced, the water requirement will be reduced to about 40–50%.

The author has traveled widely in major banana growing areas of Maharashtra and Tamil Nadu, has visited many drip irrigated banana farms and has interviewed with the farmers. Based on the data collected from the farmers and manufacturers of the drip equipments, the following conclusions are presented.

The farmers are cultivating different varieties of banana in different States. Poovan, Rasthali, Nendran, Robusta, and Karpuravalli, etc. are commonly grown in Tamil Nadu. Recently a new variety, Grand Naine, adopted from Karnataka/Maharashtra state has been introduced and its performance has been studied. This variety gives higher yield and also has good market. A case study is detailed below. This variety was culti-vated using drip irrigation in large area (>100 acres) in Jain Terra Agro farm in Ellayamuthur village of Udumalpet taluk in Coimbatore District (Figure 1.3).

1.5.1 ECONOMICS OF BANANA CULTIVATION (VAR. GRAND NAINE)

Spacing 6' × 5', i.e., 1450 plants per acre = 1500

a. Main crops (11 to 12 months)

Cost of planting materials, tissue culture plants @ Rs, 12 each = 1500 × 12 Rs./acre	18,000
Cost of cultivation (land preparation, weeding, irrigation, propping, etc.)	35,000
Cost of installation of drip irrigation including fertigation devices	25,000

b. Ratoon crop

Cost of cultivation for ratoon crop-1 (9 months), Rs./acre	35,000
For ratoon crop-2 (9 months), Rs./acre	35,000

c. Total cost (3 crops) in 30 months, Rs./acre, A   1,48,000

FIGURE 1.3   Drip irrigation in banana in Jain Terra Agro farm.

Anticipated receipts (@ 3,000 Rs. per ton of banana – average price)

Main crop (40 tons) for 11–12 months	1,20,000
Ratoon crop 1 (35 tons) for 9 months	1,05,000
Ratoon crop 2 (35 tons) for 9 months	1,05,000
Total receipts, 30 months, Rs./acre, B	3,30,000

Total yield in 2 ½ years = 110 tons/acre or 44 tons/acre/year or say
$$40 \text{ tons/ acre}$$
Net Profit per acre in 30 months, Rs./acre = $\mathbf{A} - \mathbf{B}$ = 3,30,000 – 1,48,000
$$= 1,82,000$$
Profit, Rs. per acre per year = 72,800
$$\text{or Rs. } 73,000/\text{-}$$

Water requirement for banana is about 8 to 10 liters/plant/day initially and 20 liters/plant/day later stages subject to climatic conditions and crop stage. From the above, it can be concluded that the average yield per acre was about 40 tons/acre/year or 100 tons/ha/year compared to average of 42 tons/ha or about 16.8 tons/acre/year in Tamil Nadu. This is 2.5 times more than the yield obtained at present. Water was saved up to 50%. Hence intensive cultivation of banana with less water (one of the costly inputs) will go a long way to improve the production and profit to the farmer.

1.6   INCREASING COTTON YIELD THROUGH DRIP IRRIGATION

Cotton is one of the most important cash crops in India and plays a dominant role in the industrial and agricultural economy of the country. India is one of the major producers of cotton in the world with the largest acreage, almost one-fourth of the world's area. The production share is, however, only 13.5%, ranking third after China and USA. It is estimated that about 9 million-ha are covered under cotton, achieving a production level of 17.6 million bales with a productivity of 327 kg/ha (1996–1997), as compared to the production level of 2.3 million bales from an area of 4.4 million-ha with a productivity of 88 kg/ha at the time of Indian independence. Though the area has increased to 9.2 million-ha during 1998–1999, the production and productivity are reduced to 16.5 million bales and 298 kg/ha, respectively.

The productivity of about 300 kg/ha is only around 50% of the world's average. The productivity of cotton in Israel is about 1650 kg per ha, nearly six times that of Indian average. The main reason for the low productivity is that irrigated area is only about 40% in India compared to 100% in Israel. The irrigated area was about 6% during 1947–1948 and has increased to 40% in 1994–1995. The details of irrigated area, rain-fed area, total cotton area and percent of irrigation from 1947 to 1998 are given in Table 1.12.

The average productivity has increased from 88 Kg to 300 Kg/ha and during that period the percentage of irrigation has increased from 6 to 40%. It indicates that by irrigating the crop, yield can be increased substantially by introduction of suitable irrigation methods and better water management practices. Further, it may not be possible to bring the entire area under irrigation in the country by using the present practices, however the area of irrigation can be doubled from the same quantity of water and the yield of the crop can be enhanced substantially in the coming years, by introducing micro irrigation system in the irrigated areas. By providing protective or supplemental irrigation and by adopting appropriate soil and water conservation and management practices in the rain-fed areas, the cotton yield can be increased substantially.

TABLE 1.12 Irrigated Area, Rain-Fed Area, Total Area of Cotton and Percentage of Irrigated Area (in ,000 ha), India

Year	Irrigated area	Rain-fed area	Total area	% of irrigated area	Yield, Kg/ha
1947–48	251	4062	4313	5.8	88
1950–51	465	5420	5885	8.0	85
1955–56	733	7355	8088	9.1	88
1960–61	985	6655	7640	12.9	125
1970–71	1570	6035	7605	19.5	106
1975–76	1856	5494	7350	25.3	138
1980–81	2218	5593	7811	28.4	152
1985–91	2185	5349	7534	29.0	197
1990–91	274	4866	7440	34.6	224
1995–96	3103	4822	7925	39.5	246
1998–99	-	-	9303	-	298

In the northern states as well as in major parts of Maharashtra, Andhra Pradesh and Karnataka the main source of water supply for cotton crop is canal. In UP, Gujarat and Tamil Nadu the major source is tube wells, open wells and tanks. Surface irrigation by flooding the field is the most common method practiced in the northern states. In Maharashtra, Gujarat, Tamil Nadu and Karnataka, furrow method is in vogue, which is more efficient. Drip irrigation was introduced in about 1000 ha in Maharashtra state alone. Cotton is sensitive to both excess and deficiency of soil moisture. The water should be restricted during the early growth phase to prevent excessive vegetative growth and sufficient water should be given, when the plants produce flowers and buds. However the method of flooding should be avoided to get good yield.

Compared to the world average, the yield in India is very low due. It is necessary to enhance the yield for growing population and the limited land area. The potential level of *kapas* yield in the case of irrigated cotton is about 2000–3000 kg/ha, and for the hybrid cottons it is about 4000 kg of *kapas* per ha, if all the recommended cultivation practices are adopted.

It has been reported that cotton (lint) yield of more than 2250 kg/ha using drip irrigation was obtained in Arizona, USA. This is very well suited for water scarce, shallow and sandy soils. Salt water up to 8–10 mmhos/cm can also be used for growing crops without effecting the yield. Further, improvement in quality of cotton has also been reported for fiber maturity and fineness to a great extent.

Studies conducted at Tamil Nadu Agricultural University (TNAU), Coimbatore has revealed that saving of water in drip irrigation was about 50–60% compared to furrow irrigation and at the same time, yield was also increased by about 25%. The other advantages in this system are the reduced weed growth (by 30–50%) and effective utilization of fertilizers by fertigation. It was estimated that about 25–30% of fertilizer could be saved through fertigation. The results of Drip versus traditional irrigation system are given in Table 1.13.

In order to reduce the initial cost of the system, research is underway in different places. One such method is to introduce micro tubes instead of drippers/ emitters and design the system so that each lateral can irrigate four rows of crop using micro tubes instead of two rows in the traditional system. By this, it is possible to reduce the cost by 30–40%.

TABLE 1.13 Water Use and Cotton Yield with Drip and Surface Irrigation Methods, India

Research station	Water saving, %	Increasing crop, %	Water requirement, mm/ha		Cotton yield, kg/ha	
			Surface	Drip	Surface	Drip
MPKV, Rahuri	43	40	895	511	2250	3140
TNAU, Coimbatore	60	25	856	302	2600	3260

The major limitation in adoption of drip irrigation is the higher capital cost for an average farmer. The cost of the drip system depends on the crop spacing. It is about 17,000–20,000 Rs./acre. It is ascertained from the farmers that the average yield of cotton is about 8–10×100 kg/ha in the surface method of irrigation and about 25–32 Quintals/ha in the drip irrigated field.

The cost of cultivation in drip-irrigated field is reduced by about Rs. 2500/ha. The maximum amount of water given to the crop in drip is about 8–10 liters/day/plant. This is less than 50% of the water used in surface method. From the data collected, the benefit–cost ratio is about 1.80 and the net extra income due to drip irrigation over conventional method of irrigation is about 15000 Rs./acre, and in addition more production from unit of water. The economics of drip irrigation for cotton was determined as shown in Table 1.14.

It was observed that in drip irrigated field the crop growth was uniform and vigorous. The duration of the crop was reduced by 15–20 days. The price of the *kapas* produced from drip irrigation is higher with same quality compared surface irrigated field.

From the above, it is clear that there is great potential and prospects of introducing drip irrigation for cotton in India. This is very much needed especially in water scarcity areas and in view of the fact that India requires large quantity of *kapas* for the local and export needs. The farmers who have introduced the system are very happy and they recommend drip irrigation for cotton crop in the entire country. The drip technology and the crop cultivation technology are available; therefore there is an urgent necessity to go for drip irrigation in order to save water and to increase the production of cotton.

TABLE 1.14 Economics of Drip Irrigation for Cotton

S.No	Particulars	Drip	Surface
1.	Spacing	3' × 5' × 6'	4' × 4'
2.	Cost of drip system, Rs./acre	(paired row)	-
	a) Life 5 years for Lateral, Dripper and 10 years for Main, Submain, filters.	18,000 3,060	-
	b) Depreciation, Rs./acre	1,080	-
	c) Interest, Rs./acre	860	-
	d) Repair and Maintenance 5%	5,000	-
	e) Total, Rs./acre	6,500	7,500
	f) Cost of cultivation of a crop	11,500	7,500
	Seasonal total cost = 2e+2f, Rs./acre	8–10	15–18
3.	Water used in liters/day/plant	15.0	10.0
4.	Yield of produce, 100kg/Ha	1400	1200
5.	Selling price Rs./100kg	21,000	12,000
6.	Income from produce = (6 × 5), Rs./acre	9,500	4,500
7.	Net seasonal income = (7 – 3), Rs./acre	–	-
8.	Additional area cultivated due to saving of water	11,500	-
9.	Additional expenditure due to additional area, Rs./acre	21,000	-
10.	Additional income due additional area, Rs./acre	9,500	-
11.	Additional Net income = (11 – 10), Rs./acre	23,000	7,500
	Gross cost of production, Rs./acre	42,000	12,000
12.	Gross income = (7 + 11)	14,500	—
13.	Net extra income due to drip irrigation over conventional = Drip – surface	1.83	1.60
14.	BC ratio		

1.7 INCREASING COCONUT YIELD THROUGH DRIP IRRIGATION

Coconut is an important horticultural crop in the country. This crop is grown in large areas in the Asian Sub constituent. Coconut is grown not only in the rural fields but also in urban house surroundings areas to get the daily needs of coconuts for their use. The byproducts are used for many purposes and it also generates employment opportunities for rural people.

In India, coconut is cultivated in about 3 million-ha and it produces about 14,000 million nuts/year. Over 90% of the produces are from Tamil Nadu, Andhra Pradesh, Karnataka and Kerala states. In Tamil Nadu, the area of coconut is about 0.33 million-ha (3rd rank) and the nut production is about 4300 million nuts per year. About 14,535 nuts/ha are produced that is highest in the country. The production is about 2 times of the India's average production (7781 nuts/ha). The average yield of nuts per tree is about 50 per year in India compared 100 nuts/tree in Tamil Nadu.

Karnataka ranks 2nd (about 0.4 million-ha) after Kerala and third in production (2650 million-tons). It contributes 24% of the area and 21% of the production. The crop is cultivated throughout the state, but the production is only 6700 nuts/ha, which is far below that of Tamil Nadu, Andhra Pradesh and even all India average. Though in Tamil Nadu, the area under coconut is about 16%, but the production is about 30% of the country.

Kerala has an area of three times more than in Tamil Nadu, but the production is only about 42%. The problem in cultivating this crop is only getting adequate water supply throughout the year, which is in high demand.

1.7.1 IRRIGATION FOR COCONUT

Coconut is irrigated by surface method of irrigation by flooding the entire field, which is not necessary thereby wasting the scarce water. Lot of research work is going on for years in irrigation of coconut. Advanced methods of irrigation like drip, micro sprinkler and perfo spray methods are used to save water.

Surface irrigation is very common in all places for coconut crop. The entire basin is flooded to a depth of 5–7 cm once in 5–10 days depending upon the type of soil. The quantity of water applied is >200 liters/day or about 1000–1400 liters in 5–7 days. The conveyance efficiency is less than 20–25% in surface irrigation. In surface irrigation, water/moisture is available immediately for 2 days after irrigation and after that the root has to exert more pressure to get water and this affects the growth of the tree.

1.7.2 DRIP AND MICRO SPRINKLER IRRIGATION

Drip method is an advanced method of irrigation in coconut. The maximum water saving is possible in coconut with drip irrigation. Studies have shown that drip irrigation can save water up to 50–70% and increases the yield by 30%. Further, the problem of irrigating the crop in sandy tracts can be solved through drip irrigation. The effects of drip irrigation has been demonstrated by many progressive farmers in the states of Tamil Nadu and Karnataka.

The water requirements of coconut using drip irrigation (matured crop) vary from 70 to 120 liters per day per plant depending upon the climate and season. Farmers have tried by giving water up to 150–200 liters/day, but it is found that it was not necessary. Hence the water saving in this method is substantial, by which at least 2 to 3 times more area can be irrigated by drip irrigation with the same quantity of water. Drip irrigation is practiced in about 60,000–65,000 ha in India (mainly in Karnataka, Tamil Nadu, Andhra Pradesh and Maharashtra), which is about 3.5% of the total coconut area of the country.

The cost of drip irrigation system for coconut at present is about 17,000–20,000 Rs./ha with a spacing of 27'×27' to 30'×30'. Research was carried out at Water Technology Center, Tamil Nadu Agricultural University, Coimbatore for various crops using drip system. It was found that the water saving was about 45% and the yield increase was about 30% over conventional irrigation, for coconut crop. The cost-benefit ratios were 1.41 and 5.14 taking into account excluding water saving and including water saving, respectively. From the above data, it is clear that there is tremendous scope to use drip method for coconut plantation in the country particularly in all the southern states, where more than 90% of the crop is grown and water is the constraint in growing the crop.

Coimbatore District in Tamil Nadu has about 90,000 ha of coconut crop mostly in southern part of the district, which gets more rain during southwest monsoon due to *Palghat* gap. The farmers used to pump water from their open wells till 1990. The water table in the wells were 50'–100', When the author tried to convince the farmers to go for drip irrigation for their coconut farm, their reply was that there is no problem for water and coconut need more water for good yield. Slowly the farmers are switching over to drip due to high demand of water for other uses.

Now the water table has gone up to 300–500'. Due to monsoon failure in 2001/2002, about 0.5 million-coconut trees died. There was no water even for drip irrigation in the wells. About 50% of the coconut is under drip irrigation in the district at present. To have sustainable coconut crop and to mitigate the drought, it is very essential to use drip irrigation for all coconut plantations not only in Coimbatore District but also in the entire country. This method is suited for the crop and the cost is also not much as it is wide spaced commercial crops.

1.7.3   CASE STUDIES

In 1985, a farmer in Pattiverenpatti, Dindigul District, Tamil Nadu introduced drip irrigation for his 10 acre coconut farm by installing only one HP deep well pump in his bore well (single phase) as it was not possible to get the three-phase connection paying the electricity rate on par with industrial unit. He was able to grow the crop. He was successful.

Another farmer in Kanuvai near Coimbatore has switched over from surface irrigation to drip irrigation for his 60-acre coconut farm twenty years back. He was even experimenting by giving various quantities of water (170–200 lit/tree/day). His experience was that he was able to get about 150–200 nut/tree/year under drip irrigation. He was able to irrigate the farm in a sustainable manner.

Numerous coconut growers are now using drip irrigation due to water scarcity in Tamil Nadu and Karnataka states. Time is not far away that the entire coconut farm in the country will be irrigated by drip system for its sustainability and to increase yield. The Government also should encourage the farmers to go for drip irrigation to conserve water.

1.8   SUMMARY

This success story of drip irrigation technology is narrated by the author, who has been working on drip irrigation since 1969 and has published >1200 articles in National and International journals on water resources, irrigation of which >300 are on drip irrigation. The chapter includes detailed discussion on: Historical background of drip irrigation;

components/advantages/disadvantages/limitations of drip irrigation; benefit-cost ratio for selected crops; cost analysis of drip irrigation; research advances of drip irrigation; technology to increase yield of banana, cotton, coconut; demonstration farms; and Kuppam model, etc.

KEYWORDS

- agriculture technology
- Australia
- banana
- Canada
- clogging
- coconut
- cotton
- Cyprus
- Directorate of Horticulture and Plantation Crops
- drip irrigation
- drip irrigation technology
- economics
- filtration
- flow path
- France
- fresh water
- Germany
- Government of India, GOI
- Greece
- IARI – Indian Agricultural Research Institute
- ICID
- INCID – Indian National Commission on Irrigation and Drainage
- India
- International Micro Irrigation Congress
- Iran

- Israel
- liters per hour, lph
- Malaysia
- Mexico
- micro irrigation
- National Committee on the Use of Plastics in Agriculture, NCPA
- New Zealand
- peach
- Quintal, Q = 100 kg
- relative humidity
- sandy soils
- South Africa
- sprinkler irrigation
- subsurface drip irrigation, SDI
- surface drip irrigation
- Tamil Nadu Agricultural University, TNAU
- tomato
- trickle irrigation
- United Kingdom
- USA
- water demand
- water management
- water scarcity
- World War II
- WTC

REFERENCES

1. DHPC, 2001. *Horti Statistics*. Directorate of Horticulture and Plantation Crops, Chennai.
2. GOI, 2004. *Report of the task force on micro irrigation*. G. O. I., Ministry of Agriculture, New Delhi.

3. Goyal, Megh R. (Ed.), 2013. *Management of Drip/Trickle or Micro Irrigation.* Oakville – ON, Canada: Apple Academic Press Inc., pp. 1–408.

4. Goyal, Megh R. (Ed.), 2015. *Research Advances in Sustainable Micro Irrigation, volumes 1–10.* Oakville – ON, Canada: Apple Academic Press Inc.

5. http://www.icid.org/conf_microirri_past.html

6. http://www.icid.org/sprin_micro_11.pdf

7. IARI, 2009. *Development of micro irrigation technology.* WTC, IARI, New Delhi.

8. National Committee on the Use of Plastics in Agriculture (NCPA), 1990. *Status, Potential and Approach for adoption of drip and sprinkler irrigation system* Pune, 1990.

9. Rajput, T. B. S. and Neelam Patel, 2007. *Micro irrigation manual.* WTC-IARI, New Delhi.

10. Sivanappan, R. K. and O. Padmakumari. 1980. *Drip Irrigation.* Tamil Nadu Agricultural University, Coimbatore.

11. Sivanappan, R. K., 2015. Advanced technologies for sugarcane cultivation under micro irrigation, Tamil Nadu. Pages 1:121–146, In: *Goyal, Megh R. (Ed.), 2015. Research Advances in Sustainable Micro Irrigation, volume 1.* Oakville – ON, Canada: Apple Academic Press Inc.

12. Sivanappan, R. K., 2015. Micro irrigation potential in fruit crops, India. Pages 3: 79–94, In: *Goyal, Megh R. (Ed.), 2015. Research Advances in Sustainable Micro Irrigation, volume 4.* Oakville – ON, Canada: Apple Academic Press Inc.

13. Sivanappan, R. K., 2015. Micro irrigation potential in India. Pages 1:87–120, In: *Goyal, Megh R. (Ed.), 2015. Research Advances in Sustainable Micro Irrigation, volume 1.* Oakville – ON, Canada: Apple Academic Press Inc.

14. Sivanappan, R. K., 2015. Research advances in micro irrigation in India. Pages 4: 33–42, In: *Goyal, Megh R. (Ed.), 2015. Research Advances in Sustainable Micro Irrigation, volume 4.* Oakville – ON, Canada: Apple Academic Press Inc.

15. Sivanappan, R. K. and O. Padmakumari, 2015. Principles of Drip/trickle or micro irrigation. Chapter 3, In: *Goyal, Megh R. (Ed.), 2015. Research Advances in Sustainable Micro Irrigation, volume 9.* Oakville – ON, Canada: Apple Academic Press Inc.

16. Sivanappan, R. K., 1994. *Drip Irrigation in India.* INCID, Ministry of Water Resources, G.O.I., New Delhi.

CHAPTER 2

SPREAD AND ECONOMICS OF MICRO-IRRIGATION IN INDIA— EVIDENCES FROM NINE STATES

K. PALANISAMI, KADIRI MOHAN, K. R. KAKUMANU, and S. RAMAN

CONTENTS

In this chapter, one US $ = 63.02 Rs. (Indian rupees).

2.1   INTRODUCTION

Water is becoming increasingly scarce in many parts of the world limiting agricultural development. The capacity of large countries like India to efficiently develop and manage water resources is likely to be a key determinant for global food security in the 21st century [9]. In India, almost all the easily possible ways for viable irrigation potential have already been tapped. However, the water demand for different sectors has been growing continuously [8, 12] and the demand management becomes the overall key strategy for managing scarce water resources [3]. Since agriculture is the major water-consuming sector in India, the demand management in agriculture in water-scarce and water-stressed regions would be central to reduce the aggregate demand for water to match the available future supplies [11].

Various options are available for reducing water demand in agriculture. Firstly, the supply side management practices include watershed development and water resource development through major, medium and minor irrigation projects. The second is through the demand management practices, which include improved water management technologies and practices. The micro-irrigation (MI) technologies such as drip and sprinkler are the key interventions in water saving and improving the crop productivity. Evidences show that the water can be saved up to 40 to 80% and water use efficiency (WUE) is higher in a properly designed and managed MI system compared to 30–40 per cent under conventional practice [1, 10]. The successful adoption of MI requires, in addition to technical and economic efficiency, two additional preconditions, *viz.*, technical knowledge about the technologies and accessibility of technologies through institutional support systems [4].

2.2   RESEARCH QUESTIONS AND METHODOLOGY

With regards to MI in India, much of the research has been conducted with respect to its economics and its suitability for various crops.

The available empirical evidences are comparatively limited with respect to its adoption and economics under different farm categories. Hence, the key questions are:

- Who has access to MI?
- What is the economics of MI under different farm groups (*viz.*, marginal, small and large farm groups)?
- What are the needed interventions to upscale the MI adoption?

Hence, this chapter aims to answer these questions to some extent.

The study was undertaken during 2010. All the nine states were covered for the analysis of the potential MI area and actual spread. For the farm level analysis on the costs and returns among the different farm groups, nine states were covered *viz.*, Andhra Pradesh, Gujarat, Karnataka, Kerala, Maharashtra, Orissa, Punjab, Rajasthan and Tamil Nadu (Figure 2.1). Both secondary and primary data were collected. The secondary data were collected covering the state level MI sources, cropping patterns, existing areas under MI and details for Government subsidy. The primary data was collected from 150 farmer sample from each selected state using semi-structured questionnaire covering source of irrigation, farm size, irrigated

FIGURE 2.1   Selected nine states for the study in this chapter.

area, area under MI, crops grown, subsidy availed, and income and expenditure under different crops with and without MI. Farm level constraints for adoption of MI and suggestions for better adoption were also obtained from the field surveys. The sample was post stratified into marginal, small and large farmers. The secondary data were used to work out the potential for MI in each state and the primary data were used to work out the access to and economics of MI under different farm categories as well as to document the suggestions of the farmers for better adoption of MI in the state.

The internal rate of return (IRR) due to MI was worked out using the annualized capital cost of the system (Eq. (1)), average life of the MI system and the additional crop income that will occur during the life period of the MI system in the farm.

$$\text{Annualized cost of MI} = \left[(\text{capital cost of MI}) \times (1+i)^{AL} \times i \right]$$
$$\div \left[(1+i)^{AL} - 1 \right] \quad (1)$$

Where: AL = Average life of MI system (was assumed as 8 years), and i = discount rate (was assumed as 10%). Using the farm level data, the following regression equation was fitted to study the influence of various factors on area under MI.

$$Y_i = \beta_0 + \beta_1 X_i + \beta_2 D_{1i} + \beta_3 D_{2i} + \varepsilon_i \quad (2)$$

where; Y_i = Area under MI by i-th farmer (ha), X = Farm size of i-th farmer (ha), D_{1i} and D_{2i} = Dummy variables for i-th farmer representing marginal and small farmer category respectively, $\beta_0 \beta_1 \beta_2 \beta_3$ are regression coefficients, and ε_i = error term. Analysis of variance without replications was used to test the significance of additional income earned by different categories of farmers under MI across the nine states.

2.3   RESULTS AND DISCUSSION

2.3.1   POTENTIALITY AND CURRENT STATE OF MICRO IRRIGATION

Potentiality of different MI systems in terms of drip and sprinkler was assessed using the state-wise secondary data [7]. For assessing the potentiality of MI in different states, the variables under consideration were: state-wise and source-wise irrigated area, cropped area and crop wise suitability for different MI systems. While making the assessment, the irrigated area under paddy and crop area under canal irrigation were not considered. It has been assessed that there is potentiality of bringing around 42 million-ha under drip and sprinkler in the country [7]. Out of this, about 30 million-ha are suitable for sprinkler irrigation for crops like cereals, pulses, oilseeds in addition to fodder crops. This is followed by drip with a potentiality of around 12 million-ha under cotton, sugarcane, fruits and vegetables, spices and condiments; and some pulse crops like red gram etc. In addition to drip and sprinkler irrigation, there is potentiality for bringing an area of about 2.8 million-ha under mini-sprinkler irrigation for crops like potato, onion, garlic, groundnut, cabbage, cauliflower etc., according to Raman [7].

The percentage of actual area against the potential estimated under drip irrigation in different states varied between nil in Nagaland to as high as 49.7% in Andhra Pradesh and followed by Maharashtra (43.2%) and Tamil Nadu with 24.1%. In case of sprinkler irrigation, percentage of actual area against the potential estimated was as much low as 0.01% in Bihar and highest of 51.93% in Andhra Pradesh.

Compared to the potential of 42.23 million-ha in the country, the present area under MI accounts to 3.87 million-ha (1.42 million-ha under drip irrigation and 2.44 million-ha under sprinkler irrigation), which is about 9.16% only (see, Table 2.1 and Figure 2.2). The data on the present area thus reflect the extent of MI systems covered under different Government programs as well as the own investment by the farmers. However, the actual area under MI may vary according to the extent of use of MI by the farmers.

2.3.2 MICRO IRRIGATION AND GOVERNMENT SUBSIDY

Since the introduction of MI in India, government agencies are fully aware of the fact that the system cost is high particularly for the marginal and poor farmers to adopt. Realizing this, the Central and State Governments apart from announcing subsidy schemes, at their levels, mediate with the manufacturers from time to time and try to keep the unit cost as low as possible.

TABLE 2.1   Potential (P) and Actual (A) Area Under MI in Different States

State	Area in different states of India (per 1000 ha)								
	Drip irrigation			Sprinkler irrigation			Total		
	P	A	%	P	A	%	P	A	%
Andhra Pradesh	730	363.07	49.74	387	200.95	51.93	1117	564.02	50.49
Bihar	142	0.16	0.11	1708	0.21	0.01	1850	0.37	0.02
Chhattisgarh	22	3.65	16.58	189	59.27	31.36	211	62.92	29.82
Goa	10	0.76	7.62	1	0.33	33.20	11	1.09	9.95
Gujarat	1599	169.69	10.61	1679	136.28	8.12	3278	305.97	9.33
Haryana	398	7.14	1.79	1992	518.37	26.02	2390	525.50	21.99
Himachal Pradesh	14	0.12	0.83	101	0.58	0.58	115	0.70	0.61
Jharkhand	43	0.13	0.31	114	0.37	0.32	157	0.50	0.32
Karnataka	745	177.33	23.80	697	228.62	32.80	1442	405.95	28.15
Kerala	179	14.12	7.89	35	2.52	7.19	214	16.64	7.77
Madhya Pradesh	1376	20.43	1.48	5015	117.69	2.35	6391	138.12	2.16
Maharashtra	1116	482.34	43.22	1598	214.67	13.43	2714	697.02	25.68
Nagaland	11	0.00	0.00	42	3.96	9.43	53	3.96	7.48
Orissa	157	3.63	2.31	62	23.47	37.85	219	27.10	12.37
Punjab	559	11.73	2.10	2819	10.51	0.37	3378	22.24	0.66
Rajasthan	727	17.00	2.34	4931	706.81	14.33	5658	723.82	12.79
Tamil Nadu	544	131.34	24.14	158	27.19	17.21	702	158.52	22.58
Uttar Pradesh	2207	10.68	0.48	8582	10.59	0.12	10789	21.26	0.20
West Bengal	952	0.15	0.02	280	150.03	53.58	1232	150.18	12.19
Others	128	15.00	11.72	188	30.00	15.96	316	45.00	14.24
Total	11659	1428.46	12.25	30578	2442.41	7.99	42237	3870.86	9.16

Note: P = Potential area; A= actual area. Source: Raman [7] and Indiastat, 2010.

Central Government also has launched a massive countrywide scheme to promote MI *viz.* Centrally Sponsored Scheme (CSS) on MI, which came into effect from 2005–06 financial year. However, even before the

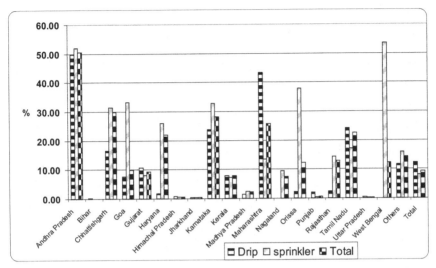

FIGURE 2.2 State wise potentiality and actual spread of area under MI, Raman [7].

start of the CSS, Andhra Pradesh and Karnataka states had MI schemes at state level. However, the subsidy levels were comparatively in low range that varied among different states from 50 to 65% depending upon the MI systems. The implementation of MI has gradually accelerated in all the states due to CSS on MI and the increase in physical performance was in the order of nearly 800% in Madhya Pradesh, 300% in Punjab and 150% in Orissa during 2006–08 [5]. In a span of 5 years (April 2005 and December 2009), an area of around 0.356 million-ha was brought under MI in the country (Figure 2.3). The level of subsidy being followed in different states and the implementing agencies are given in Table 2.2. The major crops vary from field crops (cotton, maize, groundnut, sugarcane) to vegetables, fruits (banana, papaya, mango, grapes) and plantation crops.

Many times there is time lag between the decision taken on the quantum of subsidy and its actual implementation. For example, the subsidy for drip systems for banana crop in 2010 was 65,000 Rs/ha that was based on the estimations in 2008. Any increase in the raw material prices during this time lag period will reflect on the actual cost of the system, which will be 80,000 Rs/ha thus decreasing the subsidy percent at the end users' level. Hence periodical review of the unit cost is important so that full benefit of the subsidy can be realized.

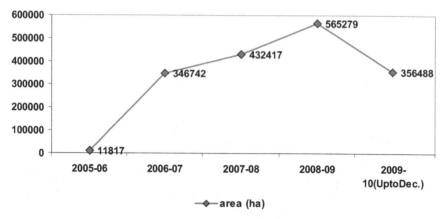

FIGURE 2.3   The trend of MI adoption in the country during 2005 to 2009. Source: NCPAH [5].

TABLE 2.2   Prevailing Subsidy Rates and Implementing Agencies in Different Indian States

State	Subsidy (%)		Major crops under MI	Implementing agency
	Drip	**Sprinkler**		
Andhra Pradesh	70	70	Chilies, mango, sweet orange, groundnut	Andhra Pradesh Micro Irrigation Project (APMIP): Autonomous body under Department of Horticulture
Bihar	90	90	Sugarcane, banana, coconut, maize, groundnut	State Horticultural Mission
Chhattisgarh	70	70	Sweet orange, vegetables	Department of Agriculture
Goa	50	50	Vegetables	Department of Agriculture
Gujarat	50	50	Cotton, vegetables, groundnut	Gujarat Green Revolution Corporation
Haryana	90	50	Orchard crops	Department of Agriculture
Himachal Pradesh	80	80	Orchard crops, cole crops	Department of Agriculture, Himachal Agro.
Jharkhand	50	50	Vegetables	Department of Agriculture
Karnataka	75	75	Grapes, vegetables, groundnut	Department of Agriculture and Department of Horticulture
Kerala	50	50	Coconut, aracanut, pepper	Department of Horticulture

TABLE 2.2   Continued

State	Subsidy (%)		Major crops under MI	Implementing agency
	Drip	Sprinkler		
Madhya Pradesh	70	70	Sweet orange, banana, vegetables	Department of Horticulture
Maharashtra	50	50	Grapes, banana, sugarcane, cotton	Department of Agriculture
Orissa	70	70	Vegetables, mango, cashew, banana	Horticultural Development Society (OHDS)
Punjab	75	75	Vegetables, orchard crops	Department of Soil and Water Conversation
Rajasthan	70	60	Groundnut, maize	Department of Horticulture
Tamil Nadu	65	50	Sugarcane, banana, coconut, maize, groundnut	Tamil Nadu Horticultural Development Agency
Uttar Pradesh	50	100	Vegetables and mango, sugarcane	Special Agricultural Department Scheme for Bundlekhand
Uttarakhand	50	50	Potato, groundnut, orchard crops	Department of Horticulture
West Bengal	50	50	Banana, maize, mango	Department of FPI and Horticulture

*Based on data by Raman [7], and field survey.

2.3.3   MICRO IRRIGATION ADOPTION BY VARIOUS FARM CATEGORIES

2.3.3.1   Farm Size and Area Under MI

Table 2.3 reveals that majority of the farmers adopting MI in the case of Kerala state (52 %) are marginal, whereas majority of farmers in Andhra Pradesh (70.67%), Karnataka (66%), Orissa (62.67%) and Punjab (55.34%) are small farmers. Only in the case of Maharashtra (63.33%) and Tamil Nadu (64.67%), majority of the farmers are large farmers (Figure 2.4). Namara [4] reported that majority of the farmers who adopted drip and sprinkler irrigation systems in case of Gujarat and Maharashtra are very rich to rich farmers. Even after providing the much-needed support for

TABLE 2.3   Farm Size and Area Irrigated by MI Systems

State	Farmer category	% of farmers	Average farm size (ha)	Average Area under MI (ha)	% of area under MI
Andhra Pradesh	Marginal	6.00	0.82	0.76	92.68
	Small	70.67	1.7	0.90	52.94
	Large	23.33	14.08	2.96	21.02
Tamil Nadu	Marginal	13.33	0.62	0.48	77.42
	Small	22.00	1.72	1.31	76.16
	Large	64.67	4.67	2.41	51.61
Kerala	Marginal	52.00	0.54	0.15	94.44
	Small	28.00	1.44	1.25	86.80
	Large	20.00	2.38	2.22	93.27
Karnataka	Marginal	6.00	1.89	1.33	70.37
	Small	66.00	5.71	1.82	31.87
	Large	58.00	18.12	6.59	36.37
Maharashtra	Marginal	20.00	1.80	0.90	50.00
	Small	16.67	3.75	2.25	60.00
	Large	63.33	6.60	3.40	51.52
Orissa	Marginal	23.33	0.51	0.07	13.72
	Small	62.67	1.74	1.23	70.44
	Large	14.00	15.52	9.56	61.60
Punjab	Marginal	5.33	0.8	0.40	50.00
	Small	55.34	2.7	1.30	48.15
	Large	39.33	8.2	4.30	52.44
Rajasthan	Marginal	14.00	0.43	0.4	93.02
	Small	35.33	1.16	0.95	81.90
	Large	50.67	3.41	2.54	74.49
Gujarat	Marginal	02.00	0.8	0.58	72.50
	Small	20.67	1.75	1.13	64.57
	Large	77.33	3.65	3.0	82.19

*The experiences of the GGRC in Gujarat indicated that in the recent years more number of small and marginal farmers is adopting the MI [7, and survey data].

promotion of MI, the percentage of area under MI is not remarkable and this has been assessed by farmer category wise in 9 states. Although the return is high under the MI, yet farmers are reluctant to expand the

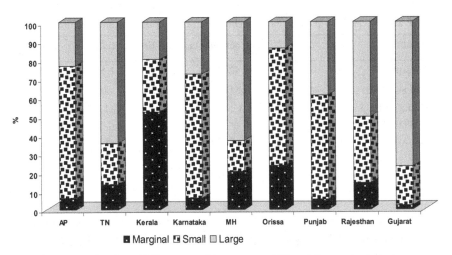

FIGURE 2.4   Adoption of MI system with respect to different farmers' categories.

area due to other constraints like high initial capital cost, lack of technical knowledge in the operation and maintenance of the systems and type of crops grown. The story is same like the SRI adoption, where the SRI results in higher yields and income, but the adoption level is much less due to operating constraints like lack of skilled labor, high management intensity, etc. [6].

2.3.3.2   Relationship Between MI Area and Farmer Categories

The regression results (Table 2.4) show that the coefficients of farm size is significant at 1% level whereas coefficient of dummy variable for small farmers is significant at 10% level and dummy variable for medium farmers is not significant. The average farm size in the 9 states was 0.91 ha, 2.41 ha and 8.51 ha for marginal, small and large farmers, respectively. On the average, each farmer was able to allot about 0.32 ha of every additional ha of land to MI, irrespective of the farm size categories.

2.3.4   *COST AND RETURNS WITH MICRO IRRIGATION*

MI system cost and farmers share after subsidy varied across the farm sizes. It is comparatively lower in the larger farms compared to the other

TABLE 2.4   Relationship Between Area Under MI, Farm Size and Category of Farms

Variables	Coefficients	Std. Error	t-stat	P-value
Intercept	1.4249	0.5686	2.5058	0.0197
Farm size (ha)	0.3152	0.0553	5.6963	0.0000
Dummy variable for marginal farmers (D1)	−1.1491	0.6161	−1.8650	0.0750
Dummy variable for small farmers (D2)	−0.8350	0.5629	−1.4834	0.1515

*Dependent Variable: Area under MI (ha); $R^2 = 0.814$, Adj. $R^2 = 0.7902$.

farms due to economies of scale (Table 2.5). In the case of Kerala, due to intercropping of the wide spaced perennial crops like rubber, coconut, areca nut, the unit cost of the system is comparatively less. In all the states, the quantum of actual subsidy is more than 30%, which is considered less

TABLE 2.5   MI Cost and Returns Across States and Farm Categories

State	Farmer's category	Average total cost of the system (Rs/ha)		Net income (Rs/ha)		IRR (%)	
		Drip	Sprinkler	Drip	Sprinkler	Drip	Sprinkler
Andhra Pradesh	M (9)	71,380	-	15,340	-	16	-
	S (91)	69,794	23282	17,612	6104	25	27
	L (50)	65,373	-	17,112	-	27	-
Tamil Nadu	M (20)	81,302	-	12,842	-	3	-
	S (33)	74,509	-	15,339	-	14	-
	L (97)	66,908	-	26,039	-	60	-
Kerala	M (78)	15,900	-	5310	-	35	-
	S (42)	18,833	-	9217	-	88	-
	L (30)	18,462	-	10,525	-	128	-
Karnataka	M (9)	57,906	-	15,699	-	29	-
	S (99)	56,950	-	15,439	-	29	-
	L (42)	56,553	-	15,331	-	29	-
Maharashtra	M (25)	42,053	-	10,026	-	22	-
	S (20)	48,085	-	13,000	-	29	-
	L (105)	45,400	-	24,360	-	115	-

TABLE 2.5 Continued

State	Farmer's category	Average total cost of the system (Rs/ha)		Net income (Rs/ha)		IRR (%)	
		Drip	Sprinkler	Drip	Sprinkler	Drip	Sprinkler
Orissa	M (15)	95,600	25,800	20,770	15,000	17	138
	S (114)	89,750	22,330	21,515	13,977	22	167
	L (21)	73,800	22,100	16,365	14,667	18	197
Punjab	M (8)	98,456	-	22,000	-	18	-
	S (83)	89,745	57,000	20,000	9500	18	5
	L (59)	86,563	42,000	18,000	9500	15	11
Rajasthan	M (25)	-	-	-	-	-	-
	S (50)		19,736		6500	-	43
	L (75)		11,765		5860	-	98
Gujarat	M (3)	61,795		14,106		19	-
	S (31)	72,482	19,300	19,683	12,617	29	188
	L (116)	73,195	10,512	19,089	10,864	27	410

*S = Small farmer; M = marginal farmer; L = large farmer; IRR = Internal rate of return.

Note: Figures in the parenthesis indicate number of farmers under each farm category.

Source: Survey data.

compared to the announced subsidy percent. Hence this may be one of the reasons for the slow spread of the MI in different states. Even though, MI was able to pay for the MI investment, farmers still expect the subsidy for MI because of following reasons:

a) MI is capital intensive, as it varies from 70,000 to 130,000 Rs per ha depending upon the crops and type of MI systems (drip or sprinkler), and farmers are reluctant to make this investment quickly;

b) Farmer's knowledge in the operation and maintenance of the MI systems is much limited, as often the systems are facing lot of problems of clogging of the filters, drippers; and the required pressure from the pumps is not always maintained due to poor conditions of the pump sets resulting in low pump discharge;

c) Except for wide spaced and commercial crops, the MI is not suitable for all crops and spacings. Except in groundwater overexploited regions, farmers in other regions do not see that MI as an immediate need.

Hence, providing incentives in terms of subsidy helps the farmers to introduce the MI in their farms and save the water.

The internal rate of return (IRR) is also varying across states and farm categories, and it ranges from 3 to 35% in case of marginal farmers, 14 to 88% for small farmers and 15 to 128% for large farmers. The IRR is higher in case of large farmers of Kerala and Maharashtra as they have a diversified inter-cropping pattern in the orchard/ plantation crops, ensuring higher rate of returns. In addition, the plantation crops are widely spaced and cost of investment is low.

The net income earned is significantly different between states (P value = 0.0.0594) at 10% level of significance (Table 2.6). The average additional income varies from Rs 8351 (Kerala) to Rs 20,000 (Punjab). However, the net incomes are not significantly different among the three categories (P value = 0.18128) of farmers. The average additional income due to drip irrigation for a marginal farmer is 14,512 Rs/ha, for a small farmer is 16,476 Rs/ha and for a large farmer is 18,353 Rs/ha.

2.3.5 FARMERS' SUGGESTIONS FOR BETTER ADOPTION OF MICRO IRRIGATION SYSTEMS

Even with the proved benefits and applicability of MI systems under different farm categories, still the overall adoption level is not high. This might be due to other constraints. This chapter further examines the suggestions from farmers and also the policy recommendations at different levels.

The major suggestions include provision of technical support for MI operation after installation, relaxation of farm size limitation in providing MI subsidies, supply of liquid fertilizers, improved marketing facilities, and access to more credit to expand the area under MI. Results indicate

TABLE 2.6 ANOVA for Net Income Under Drip Irrigation Systems

Source of Variation	SS	Df	MS	F	P-value	F critical
Rows	2.8E+08	7	39,958,830	2.618697	0.059385	2.193134
Columns	59023216	2	29,511,608	1.93404	0.18128	2.726468
Error	2.14E+08	14	15,259,048			
Total	5.52E+08	23				

that small farmers from Andhra Pradesh and Punjab, large farmers from Tamil Nadu are in need of more technical support for the adoption and management of MI. Liquid fertilizers are highly requested from Karnataka state. Market facilities of MI systems are also important in the adoption as indicated by farmers in Tamil Nadu and Punjab. At the same time farmers from these two states suggested for the provision of more credit facilities to increase the area under MI (Table 2.7).

2.4 RECOMMENDATIONS AND CONCLUSIONS

Spread of MI in India is widely noticed during 2000–2010. Even after substantial promotional efforts from government and private organizations, the rate of adoption of MI technology is still very low compared to the potential estimations. Only few states like Andhra Pradesh, Maharashtra and Tamil Nadu have expanded the area under MI. The poor adoption can be attributed to number of factors such as high cost of the MI systems, complexity of the technology and other socio-economic issues such as lack of access to credit facilities, fragmented land holdings, localized cropping pattern, etc. Majority of them are large farmers in Tamil Nadu, Maharashtra, Rajasthan and Gujarat states compared to Kerala where marginal farmers got more access to MI. The large farmers have the advantage of economies of scale compared to small and marginal farmers, whose unit cost is comparatively high thus constraining the spread of MI.

Hence reducing the capital cost and increasing the technical knowhow will help the spread of the MI in a bigger way. Keeping this in mind, discussions were held with the MI suppliers, MI experts and farmers to identify the ways and means of reducing the cost of the MI. For example, the International Development Enterprise (IDE) uses the low cost drip and sprinklers to benefit the small holders, where the cost is very low but the life period of the system is also comparatively low. Also they are not coming under the Government subsidy norms as well as under the norms of the Bureau of Indian Standards (BIS) due to the fragile structure. The Jain Irrigation is introducing now the thin walled pipes (Chapin tubes) and also helps in the economic design of the MI systems at farm level, where tubes with varying sizes are used to minimize the cost. The following cost reduction and capacity building options are also important:

Innovations in Micro Irrigation Technology

TABLE 2.7 Suggestions Rendered by the Farmers for Better Adoption and Management of MI System

State	Farmer category	Percentage of farmers surveyed					
		More Technical support	Supply Liquid fertilizers	Providing Marketing facilities	Credit to cover more area under MI	No farm ceiling	Scientific knowledge on crop production
Andhra Pradesh	M (9)	100.00	11.11	0.00	11.11	33.33	0.00
	S (91)	96.70	5.49	0.00	0.00	10.99	6.59
	L (50)	10.00	0.00	2.00	0.00	56.00	0.00
Tamil Nadu	M (20)	90.00	50.00	100.00	100.00	90.00	50.00
	S (33)	90.91	42.42	60.61	96.97	96.97	48.48
	L (97)	92.78	30.93	97.94	97.94	97.94	49.48
Kerala	M (78)	50.00	7.69	29.49	24.36	20.51	7.69
	S (42)	66.67	9.52	30.95	33.33	4.76	9.52
	L (30)	70.00	0.00	10.00	63.33	73.33	3.33
Karnataka	M (9)	11.11	88.89	11.11	66.67	0.00	0.00
	S (99)	5.05	19.19	22.22	44.44	5.05	4.04
	L (42)	4.76	21.43	23.81	50.00	59.52	0.00
Maharashtra	M (25)	20.00	24.00	16.00	32.00	8.00	88.00
	S (20)	25.00	30.00	90.00	100.00	40.00	70.00
	L (105)	5.71	21.90	54.29	53.33	55.24	50.48

TABLE 2.7 Continued

State	Farmer category	Percentage of farmers surveyed					
		More Technical support	Supply Liquid fertilizers	Providing Marketing facilities	Credit to cover more area under MI	No farm ceiling	Scientific knowledge on crop production
Orissa	M (15)	80.00	40.00	40.00	40.00	6.67	0.00
	S (114)	47.37	12.28	32.46	14.91	25.44	7.89
	L (21)	100.00	85.71	66.67	90.48	80.95	9.52
Punjab	M (8)	0.00	0.00	0.00	100.00	0.00	0.00
	S (83)	93.98	0.00	97.59	95.18	30.12	91.57
	L (59)	89.83	38.98	96.61	94.92	54.24	89.83
Rajasthan	M (25)	40.00	0.00	56.00	20.00	0.00	48.00
	S (50)	46.00	60.00	86.00	64.00	24.00	58.00
	L (75)	1.33	20.00	74.67	80.00	64.00	24.00
Gujarat	M (3)	66.67	33.33	0.00	33.33	0.00	66.67
	S (31)	19.35	19.35	25.81	19.35	12.90	38.71
	L (116)	23.28	11.21	18.10	10.34	39.66	37.07

*M = marginal; S = small; L = large farmers. Figures in the parenthesis indicate number of farmers under each farm category.

Source: Survey data.

1. **Field Level** (Affordability by the farmers-possibility for cost reduction)

   There is a good scope of reducing the system cost by slight modifications in the agro-techniques to suit small and medium farms, like paired row planting. Enough orientation needs to be given to the manufacturers/dealers/farmers so that most economic crop specific design can be made. Soil texture should be one of the important parameters in the selection of emitter spacing. This also can reduce the system cost significantly, as presently irrespective of the soil type the dripper spacing adopted is 60 cm and less. There is a need to redesign low cost drip and MI systems to suit the needs of the small and marginal farmers.

2. **State level**

   Many times there is enough time lag between the decision taken about the subsidy percent and actual implementation. Any increase in the raw material prices during this time lag period will reflect on the actual cost of the system thus decreasing the subsidy percent at the end users' level. Hence periodical review of the unit cost is important, as is done in few states.

Discussions with the MI companies and officials also indicated that the differential subsidy pattern for different crops being followed in different regions is disturbing the farmers and the implementing agencies. Hence, it is important to introduce a uniform subsidy across the state. Currently different government departments or agencies are involved in the implementation of the subsidy oriented MI schemes. Due to the variation in the norms of different schemes, which are implemented by different agencies, it is difficult to get the full details as and when required.

One of the major suggestions rendered by the farmers during the study was lack of technical support. In this connection the capacity building of the implementing team is important, which in turn can train the farmers in the use of MI systems including routine operation and maintenance. Fertigation is not done in most of the sample farms. Therefore, fertigation should be adopted in all the MI systems to increase the crop productivity and income. Capacity building units should be encouraged in each region of the state. The recent experiences indicated that in Tamil Nadu state, the introduction of the TNDRIP capacity building program in 2009 covering

100 villages and 1000 farmers has resulted in 17% yield increase and 23% water saving under different crops compared to drip farmers without capacity building activities [2]. Training to unemployed village youths to reduce time lag in installation and for entrepreneurship development is also important. A special autonomous body in each state should be created to handle the MI implementation such as the Gujarat Green Revolution Company (GGRC) Limited.

2.5 SUMMARY

Adoption of micro-irrigation (MI) has resulted in water saving, yield increase and income enhancement at farm level. However, the overall impression is that MI is capital intensive and is suited to large farms. In this context, a study was undertaken in 9 states of India during 2010, mainly to examine the actual area covered compared to the potential area and to understand the adoption level of MI as well as to analyze the cost and returns under different farm categories.

The results indicated that only about 9% of the MI potential is covered in the country (i.e., 12.2% under drip and 7.8% under sprinkler) with variation among the states. Majority of the MI farmers are small in the states of Andhra Pradesh, Karnataka, Orissa and Punjab whereas in case of Maharashtra and Tamil Nadu state, majority are large farmers. Analysis of the rate of return in MI investment across different states and different farm categories indicated that there is no significant difference in net come due to MI among the farm categories but it varied among the States. Key suggestions include: reduction in capital cost of the system, provision of technical support for MI operation after installation, relaxation of farm size limitation in providing MI subsidies, and single state level agency for implementation of the MI program.

ACKNOWLEDGEMENTS

The authors wish to thank the scientists from the nine Indian States, who did the field survey and provided the cost and returns under micro-irrigation. Dr. Ranganathan helped in the data analysis.

KEYWORDS

- analysis of variance
- Centrally Sponsored Scheme (CSS)
- cost and returns
- cropping pattern
- discount rate
- drip irrigation
- emitter spacing
- farm level constraints
- farm size
- farm size category
- field survey
- food security
- India
- Indian National Committee on Irrigation and Drainage
- Indian States
- internal rate of return
- International Water Management Institute, IWMI Sri Lanka
- irrigation potential
- large farmers
- marginal farmers
- MI adoption
- micro irrigation, MI
- mini-sprinkler
- National Committee on Plasticulture Application in Horticulture
- net income
- potential MI area
- potentiality of MI
- small farmers
- sprinkler irrigation
- SRI adoption
- subsidy

- **unit cost**
- **water**
- **water demand**
- **water management**
- **water resources**
- **water use efficiency, WUE**
- **watershed**

REFERENCES

1. INCID, 1994. *Drip Irrigation in India*. Indian National Committee on Irrigation and Drainage, Government of India. New Delhi.
2. IWMI – Tata Water Policy Program, 2011. *TNDRIP Capacity Building Program*. Evaluation Report 1/2011. International Water Management Institute, Hyderabad.
3. Molden, D. R., R. Sakthivadivel, and Z. Habib, 2001. *Basin-level Use and Productivity of Water: Examples from South Asia*. IWMI Research Report 49, International Water Management Institute (IWMI), Colombo, Sri Lanka.
4. Namara, R. E., B. Upadhyaya, and R. K. Nagar, 2005. *Adoption and Impacts of Micro-irrigation Technologies: Empirical Results from Selected Localities of Maharashtra and Gujarat of India*. Research Report 93. International Water Management Institute, Colombo, Sri Lanka.
5. NCPAH, 2009. *Evaluation Study of Centrally Sponsored Scheme on Micro-irrigation*. National Committee on Plasticulture Application in Horticulture (NCPAH), Ministry of Agriculture, Department of Agriculture and Cooperation. New Delhi.
6. Palanisami, K., and R. Karunakaran, 2010. Adoption of SRI under different irrigation sources and farm size categories in Tamil Nadu, India. Unpublished Paper IWMI-TATA Water Policy Program, South Asia Regional Office, Hyderabad, India.
7. Raman, S., 2010. State-wise micro-irrigation potential in India – an assessment. Unpublished paper, Natural Resources Management Institute, Mumbai, India.
8. Saleth, R. M., 1996. *Water Institutions in India: Economics, Law and Policy*. Commonwealth Publishers, New Delhi.
9. Seckler, D., U. Amarasinghe, D. Molden, D. Radhika and R. Barker, 1998. *World Water Demand and Supply, 1990 to 2025: Scenarios and Issues*. Research Report 19, International Water Management Institute, Colombo, Sri Lanka.
10. Sivanappan, R. K. 1994. Prospects of micro-irrigation in India. *Irrigation and Drainage Systems,* 8(1):49–58.
11. Suresh, Kumar, 2008. Promoting drip irrigation: where and why? Managing water in the face of growing scarcity, inequity and declining returns: exploring fresh approaches. IWMI TATA 7th Annual Partner Meeting Report, 1:108–120.
12. Vaidyanathan, A., 1999. *Water Resources Management: Institutions and Irrigation Development in India*. Oxford University Press, New Delhi.

CHAPTER 3

ENHANCING THE CROP YIELD THROUGH CAPACITY BUILDING PROGRAMS: APPLICATION OF DOUBLE DIFFERENCE METHOD FOR EVALUATION OF DRIP CAPACITY BUILDING PROGRAM IN TAMIL NADU STATE, INDIA

KUPPANNAN PALANISAMI, RAMA RAO RANGANATHAN,
DEVARAJULU SURESH KUMAR, and
RAVINDER PAUL SINGH MALIK

CONTENTS

In this chapter, one US $ = 63.02 Rs. (Indian rupees).

Modified and printed from "K. Palanisami, R. Ranganathan, D. Suresh Kumar, and R. P. S. Malik, 2014. Enhancing the crop yield through capacity building programs: Application of double difference method for evaluation of drip capacity building program in Tamil Nadu State, India. Open source article in Agricultural Sciences, 5(1):33–42".

3.1 INTRODUCTION

Due to water scarcity, water has limited agricultural development in many developing and developed countries across the world, because of water scarcity. By the year 2025, only 50% of the increase in demand for water can be met by increasing the effective use of irrigation water [16]. In India, almost all the easily economically viable irrigation water potential has already been developed, but the demand for water for agriculture has been growing continuously [15, 24]. The agricultural sector still consumes over 80% of water. Moreover, the water use efficiency (WUE) is only 30–40%, indicating that there is considerable scope for improving WUE. The solution to the problems of growing water scarcity and persistent degradation of water resource across regions is two-folds:

- Firstly, the supply-side management practices, such as: watershed development, and water resources development through major, medium and minor irrigation projects.
- Secondly, the demand management by efficient use of the available water both in the short-term and long-term perspectives.

Recognizing the importance of sustainable WUE in agriculture, a number of demand management strategies (like water pricing, water user's association, turnover system, etc.) have been introduced since the late seventies to increase the WUE especially in the use of surface irrigation systems. The adoption of micro irrigation technologies, such as drip and sprinkler methods of irrigation, is one of the important demand management strategies that must be vigorously promoted. The impact of drip irrigation to improve WUE has been studied by many investigators [7, 10, 11, 12, 22]. These studies have shown that the properly designed and managed drip irrigation system can increase the WUE up to 100% [4,18]. Drip method of irrigation helps to reduce the over-exploitation of groundwater. Environmental problems (like water logging and salinity) associated with the conventional irrigation are completely absent in drip irrigation [9].

Drip irrigation also helps in achieving reduction in tillage requirement, higher quality crop products, increased crop yields and higher fertilizer use efficiency [8, 14,19, 25]. In addition to these benefits, the drip irrigation generates substantial social benefits [23] in the forms of enhanced food security, women participation in agriculture (www.ide-india.org) and social status [3, 17].

Besides these potential benefits, the adoption of drip irrigation is yet need to be promoted across various Indian states. Several factors – such as huge initial investment, small size of holding, lack of technical support, nature of cropping pattern, access to water and socio-economic conditions of farmers, etc. – are major factors that affect adoption of drip irrigation [1, 2, 8,21]. In some cases, even after the adoption of drip irrigation, particularly small farmers have often discontinued the use of drip irrigation for several reasons such as: lack of proper maintenance, changes in crop pattern and unreliable water supply [6]. A study in Coimbatore district of Tamil Nadu State of India has shown that huge initial investment and small size of holding are major constraints limiting the adoption of drip technology (100% of farmers), lack of access to subsidy (46.9% of famers) and lack of technical support for follow-up actions (28.9% of farmers) [20].

To maximize adoption rate of drip irrigation by farmers, the International Water Management Institute (IWMI) under its IWMI-TATA Water Policy Program initiated a capacity building program in certain regions of Tamil Nadu in India.

This chapter discusses the impact of the capacity development program mainly on use of drip irrigation in improving crop yields. The following hypotheses were tested in this study:

- drip irrigation based on capacity building program (CBP) will enhance the crop yield compared to drip irrigation alone.
- appropriate evaluation methods will capture impact of the CBP.

3.2   CAPACITY BUILDING PROGRAM IN DRIP IRRIGATION

Realizing the significance of drip maintenance practices, drip irriga-tion capacity building and management initiative for maximizing pro-ductivity and income (TNDRIP) was developed and implemented in Coimbatore district, Tamil Nadu state of India, by International Water Management Institute (IWMI) under the IWMI-TATA Water Policy Program jointly with the Water Technology Centre (WTC) of the Tamil Nadu Agricultural University (TNAU) and Jain Irrigation Systems Limited (JISL), during 2009–2010 [20]. The objective of TNDRIP was to sustain the drip irrigation system through increased adoption and proper maintenance practices by the farmers in order to achieve increased crop yields.

The program was implemented in 100 villages covering about 1000 farmers who have adopted the drip irrigation already for various crops. A base line survey was conducted to know the current level of use of drip irrigation, inputs used and crop yields. The main part of the capacity build-ing program was one day training, which was organized in each village for farmers with different crops and farm sizes. The field-based training was given by the field technicians of the WTC and JISL through lectures, hand-on exercises with the drip systems, demonstrations of the various drip materials, and question-answer sessions.

The contents of the training mainly focused on the operation and maintenance practices of drip irrigation system besides fertigation and irrigation scheduling practices. In addition, the farmers were given hand-outs and booklets (in local language) regarding fertigation, drip demon-strations and cleaning (sand filters, PVC pipes, sub-mains and laterals, screens and disc filters, emitters) drip system using acid. The farmers were able to learn the drip and fertigation technologies and their periodi-cal maintenance.

3.3   METHODOLOGY

3.3.1   IMPACT OF TECHNOLOGY ON CROP YIELD AND IMPACT EVALUATION OF THE TECHNOLOGY

Technology is nothing but the application of improved knowledge on production relationships. Therefore, technology has the effect of raising the production function. More output per unit input (water) is possible with the new technology such as drip irrigation. This indicates that production can be increased with improved technology through the same amount of inputs that were used with traditional technology or the current production level can be reached with fewer inputs with improved technology.

In Figure 3.1, the curve AA refers the traditional irrigation technology production function, curve AB refers to the improved (drip) technology production function, and curve AC refers to the improved drip technology with capacity building program. With X units of water: traditional technology produces Y1 units of output, improved (drip) technology produces Y2 units of output, whereas improved technology with CBP produces Y3 units of output. The difference between Y2 and Y3 is the additional output due to capacity building program.

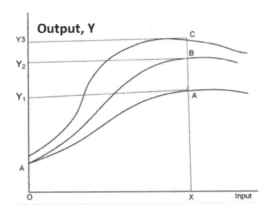

FIGURE 3.1   Effect of adoption of technology on crop yield (X – axis: Input, Y – axis: output; *Note:* AA refers the production function with traditional irrigation technology; AB refers the production function with improved (drip) technology; AC refers the production function with improved (drip) technology along with CBP).

3.3.2 IMPACT EVALUATION USING DOUBLE DIFFERENCE METHOD (DDM)

Several tools or approaches are used for impact evaluation. The most commonly used tools are the financial measures like the benefit–cost (B-C) ratio, and internal rate of return (IRR) [11, 19, 21, 22]. The major problems with this approach is that the benefits and costs are calculated using either before and after or with and without concept which ignores some of the benefits that are considered as residual, which may occur even without the intervention such as drip irrigation. Hence an approach, which considers both with and without as well as before and after situations, is important. The approach of any analysis of impact can be accomplished into two ways:

1. "with project" parameters compared to the "pre-project" situation gives the incremental benefits due to the project. But these increments in the parameters intrinsically include the changes due to state of art of technology. Thus sometimes, the benefits may be exaggerated.
2. the literature on project analysis unanimously suggests the use of comparison between the "project parameters" with the "non-project control region". This method automatically incorporates the correction for the impact of technology in the absence of the project.

For the present study, the information was collected for the pre- and post-project period and was compared with the control as well. Hence, the approach is a combination of both "with and without" and "before and after approaches", and this approach is termed as double difference method (DDM). The DDM is described in Table 3.1. For more details, the reader may consult , *"Goyal, M. R., ed., 2015. Management, Performance, and Applications of Micro Irrigation Systems. Chapters 1 and 2, pages 1 to 32. Oakville, Canada: Apple Academic Press Inc.,"*

Farm level data were collected from both types of drip farmers i.e. who have participated in the CBP and who have not participated in CBP. This enabled the use of the double difference method to study the impact of the drip capacity building program. The frame work was adopted from the program evaluation literature [5, 13].

TABLE 3.1 Double Difference Method of Impact Assessment of Drip Capacity Building Program

Particulars	Groups		
	Drip Participants	Non-participants	Difference across groups
After drip training	D1	C1	D1–C1
Before drip training	D0	C0	D0–C0
Difference across time	D1–D0	C1–C0	**Double difference** [(D1–C1)–(D0–C0)]

The resulting measures can be interpreted as the expected effect of implementing the drip CBP. In Table 3.1, the columns distinguish between groups with and without the program and the rows distinguish between before and after the program. Before the CBP, one would expect the average yield of different crops be similar for the two groups, so that the quantity (D0 – C0) would be close to zero. However, once the CBP has been implemented, one would expect differences between the groups as a result of the improvement in knowledge of the farmers about the drip maintenance, fertigation and irrigation scheduling due to the CBP. The impact of the pro- gram, however, would be better assessed considering any pre-existing observable or unobservable differences between the two randomly assigned groups. The double difference estimate was obtained by subtracting the preexisting differences between the groups (D0 – C0), from the difference after the CBP has been implemented, (D1 – C1) [23]. This is best explained in Figure 3.2.

Double Difference (DD) methodology is becoming a popular tool for studying the impact analysis, as it has the advantage to control for the time-invariant characteristics of farmers when comparing adopters and non-adopters of a technology or a CBP. In this methodology, the average impact of a capacity building program is computed by the following formula:

$$\text{Double Difference (DD)} = E(Y_1^T - Y_0^T \text{ for } T_1 = 1) - E(Y_1^C - Y_0^C \text{ for } T_1 = 0)$$

(1)

where, Y_1^T and Y_1^T denote the outcome responses for the trained and control groups at period $t(= 0, 1)$, respectively. The time period t = 0 corresponds

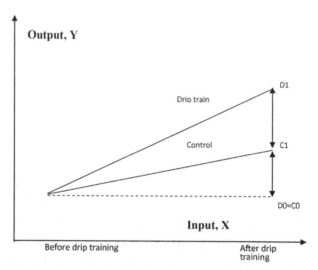

FIGURE 3.2 Impact of capacity building program by double difference method: Y-axis = Output variable; X-axis = Input variable (CBP).

to the period before the implementation of CBP and the period $t = 1$ corresponds to the period after the implementation of CBP. Further, $T_1 = 1$ implies presence of the program at time $t = 1$ and $T_1 = 0$ means absence of the program. The first term in Eq. (1) represents the average difference between before-after for the trained group and hence it is given by Eq. (2).

$$E(Y_1^T - Y_0^T \ for \ T_1 = 1) = \{1/N_T\}\sum_{i\varepsilon T}(Yi1 - Yi0) = \overline{YT1} - \overline{YT0} \qquad (2)$$

Similarly for the control group, the second term is given by:

$$E(Y_1^C - Y_0^C \ for \ T_1 = 0) = \{1/N_C\}\sum_{j\varepsilon Tc}(Yj1 - Yj0) = \overline{Yc1} - \overline{Yc0} \qquad (3)$$

Substituting these values in (1), the impact of the program can be shown as:

$$Impact = \left(\overline{YT1} - \overline{YT0}\right) - \left(\overline{Yc1} - \overline{Yc0}\right) \qquad (4)$$

The same results can be obtained by following regression approach. For each observation i, let us define a variable δ_i as $\delta_i = 0$ if the observation is from the control group; and $\delta_i = 1$ if it is from the trained group. Similarly for each observation i, let us define a variable Ti as $Ti = 0$ if the observation belongs to time $t = 0$, that is before the drip capacity building program, and $Ti = 0$ if the observation belongs to time $t = 1$, that is, after the program. Now form of the regression equation is:

$$y_i = a + b\delta_i + cT_i + (d)\delta_i(Ti) \qquad (5)$$

The following results can be easily checked in the Eq. (5):

Observation belongs to	Values of variables in Eq. (5)		
	δ	T	y_i
Control group before the program.	0	0	$\bar{y}_{C0} = a$
Control group after the program.	0	1	$\bar{y}_{C1} = a + c$
Trained group before the program.	1	0	$\bar{y}_{T0} = a + b$
Trained group after the program.	1	1	$\bar{y}_{T1} = a + b + c + d$

Therefore, using Eq. (4), we have Eq. (6):

$$\text{Impact of the program} = [\{(a + b + c + d) - (a + b)\} \\ - \{(a + c) - a\}] = d \qquad (6)$$

3.4 DATA AND VARIABLES

Out of the targeted 1000 farmers for the training, only 800 farmers actively participated in the training. Out of this, 500 farmers were selected employing random sampling procedure to study the impact of the TNDRIP during November 2011, one year after the program was introduced. In order to make a comparative study, 250 drip farmers in these villages, who have not participated in the capacity building program, were selected as control. Thus, a sample of 750 farmers was covered for the impact study. The needed information from the respondents was gathered personally administering the interview schedule. The primary information collected from the farmers included mainly: the farm size, cost of wells, cost of drip irrigation system, crops grown, maintenance costs, groundwater use, crop production including inputs used and crop yields, adoption of drip irriga-tion practices, and the constraints in using the drip irrigation. In addition, the details of the trainings attended and subject matter learnt during the training program were collected from the respondents. The base line data collected in 2009 was also used for cross checking of the inputs used and crop yields of the farmers prior to training.

3.5   RESULTS AND DISCUSSIONS

3.5.1   GENERAL CHARACTERISTICS OF THE FARM HOUSEHOLDS

Knowledge on general profile of the sample respondents will help us to understand better the impact of the capacity building program. The general characteristics were landholdings, cropped area, and irrigated area for the participants of the drip capacity building program, and these character-istics were compared with that of control farmers. The average size of holding of the drip-farmers with training was significantly higher as com-pared to control farmers. The results indicated that mostly large farmers had attended the CBP compared to small and marginal farmers, who did not took much interest due to their limited area under drip irrigation and hence the average farm size was comparatively high under participants of the drip training. However, cropping intensity and irrigation intensity are more or less same among the participants of drip training and control farmers (Table 3.2).

3.5.2   AWARENESS AND ADOPTION OF DRIP MAINTENANCE

The capacity building program has created adequate awareness about vari-ous drip management practices (Table 3.3). The perception of the trainees about the important drip management and maintenance activities before their participation in the training revealed that most of them were aware about the fertigation practice (48%) followed by the cleaning of screen/disc filters (42.7%). Many farmers (5 to 21%) were aware about the prac-tices like cleaning the sub-mains and laterals, acid treatment, protecting

TABLE 3.2   General Characteristics of the Sample Farmers

Characteristics	Units	Participants of drip training (N = 500)	Control (N = 250)
Average farm size	ha	3.41***	2.58
Net sown area	ha	3.31***	2.43
Gross cropped area	ha	3.86***	2.79
Cropping intensity [a]	%	116.6***	114.8
Net irrigated area	ha	2.97***	2.18
Gross irrigated area	ha	3.26***	2.38
Irrigation intensity [b]	%	109.8***	109.2
Percentage of area irrigated by wells to the total cropped area	%	84.5	85.3
Percentage of area irrigated under drip to gross cropped area	%	67.1	64.8
Percentage of area irrigated under drip to gross irrigated area	%	79.5	75.9

Source: Field survey.

Notes: ***indicates, values are significantly different at 1% levels from the corresponding values of control farmers.

[a]Cropping intensity is defined as the ratio of gross cropped area to net sown area and is expressed as a percentage;

[b]Irrigation intensity is the ratio of gross irrigated area to net irrigated area and is expressed as a percentage.

the drip system, valve protection and placing the laterals under shade. However, no one was aware about the important practices like cleaning the sand filter, pressure regulation in the laterals and irrigation scheduling to various crops before attending the training program.

After undergoing the training program, all the trainees were aware about all important practices and started adopting these practices. Hence, it can be inferred that the TNDRIP training has remained as the only source to provide first-hand information and created awareness about the maintenance practices, such as: pressure regulation, irrigation scheduling, acid treatment and advantages of placing the laterals under shade. However, only one-third of the trainees were aware of the practice of cleaning the sand filter even after undergoing the training. The overall observation indicated that the TNDRIP training made significant impact among farmers in adopting the drip maintenance practices in a better manner than before.

TABLE 3.3   Impact of Capacity Building on Farmers' Awareness and Adoption of Various Practices

Particulars	Awareness before training*		Awareness and adoption after training	
	No. of farmers	Percentage	No. of farmers	Percentage
Fertigation	72	48.0	78	52.0
Cleaning screen/disc filter	64	42.7	86	57.3
Cleaning sand filter	–	–	50	33.3
Cleaning sub-main	22	14.7	128	85.3
Cleaning laterals	22	14.7	128	85.3
Acid treatment	10	6.7	140	93.3
Pressure regulation	–	–	150	100.0
Thatching the drip system	32	21.3	118	78.7
Valve protection	18	12	132	88.0
Laterals in shade	8	5.33	142	94.7
Irrigation scheduling	–	–	150	100.0

*Most of the farmers were aware of these practices when the system was installed in their farms; subsequently due to lack of skills and knowledge in handling these practices, most of the farmers were not adopting these practices properly.

Source: survey.

3.5.3   BENEFITS PERCEIVED BY THE FARMERS DUE TO THE ADOPTION

The adopters of drip maintenance practices were enquired about the overall benefits they perceived due to the adoption of the drip maintenance practices in their farm. They were asked to rank their responses. The adopted farmers had altogether perceived and ranked six benefits due to the adoption of the maintenance practices taught in the training (Table 3.4). Among these, the top three benefits that were perceived and ranked were the reduction in the duration of irrigation, improved dripper discharge and achievement of uniform distribution of water in the field. Garret score in Table 3.4 was calculated as follows [25]:

$$\text{Percent position} = 100 \times [(R_{ij} - 0.5)/(N_j)] \qquad (7)$$

TABLE 3.4 Farmers' Perception About the Benefits Due to Drip Maintenance

Benefits	Garret score*	Rank
Reduction in the duration of irrigation	63	I
Improved dripper discharge	59	II
Uniform distribution of water in field	56	III
Uniform growth of plants in field	51	IV
Improvement in yield	49	V
Extended life of drip system	39	VI

*Garret scores [25], were worked out using the Eq. (7).

where, R_{ij} = Rank given for the ith factor by the jth respondent, and N_j = Number of factors ranked by the jth respondent.

The benefit namely, uniform growth of plants in the field was perceived and ranked as fourth by the adopters. All these four benefits were interlinked due to the better maintenance practices adopted by the farmers. The fact that was endorsed by majority of the adopters was that there was reduction in the duration of irrigation to an extent of up to 30% than it was observed without the adoption of the maintenance practices. Hence, this particular benefit was ranked as first by the adopters. The regular cleaning and maintenance of filters, sub-mains and laterals might rendered free flow of water through the drip system effectively. The adopters used pressure gauge to regulate the pressure in the sub-mains and laterals as per the recommendations. This particular action resulted in ensuring uniform distribution of water throughout the field thus promoting uniform growth of plants also. These benefits were observable and hence most of the farmers had perceived such benefits. The other two benefits that were perceived and ranked in the last order by the adopters were improvement in crop yield and extended life expected over the drip system due to adoption of regular maintenance practices. The adopters recorded an increase in yield up to 10% across various crops. Although this change in yield was not significant, it was perceived by some farmers and hence they ranked it fifth. Anticipation over the extended life of the drip system was perceived and ranked last, as it was not observed by the adopters but still they believed the benefit.

Sharing or recommending the learnt technologies to peers is considered as a social impact indicator of the TNDRIP project. Hence the information sharing behavior of the farmers was studied. Of the total number of 500 farmers studied, 70% of the farmers shared their information to others and 30% farmers did not share their information. Among the farmers who shared their information, 52.38% of farmers shared to their own family members and the remaining 47.62% of farmers shared to other fellow farmers (Table 3.5). This showed that the penetration of the drip fertigation technologies, their maintenance and management will be possible through capacity building programs.

TABLE 3.5 Crop Yields Under Different Farmer Samples

Type of samples	Number of farmers	Mean yield (t/ha)	Minimum yield (t/ha)	Maximum yield (t/ha)	Std. dev.
Banana 1					
Control-Before	172	45.9	37.1	57.3	3.8
Control-After	172	47.7	38.5	66.7	4.6
Trained-Before	172	65.2	55.6	85.0	3.6
Trained-After	172	69.6	55.3	94.2	4.6
Banana 2					
Control-Before	93	22.3	17.3	59.3	4.4
Control-After	93	25.0	18.0	31.9	24.2
Trained-Before	93	32.5	29.1	79.0	5.2
Trained-After	93	37.1	29.9	42.4	10.0
Sugarcane					
Control-Before	198	116.3	98.8	143.3	10.3
Control-After	198	120.0	98.8	145.7	9.9
Trained-Before	198	153.6	123.5	182.8	15.2
Trained-After	198	160.3	130.9	192.7	15.5
Turmeric					
Control-Before	91	6.3	5.2	8.3	0.8
Control-After	91	6.4	5.2	9.4	0.9
Trained-Before	91	8.7	6.9	11.4	1.4
Trained-After	91	9.0	7.1	13.6	1.4

Note: Banana-1: Variety robusta; Banana-2: Variety Nendran.

3.5.4 IMPACT OF CAPACITY BUILDING PROGRAM

Authors examined whether the capacity building program (CBP) has generated any impact on crop yield. The CBP not only created adequate knowledge on drip maintenance activities but also on irrigation scheduling and fertigation, which helped to increase crop productivity. The details of the mean yield of crops under different farmer samples indicated that crop yields were comparatively high for farmers, who were trained in drip system operation and maintenance (Table 3.6).

In the present study, the double difference method of impact assessment was employed to assess the impact due to the CBP on crop yield (Table 3.6). The two types of respondents were compared and assessed the net impacts due to CBP. For instance, the yield of banana-1 was 45.9 t/ha for the control farmers before the training period. The yield of Banana-1 for the same set of farmers after the training period was 47.7 t/ha. Therefore, there was 1.8 t/ha of increase in yield of Banana-1 even among the farmers, who did not participate in the CBP.

This increase in yield might be due to the experience, cumulative knowledge gained by the farmers, use of better quality of inputs, technological growth, and so on. Similarly, the above yields for the trained farmers (i.e., farmers who have participated in the CBP) were 65.3 t/ha and 69.6 t/ha respectively, leading to a difference of 4.3 t/ha. However, interest of authors was to assess mainly the yield increase attributed to the CBP. The double

TABLE 3.6 Impact of Capacity Building Intervention in Drip Irrigation

Observation belongs to	Banana 1	Banana 2	Sugarcane	Turmeric
Control group before the program (= a)	45.9	22.3	116.4	6.3
Control group after the program (= a + c)	47.7	32.5	153.6	8.7
Trained group before the program (= a + b)	65.3	25.0	119.9	6.4
Trained group after the program (= a + b + c + d)	69.6	37.1	160.5	9.0
Net impact due to capacity building intervention (=d)	2.5	1.8	3.3	0.3

difference method captured this yield increase as 2.5 t/ha (= 4.3 -1.8, t/ha). Similarly, the yield increase due to the training was calculated for other crops, as shown in Table 3.6.

The results for the double-difference method (DDM) using the regression analysis are presented in Table 3.7. It is seen that the drip CBP had significant impact on yield of Banana-1 and sugarcane. The technological growth indicated by time (T) had significant impact on crop yield in all the crops as evident from the coefficients of "T", which were significant for all crops. For instance, the results of Banana-1 estimates had shown interesting results. The adjusted R^2 was 0.86 indicating 86% of the variations were explained by the explanatory variables. The intercept term indicated the mean yield of the control farmers i.e. the yield for control farmers. It is evident that there was significant difference between yields in case of trained and control farmers in the base period. Similarly, there was a significant increase in yield due to time period among the control farmers. It is evident that 1.8 t/ha increase in yield was realized over time period among the control farmers. The impact of CBP was significant on the expected positive line, which showed that the CBP alone increased the crop yield by 2.5 t/ha.

TABLE 3.7 Regression Analysis for the Impact of Capacity Building Program on Crop Yield

Crops	Constant	δ	T	T × δ	Adjusted R-Squared
Banana-1	45.887	19.366***	1.847***	2.542***	0.86
	(144.75)	(43.19)	(4.12)	(4.01)	
Banana-2	22.274	2.750	10.267***	1.767	0.15
	(15.88)	(1.387)	(5.178)	(0.630)	
Sugarcane	116.404	3.574***	37.219***	3.285*	0.69
	(126.24)	(2.74)	(28.54)	(1.78)	
Turmeric	6.298	0.119	2.365***	0.265	0.53
	(51.04)	(0.68)	(13.55)	(1.07)	

Note: Figures in brackets show estimated "t" ratios.
***Significant at 1% level; *Significant at 10% level.

3.5.5 NOVELTY AND MERIT OF THE DOUBLE DIFFERENCE METHOD

The novelty of the double difference method (DDM) is that it is simple and appropriate method in capturing the yield increase due to capacity building program alone. The DDM estimated the impact of the drip CBP in terms of yield increase as 2.5 t/ha for Banana-1, 1.9 t/ha for Banana-2, 3.3 t/ha for sugarcane and 0.3 t/ha for turmeric. In the absence of the DDM, the impact of the drip CBP in terms of yield increase was 4.3 t/ha for Banana-1, 12.1 t/ha for Banana-2, 40.6 t/ha for sugarcane and 2.6 t/ha for turmeric (Table 3.8 and Figure 3.3).

Thus, the conventional approach is highly up-ward biased by overestimating the impact of the CBP. Thus the merit of the DDM is very clear from the results of this section. The results further confirmed the hypotheses that drip CBP will increase the crop yield and the use of appropriate quantification method (DDM) will capture the exact impact of the CBP. The results of the DDM further enhanced the richness of measurement methodologies in impact evaluation. Hence future impact studies can find this an interesting and valuable tool in impact evaluation of the technologies where time period is a major player along with the technologies. For example, in several cases, farmers were able to increase the crop yield over time due to better management. Thus the approach can help in decomposing the yield increase due to capacity building intervention and time period (management) as well.

TABLE 3.8 Comparison of Crop Yields Under Drip Without Training and Drip with Capacity Building Program

Crop	Increase in yield of drip farmers without training (t/ha)	Increase in yield of drip farmers with training (t/ha)	Increase in yield due to training alone (t/ha)
Banana 1	1.8	4.3	2.5
Banana 2	10.3	12.1	1.8
Sugarcane	37.2	40.5	3.3
Turmeric	2.4	2.7	0.3

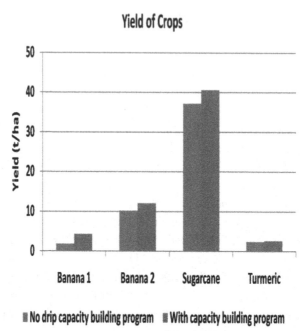

FIGURE 3.3   Impact of drip capacity building program on crop yield (t/ha).

3.5.6   *CONSTRAINTS IN ADOPTION OF DRIP MAINTENANCE ACTIVITIES*

During the field survey in this study, it was found that there are several factors that limited the adoption of maintenance practices. One of the constraints for adoption of the maintenance practices was the insufficient knowledge about the relevance of pressure regulation in the drip irrigation system. Farmers were not aware about the use of pressure gauge in the laterals to maintain uniform pressure so that the water can flow through the system effectively. The farmers believed that use of hydrochloric acid treatment in the drip system could affect the soil health and crop growth and hence feared to adopt the acid treatment technology for removing the blockage/salt encrustation in the drip system. Most of the farmers also felt that they were ignorant about the maintenance package to be adopted for the drip system, which acted as one of the constraints for its adoption. Hence the lack of knowledge and understanding of the farmers about the drip and its maintenance practices itself remained as constraints for its adoption (Table 3.9).

Farmers were asked to express the factors that remained as constraints to the adoption of the maintenance practices. The findings revealed that a majority of the trained farmers (80%) expressed high cost of the water soluble fertilizers as the major constraint. About half of them revealed that non-availability of pressure gauge was one of the constraints. It is a fact that pressure gauge apparatus was not sold commonly in the shops located in rural/town areas. This was perceived as one of the reasons for non-adoption of the maintenance practices. Some farmers (12%) were reluctant to invest in the venturi unit, which is needed for the adoption of the fertigation and acid treatment as well. About 41% of them stated that the non-availability of water soluble fertilizers in the local village/town market was as one of the reasons for non-adoption of fertigation practice, even though they were supplied with the fertilizer tank on subsidy basis by the Government of Tamil Nadu. One-fourth of the farmers (24%) expected the assistance of the extension or development staff to adopt acid treatment technology in their farms as they were not very confident in adopting the technology on their own. These constraints need to be addressed in the future training programs.

TABLE 3.9   Constraints in Adoption of the Drip Maintenance Activities

Before attending the training	Percent of farmers	After attending the training	Percent of farmers
Insufficient knowledge about pressure maintenance	94.70	High cost of water soluble fertilizers	80.00
Fear of acid treatment	90.00	Non-availability of pressure gauge	50.66
Lack of technical know-how of maintenance practices	89.30	Reluctance to invest on venturi	12.00
Misunderstanding of farmers about drip irrigation technology	88.00	Non-availability of water soluble fertilizers locally	41.33
		Lack of confidence to use correct concentration of acid	24.00

3.6   CONCLUSIONS AND RECOMMENDATIONS

TNDRIP training has brought out significant impact among farmers to adopt the drip maintenance practices in varied proportions. Practices such as cleaning the filters, sub-mains and laterals, pressure regulation, acid treatment, thatching the drip system, placing the laterals in shade etc. were adopted by a majority of the trainees as a result of their participation in the training. The study revealed that a majority of the farmers (70%) had shared the information about the drip maintenance practices to their family members and other fellow farmers.

It is found that the drip capacity building program (CBP) produced significant impact on yield of crops such as banana and sugarcane. Drip farmers with training were able to increase the crop yield by 138% of Banana-1, 19% for Banana-2, 9% for sugarcane and 13% for turmeric crops indicating that compared to drip irrigation alone, drip with CBP helped them to exploit the full potential of drip irrigation. Compared to the cost of the training (Rs 110/ha), the additional yield has resulted in an additional per ha gross margin of Rs. 14,000, Rs. 32,015, Rs. 1650, and Rs. 14,893 per year for Banana-1, Banana-2, Sugarcane and turmeric, respectively. Given this cost effectiveness of the CBP, it is suggested to introduce more number of drip CBP across regions so as to achieve more crop per drop of water. The development departments such as Department of Agriculture, Horticulture, and Agricultural Engineering may be geared up to follow up with similar kind of CBPs across the regions. This will further have multiplier impact in terms of bringing new farmers under drip irrigation as well as effective use of fertilizers. By establishing such training programs in the rural areas, local expertise in terms of entrepreneur development through public private partnership can also be developed.

3.7   SUMMARY

A capacity building program (CBM) for drip irrigation (TNDRIP) was undertaken in certain regions of the Indian State of Tamil Nadu during 2009–2010. An assessment of the impact of the program in terms of effective use of drip irrigation and increased crop yields was made in 2011 by applying double difference method (a combination of both with and without and before and after approaches). The results indicated that the drip capacity building program resulted in a yield increase of 2.5 t/ha for Banana-1, 1.9 t/ha for Banana-2, 3.3 t/ha for sugarcane and 0.3 t/ha for turmeric. The conventional method using the before and after situations showed a yield increase of 4.3 t/ha for Banana-1, 12.1 t/ha for Banana-2, 40.6 t/ha for sugarcane and 2.6 t/ha for turmeric. The conventional approach is highly upward biased in estimating the impact of the drip capacity building program and thus the double difference method will be an appropriate method to evaluate the impact of the programs that involve with and without as well as before and after situations.

ACKNOWLEDGEMENTS

This study was conducted under the IWMI– Tata Water Policy Program supported by International Water Management Institute (IWMI) and Sir Ratan Tata Trust (SRTT). Authors acknowledge the support provided by project partners: Dr. S. Chellamuthu, Director and Dr. C. Mayilswami, Professor of the Water Technology Centre of Tamil Nadu Agricultural University, and Dr. S. Narayanan, Vice President of the Jain Irrigation Systems, Coimbatore.

KEYWORDS

- agricultural sector
- assessment
- banana
- capacity building program
- crop pattern
- crop yield
- double difference method
- drip capacity
- drip irrigation

- drip technology
- fertilizer use efficiency
- food security
- impact evaluation
- India
- International Water Management Institute
- irrigation
- Jain Irrigation
- maintenance practices
- salinity
- size of holding
- small farmers
- social impacts
- subsidy
- Tamil Nadu
- Tamil Nadu Agricultural University
- Tata Water Policy Program
- TNDRIP
- water logging
- water management strategies
- water pricing
- water resource degradation
- water scarcity
- water use efficiency
- watershed development

REFERENCES

1. Cuykendall, C. H., White, G. B., Shaffer, B. E., Lakso, A. N. and Dunst, R. M., 1999. *Economics of drip irrigation for juice grape vineyards in New York State.* Department of Agricultural, Resource and Managerial Economic. College of Agriculture and Life Sciences, Cornell University, Ithaca, New York, 14853.

2. Dhawan, B. D., 2002. *Technological change in irrigated agriculture: A study of water saving methods.* Commonwealth Publishers, New Delhi.
3. Garrett, H. E. and Woodworth, R. S., 1973. *Statistics in psychology and education.* Vakils, Feffer and Simons Private Ltd, Bombay.
4. Indian National Committee on Irrigation and Drainage (INCID), 1994. *Drip irrigation in India.* INCID, New Delhi.
5. John, M. A. and Flores, R., 2005. *Impact evaluation of a conditional cash transfer program: The Nicaraguan Red Social.* Research Report 141, International Food Policy Research Institute, Washington, DC.
6. Kulecho, I. K. and Weatherhead, E. K., 2005. Reasons for smallholder farmers discontinuing with low cost micro irrigation: A case study from Kenya. *Irrigation and Drainage Systems,* 19:179–188.
7. Magar, S. S., Firke, N.N. and Kadam, J. R., 1988. Importance of drip irrigation. *Sinchan (India),* 7:61–62.
8. Namara, R. E., Upadhyay, B. and Nagar, R. K., 2005. *Adoption and impacts of microirrigation technologies: Empirical results from selected localities of Maharashtra and Gujarat States of India.* Research Report 93, International Water Management Institute, Colombo, Sri Lanka.
9. Narayanamoorthy, A., 1997. Drip irrigation: A viable option for future irrigation development. *Productivity,* 38:504–511. 5
10. Narayanamoorthy, A., 2003. Averting water crisis by drip method of irrigation: A study of two water intensive crops. *Indian Journal of Agricultural Economics,* 58:427–437.
11. Narayanamoorthy, A., 2005. Economics of drip irrigation in sugarcane cultivation: Case study of a farmer from Tamil Nadu. *Indian Journal of Agricultural Economics,* 60:235–248.
12. Palanisami, K., Raman, S. and Mohan, K., 2012. *Micro irrigation Economics and Outreach.* MacMillan Publishers India, New Delhi.
13. Pattanayak, K. S., 2009. *Rough guide to impact evaluation of environmental and development program programs.* Sandee Working Paper No. 40–09, South Asian Network for Development and Environmental Economics, Nepal.
14. Qureshi, M. E., Wegener, M. K. Harrison, S. R. and Bristow, K. L., 2001. Economic evaluation of alternate irrigation systems for sugarcane in the burdekin delta in North Queensland, Australia. In: *Brebbia,C. A., Anagnostopoulos, K, Katsifarakis, K. and Cheng, A.H.D. Eds., Water Resource Management.* WIT Press, Boston, pages 47–57. 7
15. Saleth, R. M., 1996. *Water institutions in India: Economics, law and policy.* Commonwealth Publishers, New Delhi.
16. Seckler, D., Molden, U. A. D., de Silva, R. and Barker, R., 1998. *World water demand and supply, 1990 to 2025: Scenarios and Issues.* Research Report 19, International Water Management Institute (IWMI), Colombo, Sri Lanka.
17. Shah, T., Verma, S., Bhamoriya, V., Ghosh, S. and Sakthivadivel, R., 2005. Social impact of technical innovations: Study of organic cotton and low cost drip irrigation in the agrarian economy of west Nimar Region. International Development Enterprises (India), Delhi. http://www.ide-india.org/ide/socialimpact.shtml

18. Sivanappan, R. K., 1994. Prospects of micro irrigation in India. *Irrigation and Drainage Systems*, 8:49–58.
19. Sivanappan, R. K., 2002. Strengths and weaknesses of growth of drip irrigation in India. Proceedings of the GOI Short Term Training on Micro Irrigation for Sustainable Agriculture, 19–21 June, Water Technology Centre, Tamil Nadu Agricultural University, Coimbatore.
20. Suresh Kumar, D., 2008. Promoting drip irrigation: Where and why? Managing water in the face of growing scarcity, inequity and declining returns: Exploring fresh approaches. The IWMI-Tata Water Policy Program, Seventh Annual Partners Meet, ICRISAT Campus, Hyderabad, pages 108–120.
21. Suresh Kumar, D., 2012. An analysis of economics of adoption of drip irrigation: Some experiences and evidences. *The Bangladesh Development Studies.*
22. Suresh Kumar, D. and Palanisami, K., 2010. Impact of drip irrigation on farming system: Evidences from Southern India. *Agricultural Economics Research Review*, 23:265–272.
23. Suresh Kumar, D. and Palanisami, K., 2011. Can drip irrigation technology be socially beneficial? Evidence from Southern India. *Water Policy*, 13:571–587.
24. Vaidyanathan, A., 1999. *Water resources management: Institutions and irrigation development in India.* Oxford University Press, New Delhi.
25. Verma, S., Tsephal, S. and Jose, T., 2004. Pepsee systems: Grass root innovation under groundwater stress. *Water Policy*, 6:1–16.

CHAPTER 4

UP-SCALING MODEL FOR MICRO IRRIGATION IN TAMIL NADU, INDIA

K. PALANISAMI

CONTENTS

4.1   MICRO IRRIGATION EXPANSIONS STRATEGY

Government of India has made efforts in promoting the micro irrigation (drip and sprinkler) to manage the emerging water scarcity as well as to increase the crop productivity. Subsidizing farmers' capital costs of micro irrigation systems (MI) is still seen as the key policy intervention. The importance of promoting MI adoption largely started with the *recommendations of the Micro Irrigation Task Force of India in 2004,* which recommended more financial resources for subsidies, with state governments taking up 10% of the cost, while the central funds would account for 40% of the cost; and advised greater flexibility for states to determine

In this chapter, one US $ = 63.02 Rs. (Indian rupees).

their appropriate implementation structure and institutional mechanisms for subsidy disbursement. Based on these recommendations, in 2006, the Central Sponsored Scheme (CSS) on MI was launched. The operational guidelines for *National Mission on Micro Irrigation (NMMI)* stress that "*the success of the scheme will depend on an effective delivery mechanism*". It is of utmost importance of successful implementation models, after looking at the area coverage under MI in different states and the rapid area expansion in Gujarat and Andhra Pradesh states, which witnessed more than 100% area expansion in the recently.

4.2 THE TWO SUCCESSFUL MODELS: GGRC AND APMIP

The models followed in Gujarat and Andhra Pradesh – the Gujarat Green Revolution Company Ltd (GGRC) and the Andhra Pradesh Micro Irrigation Project (APMIP), respectively – are seen as the best models in terms of "capacity and quality" of implementation. APMIP was established as a *Special Purpose Vehicle* (SPV) housed in the Directorate of Horticulture prior to the CSS in 2003 itself. GGRC was established in 2005 as a SPV in the form of a public company promoted by Gujarat State Fertilizers and Chemicals Ltd, Gujarat Narmada Valley Fertilizers Company Ltd and the Gujarat Agro Industries Corporation Ltd. Prior to these SPVs. Andhra Pradesh was one of the early adopters of MI, and in 2002, it had about 12% of the 0.5 million-ha under drip irrigation in India. Gujarat, at the same time, only had about 2.5% of the share. After the implementation of the improved implementation models, the area under MI has increased to 0.99 million-ha and 0.56 million-ha in Andhra Pradesh and Gujarat, respectively. In both these cases, there are four main actors/agents:

- funding authorities,
- implementing agency,
- MI firms, and
- farmer beneficiaries.

In addition, other players include banks and other credit agencies, third party monitoring and evaluating agencies, agriculture extension personnel, MI dealers and marketing agents (who in some cases represent multiple MI firms). The details of the key parameters of these models are given in Table 4.1. States like Maharashtra and Karnataka,

TABLE 4.1 Comparison of Different Micro Irrigation Implementation Models

Parameter	GGRC model (Gujarat)	APMIP model (Andhra Pradesh)	TANHODA model (Tamil Nadu)	Remarks
Funding source/ assistance	Government of Gujarat	NABARD assistance; GoAP	Govt. of India, Govt. of Tamil Nadu	The 'quality' of funding differs across states
Subsidy criteria	Per acre, per farmer	Per family	Per family differentiated as marginal and small farmers (100%) and "others" (75%)	No delays in release of subsidy in Gujarat and AP. Not so in TN where cycle time is on an average 200 days.
Governance	Semi-autonomous corporation supported by Gujarat state fertilizer corporation (GGSFC)	Works under the Horticulture department of GoAP	Works under Horticulture department (SPV: TANHODA). Now under NMSA, there may be change.	Decision making on operational issues faster in GGRC; APMIP suffers some administrative delays. Non-coordinated, no dedicated department and hence delays at all levels in TN.
Organizational structure	Centralized; single window operations.	Decentralized; district offices carry out key functions.	Decentralized, but no separate wing/Dept at DT level for MI. It is all part of regular, other activities.	APMIP model facilitates easy handling of huge volume of applications and smoothens monitoring & field inspection: TN suffers in getting the target achieved.

TABLE 4.1   Continued

Parameter	GGRC model (Gujarat)	APMIP model (Andhra Pradesh)	TANHODA model (Tamil Nadu)	Remarks
Subsidy: Regulated or Unregulated	Unregulated; no quotas for MI companies	Yearly quotas for drips and sprinklers fixed; MI companies allotted geographical domains.	Partial regulation. 80% of funds for small and marginal farmers and 20% for others. This may change to 60:40 area of operation, according to zones allotted to MI companies based on their operational strength.	APMIP quota systems cripples competition and distorts the MI market. Delayed release of funds from local Govt. upsets the payment cycle.
Administration and processing	Streamlined; uniform procedures	Variation between districts; ambiguous chain of command.	Streamlined-on-line registration of applications upto release of subsidy but variations in operationalization of procedures at dt level.	Administrative overlaps and non-uniformity of processes creates bottle-necks.
Transparency	Online tracking of application status	Toll-free number for enquiries about application status	On-line tracking possible, not done	Process of fixing quotas in APMIP and information about funds disbursement under GGRC lacks transparency. Information is generally freely available though there is apathy and lethargy at all levels in TN.

Source: Pullabhotla [2] and Palanisami [1].

which also witnessed increased area under MI in recent years, are following MI implementation models that are a variant of the GGRC and APMIP models

4.3   NEED FOR NEW IMPLEMENTATION MODEL FOR TAMIL NADU

Tamil Nadu state is the pioneer in introducing MI, but the area expansion for the last 10 years has been discouraging with Tamil Nadu lacking behind other states (ranking eighth position in India). Even though, water supply and availability of electric power to lift water might affect the MI expansion, the major factor is the implementation model that is followed [1, 2]. Compared to Gujarat and Andhra Pradesh states, the Tamil Nadu Horticulture Development Agency (TANHODA) model even though well planned, lacks clarity in implementing the subsidy schemes due to administrative delays in handling huge applications, fixing the MI rates and distribution of subsidies in time. Third party inspection takes months to complete the process. At time, the cycle time is on an average 180 days. Hence, an updated implementation model in line with the Gujarat's GGRC model incorporating changes (in terms of allowing more transparency, rate fixation and capacity building of the stakeholders) with adequate financial backup is highly warranted. Once this model as a SPV is in place, the area expansion is expected to reach more than one million-ha compared to the present coverage of 0.25 million-ha, where MI coverage even in canal command areas (irrigation projects) can also be achieved. IWMI-Tata Water Policy.

4.4   SUMMARY

In this chapter, *Up scaling model for implementation of micro irrigation in Tamil Nadu* state of India is compared with the two exiting models, namely: *Andhra Pradesh Micro Irrigation Project* (APMIP), and *Gujarat Green Revolution Company Ltd.*, GGRC. APMIP and GGRC models and its variants have successfully been implemented across India to promote micro irrigation.

KEYWORDS

- Andhra Pradesh
- Andhra Pradesh Micro Irrigation Project
- canal command areas
- capital cost
- Central Sponsored Scheme India
- crop productivity
- Gujarat
- Gujarat Agro Industries Corporation
- Gujarat Green Revolution Company Ltd.
- India
- International Water Management Institute
- market dynamics
- MI adoption
- Micro Irrigation Task Force of India
- micro irrigation
- National Mission on Micro Irrigation
- policy intervention
- Special Purpose Vehicle
- subsidy
- Tamil Nadu
- Tamil Nadu Horticulture Development Agency
- Tata Water Policy Program
- Up scaling model

REFERENCES

1. Palanisami, K, Kadiri Mohan, K. K. Kakumanu and S. Raman, 2011. Spread and economics of micro irrigation in India: Evidence from nine States. *Economic and Political Weekly – Review of Agriculture*, 46(26/27): June 25.
2. Palanisami, K. and S. Raman, 2012. Potential and challenges in up-scaling micro irrigation in India experiences from Nine States. Water Policy Research Highlight Report 20, *International Water Management Institute,* www.iwmi.org/iwmi-tata/apm2012.

3. Pullabhotla, K. Hemant, Chandan Kumar, and Shilp Verma, 2012. Micro-irrigation subsidies in Gujarat and Andhra Pradesh: implications for market dynamics and growth. *Water Policy Research Highlight,* Volume 43. IWMI-Tata Water Policy Research Highlight Report 43.

APPENDIX I. FIELD PHOTOS

CHAPTER 5

DESIGN OF LOW-HEAD BUBBLER IRRIGATION SYSTEM

AHMED ABDEL-KAREEM HASHEM ABDEL-NABY

CONTENTS

In this chapter: One *feddan* = 4200 m² = one ha; One LE = Egyptian unit of currency = 0.13992 US$.
Edited and abbreviated version of: "*A. A. H. Abdel-Naby, 2011. Studies on low-head bubbler irrigation system design. Thesis submitted in partial fulfillment of the requirements for the degree of Master of Science (agricultural engineering), Agricultural Engineering Department, Faculty of Agriculture at Suez Canal University, Egypt*".

5.1 INTRODUCTION

The water resources in Egypt are becoming scarce which a ninety percent of water is supplied by Nile. Egypt has 55.5 billion cubic meters according to the 1959 treaty with Sudan and there are other users who are trying to reduce Egypt's share of Nile water. With a population of approximately 76 million in 2009 and expected to increase to some 86 million by 2025, water consumption is about 730 m^3 per year (2009) to about 639 m^3 per year (2025), which is considered below the water poverty level (1,000 m^3/year/capita).

The efficient use of water in Egypt has now become a strategic goal. By law, new reclaimed lands have to be irrigated with pressurized irrigation systems. Pressurized irrigation systems (sprinkler and micro irrigation) have played an important role in improving irrigation efficiency and water application uniformity during the past two decades.

Micro irrigation applies irrigation water to the soil near the root zone. Among several advantages of micro irrigation, water and energy savings are higher than those for other modern irrigation systems [17]. Based on the hydraulic design or the method to apply water to the soil, micro irrigation systems can be broadly categorized into four types: drip, spray, bubbler and subsurface drip irrigation system. Micro irrigation can help to achieve higher irrigation efficiency and higher yields than other irrigation systems. However, one will need high-energy consumption, high capital cost and high maintenance.

The term bubbler is a genericized trademark used in some regional dialects of the United States and in Australia. A survey of US dialects undertaken between 2002 and 2004 found the word bubbler commonly used in southern and eastern Wisconsin, Rhode Island and Massachusetts. The phrase drinking fountain was common in the rest of the inland north and in the west, while water fountain dominated other parts of the country. The term bubbler is sometimes used in the Portland, Oregon region where in the late 1800s former Wisconsin resident Simon Benson installed 20 fountains, which are now known in the Portland area as "Benson Bubbler". Therefore, the name of the bubbler irrigation is derived from the fountain of water streaming out from the hoses, and from the bubbling noise made as air escapes from the pipe line when the system is turned on [39, 40].

In bubbler irrigation, water is applied to the soil surface in the form of a stream. Bubbler systems can be further sub-divided into high and low-pressure systems. Low head bubbler systems are based on gravity-flow (about 10 kPa) from a small diameter tube (1 mm to 13 mm) and high pressurize systems operate at 50 to 150 kPa. Bubbler system are restricted to slope of 1–3%, and do not require mechanical pumps or filtration systems. Therefore, the low head bubbler irrigation can help to solve problems of water scarcity and can save energy under Egyptian conditions.

This chapter discusses research results on: the design of low head bubbler irrigation system for Egypt; the effects of different operating pressures and bubbler diameters on discharge uniformity for bubbler outlets at zero land slope and parallel to the hydraulic gradient line; effects of bubbler heights on high discharge uniformity.

5.2 REVIEW OF LITERATURE

5.2.1 BUBBLER DEFINITION AND ITS APPLICATION

According to Reynolds [38], the micro irrigation system can be subdivided into four categories based on the differences in hydraulic design: drip, spray, bubbler, and subsurface irrigation systems. The design of bubbler system differs from design of other micro irrigation systems because they are based on gravity flow and do not require external energy and elaborate filtration systems. The fact that the dissemination of bubbler design has occurred largely by site visits to existing bubbler systems probably indicates that available literature does not adequately describe the simplicity of bubbler design.

Carr and Kay [14], James [26] and Lamm et al. [32] described that water is applied to the soil surface from bubbler irrigation in the form a stream or a fountain, typically from a small diameter tubes (1 mm to 13 mm) or a commercially available emitter. Because the application rates generally exceed the soil infiltration rates, small basins or furrows are needed to control the water distribution on the land to save water near the plant root zone. Two major types of bubbler irrigation systems are high and low pressurized systems. The low head bubbler systems are based on gravity flow (about 10 to 50 kPa) and pressurized systems operate at 50 to 150 kPa of pressure. Hull [24] stated that bubbler system is restricted to land slope of 1–3%. According to Rawlins [37], Behoteguy and Thornton [8], Carr and Kay [14] and Hull [24], low head bubbler irrigation system is defined as the one that reduces energy requirement. This is a type of micro irrigation system that typically delivers flow rates of 0.032 to 0.063 lps to each tree through a small diameter polyethylene (PE) tubing (delivery hose) attached to a large diameter lateral of corrugated plastic pipe which is buried between two tree rows by using 38.1 to 120 mm diameter of lateral PE pipe.

Awady et al. [7] developed first trickle irrigation system, installed and tested in Egypt as early as 1973. The system was operated on a very low head of 40 cm, being close to bubbler and it proved to reduce clogging problem.

Yitayew et al. [49] mentioned that the distinguishing feature of low-head bubbler systems is the use of flexible delivery hoses. Water is distributed to the bubbler tubes by adjusting the elevations of the tube outlets along the lateral so that water flows out from all hoses at approximately same rates. Despite this early experimental success, the bubbler concept has not been widely adopted in agriculture. Perhaps one of the main reasons is lack of interest in design criteria and recommended operating procedures.

Hull [24] illustrated that the bubbler irrigation is very sensitive to changes in pressure head, and a constant head source is essential for a commercial orchard or plantation. A change in pressure head at the inlet results in non-uniformity of water application at each outlet. A pressure head of one meter is very small, and small changes in head can thus have a marked effect on the flow rate, which is fixed once the system is installed.

Bubbler systems are well suited for perennial crops, particularly orchards and vines, because the irrigation system typically includes buried pipes and small earthen basins around the plantings. Bubbler systems can also be adapted to row crops that utilize furrows. The laterals are placed along the furrows after planting and are removed from the field after harvest. A fine textured soil is preferred. Bubbler systems can readily utilize low-head water supplies, similar to surface irrigation systems [32].

5.2.2 ADVANTAGES AND DISADVANTAGES OF BUBBLER IRRIGATION

Behoteguy and Thornton [8], Hull [24], Phocaides [36] and Lamm *et al.* [32] indicated that bubbler systems have some advantages and disadvantages compared to other micro irrigation systems.

Advantages	Disadvantages
Energy requirements are low due to gravity flow.	Very few agricultural bubbler systems have been installed.
Maintenance is low as a result of few equipments (filters, pumps).	Design criteria and recommended operating procedures are not well documented.
Susceptibility to emitter clogging is low due to large diameter delivery hose.	Entrapment of air in the pipe network can lead to blockages.
Water with high-suspended solids concentration can be used.	Farm topography needs to be nearly level.
Operating costs are low because of the lower energy and maintenance requirements.	Bubblers are not suitable for sandy soils due to high infiltration rate.
Intervals between irrigations are long.	Small earthen basins are typically required around plants to hold the water near the root zone.
Duration of an irrigation event is short because of high discharge rates.	Cultural practices are more difficult to perform around earthen basins.

Advantages	Disadvantages
Accumulated salts are uniformly leached.	Small water flow cannot be used as in other micro irrigation systems.
Bubbler basins increase catchment's of rainfall.	Limited to orchard and plantation type crops because of costs.
High irrigation application uniformity up to 75%.	Possibly more leaching and evaporation losses than with trickle irrigation system.
The entire piping network is buried therefore few problems in field operations.	Usually greater water consumption than trickle system.
The technology is simple and no highly sophisticated equipment is needed.	The bubbler concept has not been widely adopted in agriculture,
The initial cost and maintenance costs are low compared to other micro irrigation systems.	
Reduced tail water.	
The ability to more precisely apply nutrients to the tree.	
The system can be operated by unskilled farmers and laborers.	

5.2.3   DESIGN OF BUBBLER IRRIGATION SYSTEM

Design procedures for gravity systems have been developed over the last several years and are relatively unique to this type of irrigation [22, 23]. Rawlins [37] reported that to ensure equal discharge from all delivery hoses, the elevation of each delivery hose was calculated by subtracting from the static head, the friction losses in the pipes and the changes in elevation. After the delivery hoses were installed at these computed elevations, the outflows of the delivery hoses were adjusted to be approximately equal by dynamically calibrating the system.

Dynamic calibration is a procedure by which errors in friction loss calculations can be evenly distributed along the lateral by adjusting the elevation of each delivery hose. Dynamic calibration is performed after the delivery hose elevations have been set at the calculated elevations. The discharge uniformity values were 89.2% before dynamic calibration at Tacna, Arizona and 97.3% uniformity at Riverside – CA after dynamic calibration. Designing a bubbler irrigation system includes designing the lateral line, delivery tube and determining the height of bubbler outlet [8].

5.2.3.1 Delivery Tube Design

Emitters for gravity flow bubblers are unique in that they are not designed to dissipate energy, unlike those associated with the other types of micro irrigation systems. Bubbler emitters are essentially delivery tubes for transferring water from irrigation laterals to the plants. The delivery hose length was calculated by using Eq. (1) by Lamm et al. [32].

$$L_{dh} = 0.5\ S_r + d_l + H_{max} \tag{1}$$

where: L_{dh}: delivery hose length, m; S_r: plant spacing, m; d_l: burial depth of lateral line, m; H_{max}: maximum height of delivery hose, m.

Small changes in elevations throughout the system have a large impact on discharge rates. Additionally, friction losses within the pipes and tubes affect water pressures within the system, and therefore affect the discharge rates. Although discharges are usually less than 225 lph, yet friction losses in the delivery tubes do affect the flow rates. These losses must be esti-mated and taken into account to select proper tube diameter. For small diameter and smooth pipes, the Darcy-Weisbach and Blasius equations can be combined to predict friction head loss, $h_f(m)$, accurately in bubbler tubes [31]:

$$h_f = K_{fdw} \times [Q^{1.75}/D^{4.75}] \times L \tag{2}$$

where: h_f: friction head losses, m; D: inside diameter, mm; Q: flow within tube, lps; L: length of tube, m; and K_{fdw}: a constant = 7.89×10^5 for SI units at a water temperature of 20°C. The head loss gradient (pipe friction loss as a function of length) for a variety of pipe diameters and flows are given in Figure 5.1. Generally, bubbler laterals are centrally situated between plant rows with delivery tubes placed on both sides of the lateral. Tube lengths can range from less than 1 m in row crops to more than 5 m for orchards.

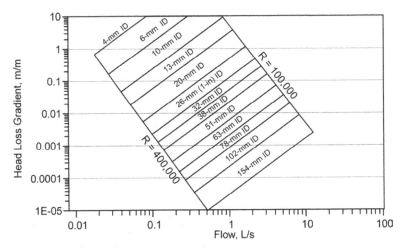

FIGURE 5.1   Head loss gradient versus flow rate for smooth (*PE* and *PVC*) pipes for Reynold's number (R) between 100,000 and 400,000 and for a water temperature of 20°C [39, 40].

Discharge rate as a function of tube length can be derived from fundamental hydraulic principles. Energy conservation within the bubbler tube can be described by following Bernoulli's equation (3):

$$p_1/\gamma + Z_1 + (V_1)^2/2g = p_2/\gamma + Z_2 + [(V_2)^2/2g] + h_f + h_{ml} \qquad (3)$$

where: h_f: total friction head loss in pipes, m; h_{ml}: total minor loss at pipe fittings, m; V_1 and V_2: flow velocities in the pipe at locations 1 and 2, respectively, m/s; P_1 and P_2: pressures within the pipe at locations 1 and 2, respectively, kPa; Z_1 and Z_2: elevations of pipe at locations 1 and 2, with respect to a reference datum, m; γ: specific weight of water = 9790 N/m³ at 20°C; and g: acceleration due to gravity = 9.81 m/s².

When applying equation (3) to a bubbler tube, points 1 and 2 can be set at the entry and outlet of the tube. Several following assumptions can then be made to simplify the equation:

- Minor losses (h_{ml}) can be neglected.
- No elevation change along tube, $Z_1 = Z_2$.
- Continuity equation applies for a same diameter of tube, $V_1 = V_2$.
- $P_2 = 0$, atmospheric pressure.

Based on the above assumptions and using Eq. (2) for the head loss (h_f), we get following Eq. (4) for the bubbler tube discharge:

$$Q_b = K_b \times [P/L_b]^{0.57} \times D^{2.71} \qquad (4)$$

where: q_b: bubbler tube discharge, lph; P: operating pressure, kPa; L_b: length of bubbler tube, cm; D: diameter of bubbler tube, mm; and K_b: a constant = 5.52. Equation (4) can be rearranged to solve for L_b:

$$L_b = K_l \times [(D)^{4.75}/(q_b)^{1.75}] \times P \qquad (5)$$

where: $K_l = 19.88$ defined by Lamm et al. [32].

If the delivery tubes are cut to the same length, the flow to the tree basin will be controlled only by the height of the outlet at delivery tube. Each hose can be attached conveniently to the tree by stapling the delivery tube from the lateral to the trunk of the tree [37] and installing a barbed tee with its horizontal arm at the desired elevation as shown in Figure 5.2.

El-meseery [16] used the equations for pressurized irrigation systems to derive design equation for bubbler irrigation system. When the delivery outlets were parallel to the hydraulic gradient line, the uniformity coefficient (Cu) of discharge was about 99%, but for the outlets at the same

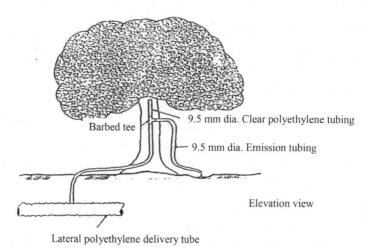

FIGURE 5.2   Typical installation of bubbler irrigation system [8].

elevation, the discharge uniformity coefficient (Cu) was increased with decreasing of the initial operating pressure.

5.2.3.2 Elevation of a Delivery Tube (Height of Bubbler Outlet)

Rawlins [37] described two procedures to determine the proper elevation of the supply hose at each tree to provide same flow rate. First method consisted of allowing water to pond at a fixed static head in the lateral; a reference level was found and marked on each tree by lowering each supply hose until the water level was at its opening. During the procedure, all other hoses were kept elevated above this level so that water did not flow through these, causing a pressure head gradient within the lateral. All subsequent elevation measurements were made relative to this reference elevation. In the second method, he estimated the head losses that will occur in the lateral between each pair of connections when the system was in operation. This head loss in the lateral was then compensated by lowering the point of attachment of the supply hose from one tree to another by a distance equal to it. Hydraulic head, delivery hose outlet, and ground levels as a function of distance from the water source are shown in Figure 5.3.

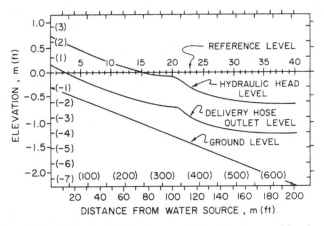

FIGURE 5.3 Hydraulic head, delivery hose outlet level and ground level as a function of distance from the water source. The lateral pipe *ID* changes from 102 mm (4 inches) to 76 mm (3 inches) at 100 m (330 ft) according to Rawlins [37].

Rawlins [37] added that because the bubbler irrigation system operates at low pressure, the existing elevation of pipes used for furrow or flood irrigation should often be sufficient to provide this information. There is, of course, a minimum elevation required, either to keep the lateral pipe size within economic limits, or in some cases to maintain flow velocities high enough to prevent siltation or to allow periodic flushing. Hull [24] gave the following procedure for adjusting the level of bubbler:

- Find the reference level by raising all delivery hoses, until no flow occurs in the system and water pond at the exit of delivery hose.
- Calculate the total head losses in the lateral to each tree, making sure that this does not exceed the total head available at any tree.
- If it does, then larger pipe size will have to be used to reduce head losses. Measure a distance downwards from the reference level at each tree
- With the system operating, at each delivery hose in turn, rise the delivery hose downwards from the reference level. This refines the system and allows for any discrepancies in lateral head from reference level.
- Measure the discharge at each tree to confirm the discharge expected.

Generally, all delivery hoses in a system are assumed to have same length and the maximum and minimum delivery hose heights are assumed to equal 1.0 m and 0.3 m (3.3 ft and 1 ft), respectively. Delivery hoses set at elevations lower than 0.3 m (1 ft), there is a risk damage from ponded water or trampling by workers or animals. Delivery hose heights can be set at heights higher than 1 m (3.3 ft), but the flowing water will increase soil erosion at the point of impact. One way to increase delivery hose heights without increasing soil erosion is to place a tee at the point of discharge of the hose, and run a delivery tube from the side of the tree down to the basin [39, 40]. Abozaid et al. [1] derived following equation to determine the bubbler height to achieve high uniformity of discharge:

$$hbn = H_i - h_e - h_{ln} \qquad (6)$$

$$hbn = H_i - [q/a]^{(1/b)} - [61111q^{1.75}D^{-4.75}(s+cl)]\left\{\sum_{n=1}^{N}(N-n+1)^{1.75}\right\} \quad (7)$$

where: hbn: bubbler height at location "n", cm; H_i: initial head, cm; h_e: effective head, cm; h_{ln}: total head loss at bubbler location, cm;

q: bubbler discharge, liters/min; D: lateral line inside diameter, mm; S: distance between bubblers, m; cl: barbed length, m; and N: total number of bubblers.

5.2.3.3   Design of Lateral Line and Manifold for Bubbler Irrigation

Lamm et al. [32] indicated that laterals and manifolds for bubbler systems are typically constructed from smooth PVC and/or corrugated PE pipe. Due to relatively high emission discharge rates, the diameters of laterals and manifolds are generally larger and/or their lengths are shorter than those in other micro irrigation systems. For typically sized lateral and manifold PVC pipes used in bubbler systems, Hazen-Williams equation is used for predicting friction head loss, h_f (m), as a function of flow rate, pipe length, and pipe diameter. The following Hazen-Williams equation (8) is very similar to the Darcy-Weisbach equation (2) for small diameter of bubbler tubes:

$$h_f = K_{fhw} \ [(Q^{1.85})/(D^{4.87})][L] \tag{8}$$

$$h_f = K_{fhw} \ [(Q^{1.85})/(D^{4.87})][L] \tag{9}$$

$$h_f = K_p \ [(Q^2)/(D^7)][L] \tag{10}$$

where: h_f: friction head loss, m; K_{fdw}: $= 1.135 \times 10^6$, a constant for SI units at 20°C; K_p: a constant $= 5.78 \times 10^6$; Q: inlet flow rate, lps; D: inside pipe diameter, mm; and L: length of pipe, m.

The Christiansen reduction coefficient, F, can be applied to Eq. (8) to account for head loss in pipes for discharge flow uniformly along the length of pipe via laterals and manifolds. Reduction coefficients are listed in Table 5.1. Depending on the location of the first outlet relative to the inlet of lateral, F_1, F_2, or F_3 is selected. F_1 is used when the distance from the lateral inlet to the first outlet is Sb. F_2 is used when the first outlet is adjacent to the lateral inlet. F_3 is used when the distance from the lateral inlet to the first outlet is $Sb/2$. With minor modification to Eq. (8), taking into account the outlets for the bubbler tubes, the Eq. (9) gives the total head loss for a lateral or manifold. Because of its relatively low cost,

TABLE 5.1   Values of Reduction Factor (F) for Plastic Pipe [9]

Number of outlets	F_1 [1]	F_2 [2]	F_3 [3]
5	0.469	0.337	0.410
10	0.415	0.350	0.384
12	0.406	0.352	0.381
15	0.398	0.355	0.377
20	0.389	0.357	0.373
25	0.384	0.358	0.371
30	0.381	0.359	0.370
40	0.376	0.360	0.368
50	0.374	0.361	0.367
100	0.369	0.362	0.366

[1] F_1 is used when the distance from the lateral inlet to the first outlet is S_b.
[2] F_2 is used when the first outlet is adjacent to the lateral inlet.
[3] F_3 is used when the distance from the lateral inlet to the first outlet is $S_b/2$.

corrugated PE pipe can also replace PVC pipe for low-pressure systems. Friction head loss, however, is greater for the corrugated PE, and the values presented in Figure 5.1, which were established for smooth pipes, are not applicable here.

According to Hermsmeier and Willardson [19], the friction head loss for corrugated plastic pipe for a water temperature of 20°C is defined in Eq. (10). Based on this equation, the friction loss gradient for corrugated plastic pipe, (h_f/l), is presented in Figure 5.4 for pipe diameters between 51 and 204 mm and flow rates between 0.2 and 100 lps. Laterals and manifolds are sized according to the allowable friction loss in the system, by taking into account the reduction coefficient, F, as described in Eq. (8) and Table 5.1.

Selection of pipe size for the manifold is to a large extent an economic decision, which involves balancing friction losses against various economic factors. One common method of pipe size selection is the "percent head loss method," where the allowable friction loss in the manifold is limited to (5 to 20%) of the irrigation system's design head, (H_d). In practice, both the (5 and 20%) conditions are often calculated, and the final decision is based on the calculated results and on additional factors such

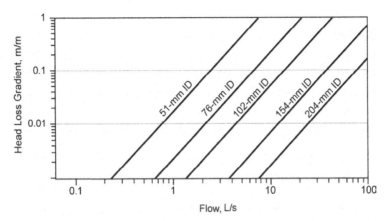

FIGURE 5.4 Head loss gradient for corrugated PE pipe for water temperature at 20°C [39, 40].

as price differences, availability, installation, maintenance requirements, and end-user preferences.

The allowable friction loss in the manifold, (h_{fam}), may then be expressed by the Eq. (11) for either 5% or 20% of design head of the irrigation system. The allowable head losses gradient in the manifold is then expressed in Eq. (12).

$$h_{fam} = \{[5\% \ or \ 20\%]/100\} \ H_d \qquad (11)$$

$$h_f/L = (h_{fam})/(F \times L_m) \qquad (12)$$

where: h_{fam}: allowable friction loss in the manifold, m; H_d: design head, m; L_m: length of the manifold, m; and h_f/L: allowable head losses gradient in the manifold.

In addition to friction loss in pipes, the slope of the field is a variable that must be considered in designing laterals and manifolds. Elevation differences are especially critical in gravity irrigation system, because minor changes in elevation head may have a significant effect on pressures within the system. Additional considerations for bubbler systems include equipment in the control head and air release valves in gravity flow networks. Clogging of bubbler tubes in low-pressure systems is usually not a concern, because tube openings are relatively large. Chemical injectors for fertilizers and other chemicals may also be incorporated in bubbler irrigation systems.

For gravity bubbler systems, a constant head device is required when the water source (reservoir or canal) is not maintained at a constant elevation. A constant head device (e.g., standpipe and gate valve) can be installed near the water source or elsewhere along the mainline to maintain a constant design head during operation of bubbler system.

5.2.3.4 Minor Losses

Keller and Karmeli [30] and James [26] calculated the head losses due to the emitter connection by an equivalent length method, and this length was added to the length of lateral line. The typical equivalent length for various emitter connections (fitting) to the lateral line are as follows:

- in-line with barbed or layout connection: from 1.0 to 3.0 m,
- on-line with barbed connections: from 0.1 to 0.6 m,
- in-line with smooth connection, which does not appreciably restrict the flow: from 0.3 to 1.0 m.

Watters and Keller [45] presented that the barbed friction loss (*cl*) in terms of a length of lateral that produces a friction loss of the same magnitude of the localized loss produced by the barb. They presented graphic data on emitter barb losses for various pipe diameter and barb dimensions. The following equation (with a correlation coefficient of $R = 0.99$) was based on their results.

$$Cl = 0.25\ W\left[19\ D^{-1.9}\right] \tag{13}$$

where: Cl: equivalent length of pipe, m; W: emitter barb diameter, mm; and D: diameter of lateral, mm.

5.2.3.5 Elimination of Air-Locks

Air locks are often found in low-pressure gravity flow irrigation systems, where pockets of air may accumulate at the crest of pipe undulations. These air pockets absorb a significant amount of energy and may partially block or reduce the flow of water. When the flow is entirely blocked by air, no water will be discharged until the air is removed.

Installation of air relief valves or standpipes just downstream from the crest of pipe undulations is the most common method to release air accumulations in water lines. However, installing air valves in bubbler systems is not a practical or an economical solution. Although air relief valves may be installed throughout the system, yet a cost effective procedure is to maintain pipe velocities greater than 0.3 m/s. At these velocities, water turbulence prevents air accumulation in the pipes. Therefore, emission tubes less than (13 mm) in diameter are recommended for these hydraulic conditions under low-pressure operation. From empirical data, the following equation can be used to calculate the minimum pipe flow rate to prevent air locks in both types of bubbler irrigation systems:

$$Q = K_a D^{2.45} \tag{14}$$

where: Q: flow within pipeline, lps; D: inside diameter of the pipeline, mm; $K_a = 0.0001$, a constant defined by Reynolds and Yitayew [39].

Jordan [27] gives a good analysis of air locks and how to avoid air pockets in the design of gravity-flow water supply systems. However, his analysis is not directly applicable to bubbler irrigation systems since his analysis is for water supply system with large elevation differences and long lengths of pipes. To prevent air locks from occurring in small-diameter pipes. Harrington [18] suggested the following:

a. Avoid air locks by:
  • Eliminating pipe undulations,
  • Keeping the hydraulic gradient line above the pipeline.
  • Ensuring air does not enter at the pipeline inlet.
  • Ensuring that pipe flow will be sufficient to flush out air in the pipeline.

b. Relive air locks by:
  • Providing outlets, air valves or standpipes, at critical locations along the pipeline.
  • Arranging the water supply so that higher pressure can be introduced at the start of operation, and then cut back to normal pressure after all air has been flushed from the line.

Waheed [44] revealed that the undulations, which are created during field installation, are the primary cause of air locking. The head needed to flush out the trapped air is independent of tube diameter, shape of the undulations and presence of water in the lower portions of undulations, but depends on the sum of heights of successive undulations. It was concluded that if the sum of heights of all the undulations exceeds the maximum allowable head loss in the tubing, water will not be able to flow out of the tubing.

5.2.4 *HYDRAULIC EVALUATION OF BUBBLER IRRIGATION SYSTEM*

The hydraulic performance is used to determine the characteristics of the bubbler irrigation systems; and also to verify and compare the published data by researchers and manufacturers. Hydraulic evaluation can be determined on the basis of parameters, such as: Coefficient of manufacturing variation (*Cv*); Coefficient of uniformity (*Cu*); and (*k, x*) parameters. Irrigation system performance can be expressed in terms of the determined coefficient of manufacturing variation and coefficient of uniformity. The more uniformly water is applied, potentially the more efficient is the irrigation.

5.2.4.1 Bubbler Discharge

The bubbler discharge is characterized by the relationship between discharge, pressure and a bubbler discharge exponent. The Eq. (15) for bubbler flow can be expressed as:

$$q = k\, h^x \tag{15}$$

where: *q*: bubbler discharge rate, lph; *k*: dimensionless constant of proportionality that characterizes each bubbler; *h*: pressure head at the bubbler, *m;* and *x*: dimensionless bubbler discharge exponent that is characterized by the flow regime. The sensitivity to *h* of a bubbler discharge depends mainly on the value of *x*, which determines how sensitive the discharge is to the pressure. The lower the value of x, the discharge will be less affected by variations in pressure. The value of x typically ranges from 0.1 and 1.0 depending on the make and design of the bubbler, i.e., hydraulic characteristics (Table 5.2).

The flow from non-compensating orifices and nozzle bubblers are always fully turbulent with $x = 0.5$. However, the exponent of long path bubblers may range between 0.5 for fully turbulent flow and $x = 1$ for laminar flow [28].

5.2.4.2 Coefficient of Manufacturing Variation, *Cv*

Lamm et al. [32] mentioned that the manufacturer's coefficient of variation for five models tested ranged from 8 to 21%, which is relatively high for micro irrigation emitters. ASAE [4] recommends values less than 11% and suggests that values greater than 15% are unacceptable. ASABE Standards [2] classified emitters based on coefficient of manufacturer's variation (*Cv*) as shown in Table 5.3 for point source emitters.

Wu et al. [47] reported that the total emitter flow variation is mainly affected by manufacturer's variation, temperature changes, and bulging. Assuming the temperature variation is small and the plugging problem is under control, the total emitter flow variation will be affected by the variation caused by the manufacturer. Wu et al. [48] showed that hydraulic design of drip irrigation lateral line is usually based on a design criterion using an emitter flow variation (q_{var}) of either 10 or 20%, which is equivalent to coefficient of variation (s/q) of 3 or 6%, where: s is the standard

TABLE 5.2 Classification of Flow Regime According to the Value of x

x	Classification*
0.00	Fully pressure compensating
0.25	Partially pressure compensating
0.50	Fully turbulent flow regime
0.75	Partially turbulent or unstable flow regime
1.00	Laminar flow regime

*According to Howell and Hiler [20, 21]; Wu and Gitlin [46]; Karmeli [28]; Solomon and Bezdek [42]; Braud and Soon [12]; and Boswell [10].

TABLE 5.3 Classification Based on Manufacturer's Coefficient of Variation (*Cv*), According to ASABE Standards [2]

Cv range	Classification
<0.05	Excellent
0.05 to 0.07	Average
0.07 to 0.11	Marginal
0.11 to 0.15	Poor
>0.15	Unacceptable

deviation of emitter flow; and *q* is the mean emitter flow. Bralts [11] and Solomon [41] indicated that the manufacturer's variation significantly attributes to the total flow variation than the variation caused by hydraulics, if the design is based on 10 or 20% of emitter flow variation.

5.2.4.3 Coefficient of Discharge Uniformity, *Cu*

The uniformity coefficient (*Cu*) for bubbler irrigation system was estimated by Perold [35], as shown in Eq. (16).

$$Cu, \% = [1 - Abs\,(\sigma_{md})] \times 100 \tag{16}$$

where: *Cu*: coefficient of uniformity, %; and *Abs* (σ_{md}): absolute mean deviation of discharge of lateral line. *Abs* (σ_{md}) was calculated by using the Eq. (17a).

$$Abs(\sigma_{md}) = [1/n][\Sigma(q - qmean)] \tag{17a}$$

$$qmean = [1/n][\Sigma q] \tag{17b}$$

where: *n*: number of bubblers; *qmean*: mean discharge mean defined in Eq. (17b), lps; and *q*: discharge from bubbler, lps.

The *Cu* is a better way of expressing the variation in discharge of lateral lines. Nakayama and Bucks [33] studied the relationship between emitter flow variation and uniformity coefficient, and reported that a uniformity coefficient of about 98% is equivalent to an emitter flow variation of 10% and a uniformity coefficient of about 95% is equivalent to an emitter flow variation of 20%.

Benami and Ofen [9] recommended that allowable variation in pressure head should be limited to 15% for lateral line design in drip irrigation system. Due to the lack of well-defined design procedure for bubbler irrigation system and difficulties associated with the change of height of bubbler tube along lateral line, this study was carried out to get an appropriate system for bubbler irrigation by changing diameters of outlets along lateral line.

Awady and Habib [6] stated that the discharge uniformity of bubbler irrigation system is controlled by varying the tube diameter and/or length and/ or using valve for each bubbler along lateral line as shown in Figure 5.5.

This chapter compares the results to standards of field performance of micro irrigation performance by the ASAE [3]. The general evaluation standards for (EU) values are: >90%, excellent; 80–90%, good; 70–80%, fair; and <70%, poor. Table 5.4 shows recommended range of (EU) values. In fact, this statement is not only for (EU), but it also applies to all other uniformity expressions. For micro irrigation, which has a relatively high uniformity in design, all the uniformity expressions can be converted and used for other uniformity expressions.

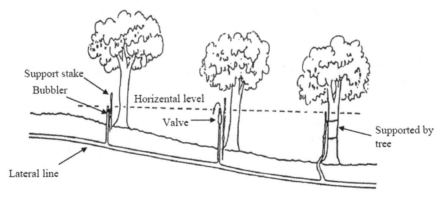

FIGURE 5.5   Bubbler irrigation system by Awady and Habib [6].

TABLE 5.4 Recommended Ranges of Design Emission Uniformity (*EU*), [2]

Emitter type	Spacing, *m*	Topography	Slope,%	EU range,%
Point source on	>4	Uniform Steep or undulating	<2	90 to 95
Perennial crops			>2	85 to 90
Point source on	<4	Uniform Steep or undulating	<2	85 to 90
Perennial or semi-permanent crops			>2	80 to 90
Line source on annual or	All	Uniform Steep or undulating	<2	80 to 90
Perennial crops			>2	70 to 85

5.3 MATERIALS AND METHODS

The study was conducted at the Farm of Agriculture Faculty, Suez Canal University, Ismailia, Egypt, during November, 2008 through August, 2009. The study consisted of laboratory and field experiments. The laboratory experiment was carried out in the Hydraulics Laboratory of Agricultural Engineering Department to determine bubbler discharge exponent constants and manufacturer's coefficient of variation for three bubbler tube outlets. The field experiment was carried out to: (i) Evaluate the effects of different initial operating pressures and bubbler tube diameters on bubbler discharge uniformity; (ii) Determine the optimum height for each bubbler diameter, which will give highest bubbler uniformity; (iii) Calibrate equation for bubbler height.

5.3.1 INSTRUMENTS FOR LABORATORY AND FIELD EXPERIMENTS

- Graduated cylinder of one-liter capacity with an accuracy of 10 cm³ was used to measure the water volume. A stopwatch was used to measure the elapsed time in different operations.
- A steel tape of one meter length was used to determine the bubbler height.
- Electronic digital caliper with accuracy of 0.01 mm was used for measuring the inside diameter of the bubbler tubes.

• Pressure gage range (0.6 bar) with 0.02 bar increment scale.
• Electrical Drill and pincer were used to perforate the lateral pipes to mount the bubbler tubes.

5.3.2 LABORATORY EXPERIMENTS

Laboratory experiments were carried out to find the volume of water from different diameters of bubbler at varying operating pressures to determine the bubbler constants. The discharge through bubblers was measured along lateral pipe at different pressures. Pressure head was measured in laboratory experiment by piezometric tube with 1 cm increment scale. The tested pressures ranged from 11 to 20 kPa with an increment of 1 kPa.

5.3.3 FIELD EXPERIMENTS

The completely randomized factorial design was used for 3 bubbler tube diameters, 3 initial operating pressures with 3 replications with as shown in Figure 5.6. Before starting the experiments, air in the lateral lines was

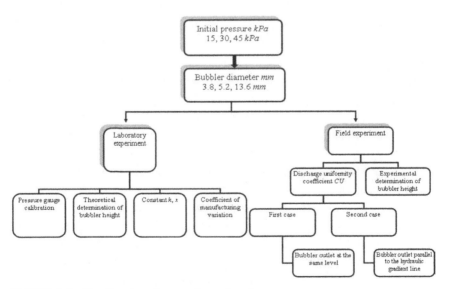

FIGURE 5.6   The flowchart for experimental study.

flushed out by opening the tubes at downstream end. Pressures were set at 15, 30 and 45 kPa. The bubbler discharge was measured by collecting the volume of water in plastic container in 5 minutes. The experiment was executed at the level ground surface (0% land slope). Specific bubbler flow functions were determined, such as: Pressure flow relationship, manufacturing coefficient of variation, coefficient of uniformity, and bubbler heights.

5.3.3.1   The Experimental Setup

The experimental bubbler irrigation systems are shown in Figures 5.7 and 5.8), and are described as follows:

- The water is pumped from the water source by using self-priming centrifugal pump, with suction tube diameter of 38.1 mm and delivery pipe diameter of 31.8 mm; powered by an electric motor of 3 horse Power (2.2 KW) at 220 volts.
- The water was pumped to a cylindrical plastic tank with dimensions: height = 0.9 m, diameter = 0.49 m, with a 0.17 m³ capacity.

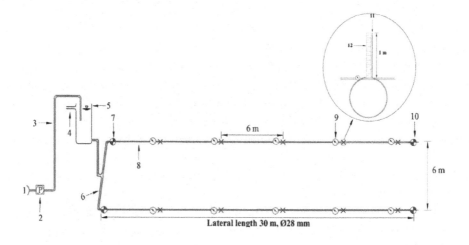

1- Water source	5- Tank	9- Pressure gauge
2- Centrifugal pump	6- Sub main pipe	10- Flushing valve
3- Delivery tube	7- Valve	11- Bubbler tube
4- Over flow tube	8- Lateral pipe	12- Steel tape

FIGURE 5.7   The experimental setup.

• The water level was kept constant in the tank by using an over flow tube with diameter 50 mm.
• The main pipe was branched into two submains with one lateral mounted in each submain. Two valves were mounted at entrance and end of each lateral to control and flush out the air from it. The lateral pipe was a smooth polyethylene with 30 m length and 28 mm internal diameter. The lateral pipe slope was zero.
• Five delivery tubes (bubblers) were mounted on each lateral pipe at 6 *m* spacing. The bubbler tubes were smooth polyethylene with nominal diameter of 4.5, 6 and 16 mm (ID were 3.8, 5.2 and 13.6 mm), respectively. The length of each bubbler was 5 m as shown in Figure 5.7. The bubbler was tide to wooden stakes.
• Pressure gages were mounted before each bubbler inlet to measure the pressure.

5.3.3.2   Performance and Evaluation of Bubblers

5.3.3.2.1   Pressure-Flow Relationships

For studying the hydraulic performance, bubbler discharge was measured at three initial operating pressures (P_i: 15, 30 and 45 kPa). Bubbler flow as a function of pressure can be expressed as below [5, 21, 29, 46]:

$$q = k\,h^x \qquad (18)$$

where: q: bubbler discharge rate, lph; k: the constant of proportionality that characterizes each bubbler; h: working pressure head at the bubbler, m; and x: the bubbler discharge exponent that is characterized by flow regime. The magnitude of k is a size or capacity parameter for a bubbler, since its value is equal to the bubbler flow rate for $h = 1.00$ [21]. The suggested criteria for (x) values were presented in Table 5.2.

A different effective pressure (P_e) from 11 to 20 kPa with an increment 1 kPa was used for the bubbler system under investigation. The effective pressure was obtained by changing the bubbler height. The discharge was measured at each effective pressure by collecting the water from the bubbler in plastic container for same time duration, and then the discharge was calculated. The values of k and x were determined by nonlinear regression analysis power between measured q and effective pressure (P_e). The following equation was used to calculate the percentage difference between the discharge rates.

$$q_c = 100 \times [(q_{pe} - q_{med})/q_{med}] \qquad (19)$$

where: q_c: percentage of discharge variation from medium value, %; q_{Pe}: bubbler discharge at any effective pressure P_e, liters/min; and q_{med}: bubbler discharge at medium value of effective pressure and the same water temperature, liters/min. The percentage difference of uniformity coefficient was calculated as follows:

$$CU_c = 100 \times [(CU_{pe} - CU_{med})/CU_{med}] \qquad (20)$$

where: CU_c: percentage of uniformity coefficient variation from medium value, %; CU_{Pe}: bubbler uniformity coefficient at any effective pressure P_e; and Cu_{med}: bubbler uniformity coefficient at medium value of effective pressure and the same water temperature.

5.3.3.2.2 Bubbler Manufacturer's Coefficient of Variation, Cv

The manufacturer's coefficient of variation (Cv) was calculated for the bubbler inside diameter of 3.8, 5.2 and 13.6 mm by measuring the bubbler discharge according to ASABE Standards [2]:

$$Cv = S/\overline{x} \qquad (21)$$

$$S = [1/(n-1)]^{0.5} \left[\sum_{i=1}^{n} (xi - \overline{x}) \right]^{0.5} \qquad (22)$$

where: Cv: manufacturer's coefficient of variation, dimensionless; S: standard deviation of bubbler discharge (lph) in the sample, according to equation (22); $\overline{x}$: mean discharge of bubblers, lph; x_i: discharge of an bubbler; and n: number of bubblers.

The experimental study was done with no plugging at the same hydraulic design and temperature, so that the average bubbler discharge variation was caused only by bubbler manufacturing variation. The manufacturing coefficient of variation ranged from 0.05 to 0.2 for different bubbler and lateral lines [11, 41].

Ten bubbler tubes were tested for each bubbler diameter to determine Cv. Three piezometer tubes were used to monitor the pressure in the lateral line at the beginning, middle and end of lateral pipe as shown in Figure 5.8.

5.3.3.2.3 Bubbler Discharge Uniformity Coefficient, Cu

The uniformity of irrigation water was calculated in this study in two different cases:

- Case 1: when the bubbler outlets were at same elevation; and
- Case 2: when the bubbler outlets were parallel to the hydraulic gradient line.

The Christiansen uniformity coefficient (Cu) was calculated by Perold [35] for bubbler irrigation system as follows: (23)

$$Cu = 100 \times \left[1 - |\bar{\sigma}|\right] \tag{23}$$

$$|\bar{\sigma}| = [1/n]\left[\Sigma[q - \bar{q}]\right], \text{ where } \bar{q} = [1/n]\Sigma q \tag{24}$$

where: Cu: coefficient of uniformity, %; $|\bar{\sigma}|$: absolute mean deviation of discharge of lateral pipe; $\bar{q}$: mean discharge; q: discharge from bubbler; and n: number of bubblers. The absolute mean deviation $|\bar{\sigma}|$ is calculated from equation (24). In this study, the discharge uniformity was calculated

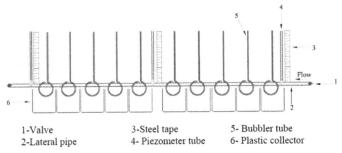

| 1-Valve | 3-Steel tape | 5- Bubbler tube |
| 2-Lateral pipe | 4- Piezometer tube | 6- Plastic collector |

FIGURE 5.8 Experimental setup for manufacturer's coefficient of variation.

for each bubbler along the lateral pipe at different initial operating pressure of 15, 30 and 45 kPa.

5.3.3.2.3.1   Bubbler outlets at same elevation, case one

This experimental study for bubbler outlet at same elevation (Figure 5.9) was conducted to evaluate the effects of different initial and operating pressures (P_o) on bubbler discharge and discharge uniformity. Three initial operating pressures $(P_i$, between the distance between the water level in the tank and ground level) were 15, 30 and 45 kPa. Six operating pressure $(P_o$, the distance between the bubbler outlet height and the water level in the tank) were determined at bubbler height of 0.0, 0.2, 0.4, 0.6, 0.8 and 1.0 m for each initial operating pressure. A completely randomized factorial design $6 \times 3 \times 3 \times 1$ was used with three replications. Figure 5.10 illustrates the experimental design of the uniformity in case one which Five plastic collectors with a capacity of 60 liters were located under the bubblers to collect the volume of water from each lateral pipe. The bubbler discharge was calculated by volume of water in a specified time by stop watch.

5.3.3.2.3.2   Bubbler outlets parallel to hydraulic gradient line: Case 2

The hydraulic gradient line was determined by measuring the pressure at each bubbler inlet and knowing the friction losses along the lateral pipe.

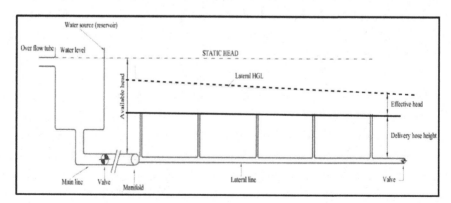

FIGURE 5.9   Bubbler systems (case one).

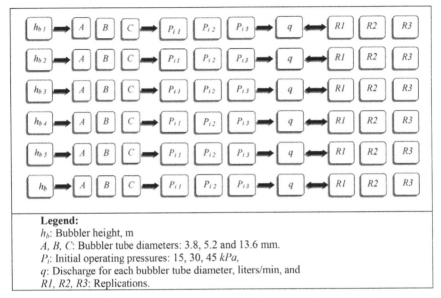

FIGURE 5.10   The experimental design for the discharge uniformity: Case one.

Then the height of bubbler outlet was calculated so that line joining these heights was parallel to the hydraulic gradient line as shown in Figure (11). Three effective pressures or operating pressures (the distance between hydraulic gradient line and level of bubbler outlets) were chosen. Figure (12) illustrates the completely randomized factorial design to evaluate discharge uniformity with three replications: $3 \times 3 \times 3 \times 1$. The predetermined effective pressures were chosen depending on the highest

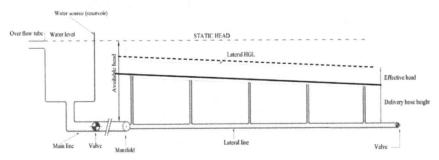

FIGURE 5.11   Bubbler irrigation system: Case 2.

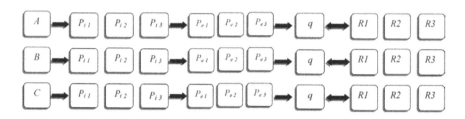

A, B, C: Bubbler tube diameters, (3.8, 5.2 and 13.6 *mm*)

P_i: initial operating pressures, (15, 30, 45 *kPa*)

P_e: Effective pressure, *kPa*

q: Discharge for each bubbler tube diameter, ℓ/min and

R1, R2, R3: Replications.

FIGURE 5.12   The experimental design for the discharge uniformity: Case 2.

values of discharge uniformity in the experimental study of the bubbler outlets at same elevation (case 1).

5.3.3.3   Determination of Bubbler Height for a Specified Bubbler Discharge

To achieve high uniformity of bubbler discharge on lateral pipe, two methods were used:

- First: by controlling the cross section area of bubbler.
- Second: by adjusting the bubbler height on the lateral pipe.

In order to determine the bubbler height on lateral pipe for specified bubbler tube diameter, a generalized equation was derived from several equations for design of micro irrigation system. To validate the theoretical equation, an experiment was conducted. The analysis of variance using t-test was used to determine the significance between theoretical and experimental results. A completely randomized factorial design ($3 \times 3 \times 3 \times 1$) with three replications, Figure 5.13, was used to compare the theoretical and experimental bubbler heights.

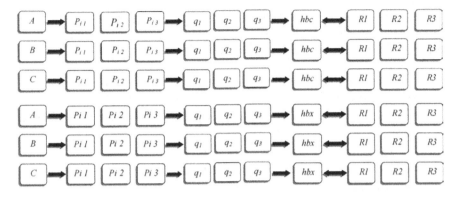

A, B, C: Bubbler tube diameters, (3.8, 5.2 and 13.6 *mm*)

P_i: Initial operating pressures, (15, 30, 45 *kPa*)

q: Bubbler tube discharge, ℓ/min

hbc: Calculated bubbler height, and

hbx: Experimental bubbler height.

R1, R2, R3: Replications.

FIGURE 5.13 The experimental design for bubbler height determination.

5.3.3.3.1 *Theoretical Method to Determine Bubbler Height for a Specified Bubbler Discharge*

The bubbler height (h_b) parallel to the hydraulic gradient line depends on several variables, such as:

- Design parameters: Length (*L*), diameter (*D*) and coefficient of friction (*F*) of lateral pipe; and length (*l*), location (l_l), diameter (*d*), coefficient of friction (*f*) of bubbler tube; and spacing between bubblers (*s*).
- Operating parameters: Lateral pipe discharge (*Q*), bubbler discharge (*q*) and initial operating pressure (P_i).

Therefore, the bubbler height (h_b) can be expressed as function of these variables as shown below:

$$h_b = f_l [L, D, F, l, s, Q, H_i, q, l_i, d, f_i] \qquad (25)$$

In order to define the function, f_i, several following assumptions were considered for the bubbler irrigation system as shown in Figures 5.11 and 5.12:

- An equal bubbler discharge (q), which is the objective of any bubbler irrigation system.
- An equal bubbler discharge can be achieved by an equal effective pressure (P_e).
- Drawing a curve parallel to the bubblers outlet line by a distance equal to the (P_e) leads to the gradient of bubbler height as shown in Figure 5.11.
- The regime flow in lateral pipe is turbulent.
- Materials of lateral pipe and bubbler tubes are polyethylene.

For the system under study, the distance between bubblers (s) was equal to the distance from water source to the first bubbler. Therefore, the final equation for the bubbler height can be expressed as follows, according to Abozaid et al. [1]:

$$hbn = H_i - [q/k]^{(1/x)} - [61111q^{1.75}D^{-4.75}(s + cl)]^{1.75}\}$$ (26)

where: hbn: bubbler height at location "n", cm; H_i: initial head, cm; q: bubbler discharge, liters/min; D: lateral line inside diameter, mm; S: distance between two successive bubblers, m; cl: bubbler inlet barb equivalent length, m; k and x: bubbler constants; N: total number of bubblers; and n: nth location of bubbler.

For the system under study, minor losses were found for bubblers inlet barbs. The bubbler inlet losses should then be substituted by equivalent length (cl) as indicated by James [26] as follows:

$$cl = 1.729 \, D^{-1.935}$$ (27)

where: cl: bubbler inlet barb equivalent length, m; and D: diameter of lateral pipe, cm.

Therefore, the distance between bubblers in the final equation for bubbler height was substituted by ($s + cl$).

5.3.3.3.2   Validation of Theoretical Equation for Bubbler Height

To validate theoretical bubbler height by Eq. (26), the hydraulic gradient line was drawn by using pressure gages at bubbler inlets with same bubbler outlet levels. Then, the effective pressure was calculated and the bubbler discharge was measured at this effective pressure. Two additional values of bubbler discharge were measured at effective pressure above and below the estimated effective pressure.

5.3.3.4   Calibration of Bubbler Height with Pressure Gages

Pressure gages were laboratory calibrated by using piezometric tubes. The experiment setup consisted of 32 mm nominal diameter (ID 28 mm) lateral pipe with two valves at the inlet and outlet to control the water pressure. The piezometric tube 3 m height was mounted on the center of lateral pipe with steel tape (0.5 cm increment scale) was used to measure the pressure head. Calibrated pressure gage was connected next to piezometric tubes as shown in Figure 5.14. The actual reading was obtained from the

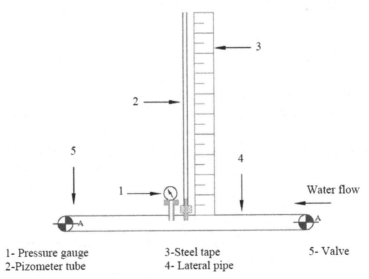

| 1- Pressure gauge | 3-Steel tape | 5- Valve |
| 2-Pizometer tube | 4- Lateral pipe | |

FIGURE 5.14   Experimental setup for pressure gage calibration.

piezometric tube and the indicated reading was obtained from the pressure gage. Then the relationship between indicated and actual value was established to get the percentage of error for each pressure gage.

The percentage error was calculated for each pressure gage by the following equation:

$$Percentage\ error = 100 \times [indicated\ value - actual\ value] \div [maximum\ value\ on\ the\ scale] \qquad (28)$$

5.3.3.5   Software and Programs Used

The statistical analysis was conducted to find significant differences with a T-Test (in groups), according to Steel et al. [43]. AutoCAD is a CAD (Computer Aided Design) software application for *2D* and *3D* design and was used draw the field diagrams [13].

5.4   RESULTS AND DISCUSSION

5.4.1   *LABORATORY EXPERIMENTS*

5.4.1.1   *Effects of Operating Pressure on Discharge*

Three bubbler tube internal diameters, in this study, were 3.8, 5.2 and 13.6 mm. Each tube was tested at an operating pressure of 11 to 20 kPa. Table 5.5 shows the discharge through each bubbler tube diameter for all operating pressures. Bubbler discharge was proportionally increased with increasing the operating pressure. By increasing the operating pressure from 11 to 20 kPa, the discharge (liters/min) was increased from 0.57 to 0.65, 0.97 to 1.29 and 7.12 to 9.53 for 3.8, 5.2 and 13.6 mm bubbler tube diameter, respectively.

The Figure 5.15 shows the power relationships between bubbler discharge and effective pressures. All coefficients of determination were above 0.95. Two bubbler diameters 5.2 and 13.6 mm gave fully turbulent with bubbler discharge exponents of 0.5 and 0.45, respectively. The third diameter 3.8 mm was partially pressure compensating with bubbler discharge exponent of 0.23, according to Boswell [10]. Table 5.6 gives he

TABLE 5.5   The Bubbler Discharge and Manufacturer's Coefficient of Variation for Different Effective Pressure and Bubbler Tube Diameters

Mean effective pressure, P_e	Internal diameter of bubbler tube, ID, mm					
	3.8		5.2		13.6	
	Mean discharge	Cv	Mean discharge	Cv	Mean discharge	Cv
kPa	liters/min	—	liters/min		liters/min	–
11	0.57	0.006	0.97	0.007	7.12	0.006
12	0.58	0.005	1.00	0.005	7.76	0.008
13	0.59	0.005	1.03	0.005	8.19	0.008
14	0.60	0.004	1.06	0.004	8.47	0.010
15	0.61	0.003	1.11	0.004	8.70	0.009
16	0.62	0.003	1.16	0.003	8.90	0.008
17	0.63	0.004	1.19	0.006	9.10	0.010
18	0.63	0.004	1.23	0.005	9.24	0.009
19	0.64	0.004	1.27	0.004	9.39	0.011
20	0.65	0.004	1.29	0.004	9.53	0.009

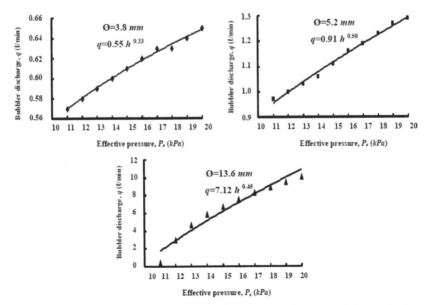

FIGURE 5.15   Relationship between effective pressure (kPa) and bubbler discharge (liters/min.) for three bubbler tube diameters (mm).

TABLE 5.6 The Nonlinear Regression Coefficients and Coefficient of Determination for the Bubbler Discharge Equation Under Different Bubbler Tube Diameters: $q = k\,P^x$

Bubbler diameter, (*mm*)	*k*	*x*	*R²*
3.8	0.55	0.23	0.99
5.2	0.91	0.50	0.99
13.6	7.12	0.45	0.95

nonlinear regression coefficients and coefficient of determination for the bubbler discharge equation under different bubbler tube diameters.

5.4.1.2 Bubbler Manufacturer's Coefficient of Variation, *Cv*

Table 5.5 and Figure 5.16 show the manufacturing coefficient of variation (*Cv*) for each bubbler diameter. The *Cv* values for three bubbler diameters ranged between 0.003 to 0.011 at 11 to 20 *kPa* effective pressure, respectively, which was considered excellent according to the classification of manufacturing variation coefficient for point source emitter [4]. The *Cv* values indicated fluctuations for three bubbler tube diameters with increasing effective pressures. These results have a good agreement with El-Lithy [15] and Hussein [25].

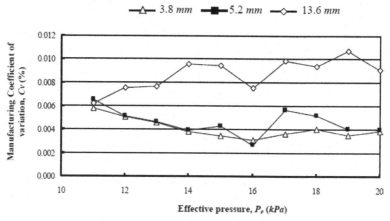

FIGURE 5.16 The relationship between effective pressure and manufacturing coefficient of variation for three bubbler diameters.

5.4.2 FIELD EXPERIMENTS

Two field experiments were carried out to: (i) Find the highest value of discharge uniformity coefficient (*Cu*) for different heights with several pressures, in each bubbler tube diameter; (ii) Test the highest uniformity at equal outlet elevations, and then apply this highest uniformity to the case 2 of bubbler outlets parallel to the hydraulic gradient line. Also, a theoretical bubbler height equation was validated.

5.4.2.1 Discharge Uniformity Coefficient, Cu

5.4.2.1.1 The Outlets at Same Elevation (Case 1)

Table 5.7 and Appendix I show the Christiansen uniformity coefficient (*Cu*) for three bubbler tube diameters at same outlet elevations.

a. Bubbler height of 0.0 m

The mean effective pressure (P_e) and the mean discharge were proportionally increased with increase in initial operating pressure (P_i), for all bubbler tube heights and diameters. For 3.8 mm bubbler tube diameter, the P_e values were 7.02, 24.38 and 40.46 kPa at P_i of 15, 30 and 45 kPa, respectively. At initial operating pressures 15, 30 and 45 kPa, Figures 5.17a, 5.18a, and 5.19a show that the values of bubbler discharge (q) were 0.51, 0.68 and 0.76 liters/min; the discharge uniformity coefficients (Cu) were 98.8, 98.8 and 98.2%, respectively. It was found that the discharge uniformity coefficient (*Cu*) values were relatively constant at different initial operating pressures for all bubbler tube heights.

Meanwhile for 5.2 *mm* bubbler tube diameter, (P_e) values were 12.96, 24.82 and 35.58 kPa at (P_i) 15, 30 and 45 kPa, as shown in Figure 5.17b. The discharge (q) values were 1.03, 1.43 and 1.72 liters/min; and the discharge uniformity (Cu) values were 94.4, 99.2 and 96.8% at initial operating pressures 15, 30 and 45 kPa, respectively, as shown in Figures 5.18b and 5.19b. The discharge uniformity coefficient values were increased with increasing the initial operating pressure from 15 to 30 kPa and were decreased with increasing the initial operating pressure from 30 to 45 kPa for all bubbler tube heights.

TABLE 5.7 Bubbler Mean Effective Pressures, Bubbler Discharge, Discharge
Uniformity Coefficients at Different Initial Operating Pressures for Internal Bubbler
Diameters at the Same Bubbler Heights (Case 1)

h_b	ID, Ø	P_i	Mean effective pressure, P_e	Mean discharge q	Cu
m	mm	kPa	kPa	liters/min	%
0	3.8	15	7.02	0.51	98.8
		30	24.38	0.68	98.8
		45	40.46	0.76	98.2
	5.2	15	12.96	1.03	94.4
		30	24.82	1.43	99.2
		45	35.58	1.72	96.8
	13.6	15	9.12	6.83	65.8
		30	15.72	8.70	56
		45	21.04	9.93	54.2
0.2	3.8	15	6.50	0.50	98.8
		30	23.40	0.68	98.8
		45	38.30	0.75	98
	5.2	15	12.12	1.00	94.4
		30	24.18	1.42	99.2
		45	34.66	1.70	96.8
	13.6	15	8.64	6.66	66.2
		30	15.00	8.53	56.8
		45	19.90	9.69	54.4
0.4	3.8	15	6.16	0.49	98.8
		30	22.76	0.67	98.8
		45	36.34	0.74	98.2
	5.2	15	11.84	0.99	94.6
		30	23.70	1.40	99.2
		45	33.86	1.68	96.8
	13.6	15	8.10	6.48	66.8
		30	14.4	8.39	57.6
		45	18.80	9.45	55.4

TABLE 5.7 Continued

h_b	$ID, \emptyset$	P_i	Mean effective pressure, P_e	Mean discharge q	Cu
m	mm	kPa	kPa	liters/min	%
0.6	3.8	15	5.78	0.49	98.8
		30	22.06	0.67	98.8
		45	34.10	0.73	98
	5.2	15	11.44	0.97	94.6
		30	23.02	1.38	99.2
		45	33.06	1.66	96.8
	13.6	15	7.76	6.35	68.4
		30	14.06	8.30	58
		45	17.60	9.17	55.4
0.8	3.8	15	5.50	0.48	98.8
		30	21.30	0.66	98.8
		45	32.46	0.73	98.4
	5.2	15	11.02	0.95	94.8
		30	22.52	1.37	99.4
		45	32.20	1.64	96.8
	13.6	15	7.24	6.16	69.6
		30	13.68	8.17	58.6
		45	16.28	8.85	55.8
1.0	3.8	15	5.16	0.47	98.8
		30	20.74	0.66	98.8
		45	30.92	0.72	98.4
	5.2	15	10.56	0.94	95.6
		30	21.80	1.34	99.4
		45	31.18	1.61	97
	13.6	15	6.84	6.00	72.8
		30	13.08	8.02	62.2
		45	15.40	8.61	61.8

h_b: bubbler height; P_i: initial operating pressure; Cu: discharge uniformity coefficient.

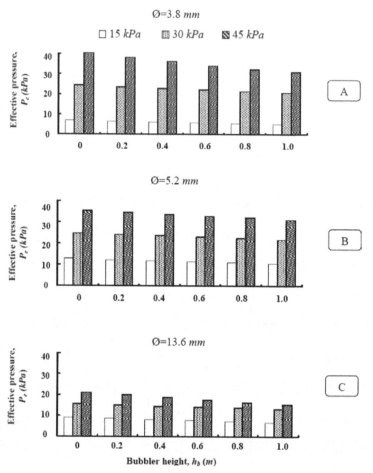

FIGURE 5.17 The relationship between bubbler height and effective pressure for different bubbler diameters and initial operating pressures (case 1).

Likewise for 13.6 *mm* bubbler tube diameter, P_e values were 9.12, 15.72 and 21.04 kPa, at (P_i) 15, 30 and 45 kPa, as shown in Figure 5.17c. At initial operating pressures of 15, 30 and 45 kPa, the discharge (q) values were 6.83, 8.7 and 9.93 liters/min, and the discharge uniformity (Cu) were 65.8, 56 and 54%, respectively as shown in Figures 5.18c and 5.19c. The discharge uniformity coefficient values were decreased with increase in the initial operating pressure from 15 to 45 kPa for all bubbler tube heights.

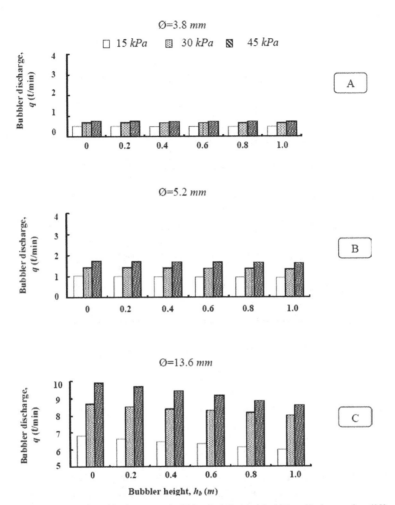

FIGURE 5.18 Relationship between bubbler height and bubbler discharge for different bubbler diameters and initial operating pressures (case 1).

b. Bubbler height of 0.2 m

For 3.8 *mm* bubbler tube diameter, the mean effective pressure (P_e) values were 6.5, 23.4 and 38.3 kPa, at P_i of 15, 30 and 45 kPa, as shown in Figure 5.17a. At initial operating pressures of 15, 30 and 45 kPa, the discharge (q) values were 0.50, 0.68 and 0.75 liters/min and the discharge uniformity (Cu) were 98.8, 98.8 and 98.0%, respectively as shown in Figures (4.4.A) and (4.5.A).

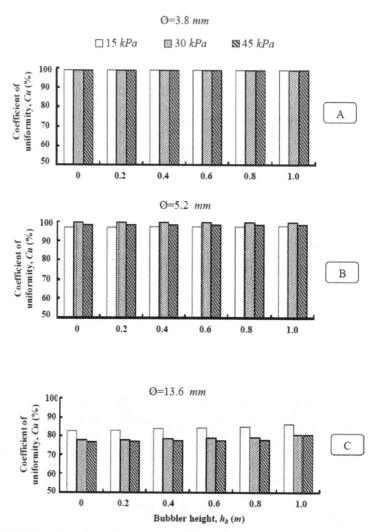

FIGURE 5.19 The relationship between bubbler height and coefficient of uniformity for different bubbler diameters and initial operating pressures (case 1).

Similarly, for 5.2 *mm* bubbler tube diameter: the mean effective pressure (P_e) values were 12.12, 24.18 and 34.66 *kPa*, at P_i 15, 30 and 45 kPa, as shown in Figure 5.3b. At initial operating pressures 15, 30 and 45 *kPa* the discharge (q) values were 1.0, 1.42 and 1.7 liters/min, and the discharge uniformity coefficients (Cu) were 94.4, 99.2 and 96.8%, respectively as shown in Figures 5.18b and 5.19b.

Meanwhile, for 13.6 *mm* bubbler tube diameter: the mean effective pressure (P_e) values were 8.64, 15.0 and 19.9 kPa, at P_i 15, 30 and 45 kPa, as shown in Figure 5.17c. At initial operating pressures 15, 30 and 45 kPa: the discharge (q) values were 6.66, 8.53 and 9.69 liters/min; and the discharge uniformity coefficients (Cu) were 66.2, 56.8 and 54.4%, respectively as shown in Figures 5.18c and 5.19c.

c. Bubbler height of 0.4 m

For 3.8 *mm* bubbler tube diameter: the mean effective pressure (P_e) was proportionally increased with increase in initial operating pressures (P_i). The P_e values were 6.16, 22.76 and 36.34 kPa, at P_i of 15, 30 and 45 kPa, as shown in Figure 5.17a. At initial operating pressures 15, 30 and 45 kPa: the discharge (q) values were 0.49, 0.67 and 0.74 liters/min and the discharge uniformity coefficients (Cu) were 98.8, 98.8 and 98.2%, respectively as shown in Figures 5.18a and 5.19a.

Meanwhile, for 5.2 mm bubbler tube diameter: the mean effective pressure (P_e) values were 11.84, 23.7 and 33.86 kPa, at P_i of 15, 30 and 45 kPa, as shown in Figure 5.17b. At initial operating pressures 15, 30 and 45 kPa: the discharge (q) values were 0.99, 1.40 and 1.68 liters/min; and the discharge uniformity coefficients (Cu) were 94.6, 99.2 and 97.0%, respectively as shown in Figures 5.18b and 5.19b.

For 13.6 *mm* bubbler tube diameter: the mean effective pressure (P_e) values were 8.1, 14.4 and 18.8 kPa at P_i of 15, 30 and 45 kPa, as shown in Figure 5.17c. At initial operating pressures of 15, 30 and 45 kPa: the discharge (q) values were 6.48, 8.39 and 9.45 liters/min and the discharge uniformity (Cu) were 66.8, 57.6 and 55.4%, respectively as shown in Figures 5.18c and 5.19c.

d. Bubbler height of 0.6 m

For 3.8 *mm* bubbler tube diameters: the mean effective pressure (P_e) values were 5.78, 22.06 and 34.1 kPa, at P_i of 15, 30 and 45 kPa, as shown in Figure 5.17a. At initial operating pressures 15, 30 and 45 kPa: the discharge (q) values were 0.49, 0.67 and 0.73 liters/min; and the discharge uniformity (Cu) were 98.8, 98.8 and 98.0%, respectively as shown in Figures 5.18a and 5.19a.

For 5.2 mm bubbler tube diameters, the mean effective pressure (P_e), was proportionally increased with increase in initial operating pressures (P_i). The P_e values were 11.44, 23.02 and 33.06 kPa at P_i of 15, 30 and 45 kPa, as shown in Figure 5.17. At initial operating pressures 15, 30 and 45 kPa: the discharge (q) values were 0.97, 1.38 and 1.66 liters/min; and the discharge uniformity (Cu) were 94.6, 99.2 and 96.8%, respectively as shown in Figures 5.18b and 5.19b.

Meanwhile, the mean effective pressure (P_e) values were 7.76, 14.06 and 17.6 kPa for 13.6 mm bubbler tube diameter, at P_i of 15, 30 and 45 kPa, as shown in Figure 5.17c. At initial operating pressures 15, 30 and 45 kPa: the discharge (q) values were 6.35, 8.3 and 9.17 liters/min; and the discharge uniformity coefficients (Cu) were 68.4, 58 and 55.4%, respectively as shown in Figures 5.18c and 5.19c.

e. Bubbler height of 0.8 m

For 3.8 mm bubbler tube diameter: The mean effective pressure (P_e) values were 5.5, 21.3 and 32.46 kPa, at P_i of 15, 30 and 45 kPa, as shown in Figure 5.17a. At initial operating pressures of 15, 30 and 45 kPa, the bubbler discharge (q) values were 0.48, 0.66 and 0.73 liters/min and the discharge uniformity coefficients (Cu) were 98.8, 98.8 and 98.4%, respectively as shown in Figures 5.18a and 5.19a.

For 5.2 mm bubbler tube diameter: the mean effective pressure (P_e) values were 11.02, 22.52 and 32.2 kPa at P_i of 15, 30 and 45 kPa, as shown in Figure 5.17b. At initial operating pressures of 15, 30 and 45 kPa the discharge (q) values were 0.95, 1.37 and 1.64 liters/min and the discharge uniformity coefficients (Cu) were 94.8, 99.0 and 96.8%, respectively as shown in Figures 5.18b and 5.19b.

Meanwhile for 13.6 mm bubbler tube diameter: the mean effective pressure (P_e) values were 7.24, 13.68 and 16.28 kPa, at P_i of 15, 30 and 45 kPa, as shown in Figure 5.17c. At initial operating pressures of 15, 30 and 45 kPa, the discharge (q) values were 6.16, 8.17 and 8.85 liters/min and the discharge uniformity coefficients (Cu) were 69.6, 58.6 and 55.8%, respectively as shown in Figures 5.18c and 5.19c.

f. Bubbler height (1.0 m)

For 3.8 *mm* bubbler tube diameter: mean effective pressure (P_e) values were 5.16, 20.74 and 30.92 kPa, at P_i of 15, 30 and 45 kPa, as shown in Figure 5.17a. At initial operating pressures of 15, 30 and 45 kPa, the discharge (q) values were 0.47, 0.66 and 0.72 liters/min and the discharge uniformity coefficients (Cu) were 98.8, 98.8 and 98.4%, respectively as shown in Figures 5.18a and 5.19a.

Similarly, for 5.2 mm bubbler tube diameter: the mean effective pressure (P_e) values were 10.56, 21.8 and 31.18 kPa at P_i of 15, 30 and 45 *kPa*, as shown in Figures 5.17b. At initial operating pressures of 15, 30 and 45 kPa, the discharge (q) values were 0.94, 1.34 and 1.61 liters/min and the discharge uniformity coefficients (Cu) were 95.6, 99.4 and 96.2%, respectively as shown in Figures 5.18b and 5.19b.

Meanwhile for 13.6 mm bubbler tube diameter, the mean effective pressure (P_e) values were 6.48, 13.08 and 15.4 kPa for at P_i of 15, 30 and 45 *kPa*, as shown in Figure 5.17c. At initial operating pressures of 15, 30 and 45 kPa, the discharge (q) values were 6.0, 8.02 and 8.61 liters/min and the discharge uniformity coefficients (Cu) were 72.8, 62.2 and 61.8%, respectively as shown in Figures 5.18c and 5.19c.

It can be observed that the mean effective pressure (P_e) was decreased due to the increase in bubbler height. The bubbler discharge (q) was consequently decreases for all bubbler tube heights (h_b) from 0.0 to 1.0 *m* at three initial operating pressures, for the three bubbler tube diameters. The mean bubbler outlet heights, according to outlet elevation, gradually rise up from the datum and with variation in velocity head and pressure head, as shown in Figures 5.20 and 5.21. The mean effective pressure variation changed with bubbler heights and initial operating pressures. But the variation had a same pattern with different bubbler heights at each initial operating pressures, whereas it was had a different pattern for different initial operating pressures. The changes in pattern were due to the interaction between velocity head and pressure head.

Subsequently, the mean effective pressure (P_e) along the lateral pipe was decreased with increase in bubbler distance from inlet. As a result, the bubbler discharge (q) was decreased for all bubbler tube heights (h_b) from 0 to 1.0 m, at three initial operating pressures for the three bubbler tube

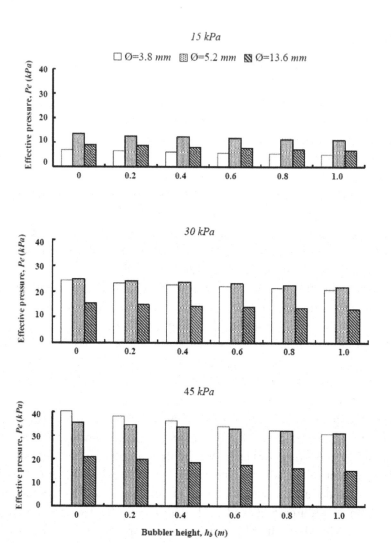

FIGURE 5.20 The relationship between bubbler height and effective pressure for different initial operating pressures and bubbler diameters (Case 1).

diameters as shown in Appendix I. This decrease in effective pressure (P_e) was normal due to the friction losses along the lateral pipe.

On the whole, for all bubbler tube diameters (*ID*), at all bubbler tube heights (h_b) from 0 to 1.0 m: The discharge uniformity coefficient (*Cu*) was relatively constant at the same initial operating pressure (P_i) for 3.8 mm bubbler diameter.

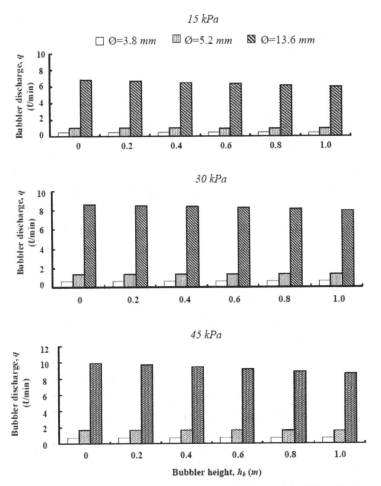

FIGURE 5.21   The relationship between bubbler height and bubbler discharge, for different initial operating pressures and bubbler diameters (Case 1).

While for 5.2 *mm* bubbler diameter: the uniformity coefficient (Cu) was increased with initial operating pressure (P_i) increasing from 15 to 30 kPa and was decreased with P_i increasing from 30 to 45 kPa.

But for 13.6 *mm* bubbler diameter: the discharge uniformity coefficient (Cu) was decreased with initial operating pressure (P_i) increasing from 15 to 45 kPa as shown in Figure 5.19. The highest values of discharge uniformity coefficient (Cu) were recorded with bubbler diameters of 5.2 and 3.8 mm, while Cu value was considered marginal for 13.6 mm.

These results agree with those by Reynolds et al. [39, 40], who indicated that bubbler hose diameters greater than 10 mm are not recommended for low-head bubbler system due to poor water distribution uniformity.

(i) Initial Operating Pressure (15 kPa)

For 3.8 mm bubbler tube diameter: discharge uniformity coefficient (Cu) was high and constant 98.8% at all bubbler tube heights (h_b). On the other hand at 5.2 mm, the discharge uniformity coefficient (Cu) was relatively constant value $94.5 \pm 0.1\%$ with h_b increase from 0.0 to 0.6 m; and was slightly increased from 94.8 to 95.6% with h_b increase from 0.8 to 1.0 m. However at 13.6 mm: the discharge uniformity coefficient (Cu) was relatively constant value $66 \pm 0.2\%$ with h_b 0.0 and 0.2 m, and (Cu) was increased from 66.8 to 72.8% with h_b increase from 0.4 to 1.0 m as shown in Figures 5.22a.

In conclusion, the discharge uniformity coefficient was more sensitive to increasing bubbler height with 13.6 mm diameter than Cu at 5.2 mm. Generally for the two previous diameters, the uniformity was increased with bubbler height increase from 0.4 to 1.0 m.

(ii) Initial Operating Pressure (30 kPa)

For 3.8 mm bubbler diameter: discharge uniformity coefficient (Cu) was high and constant 98.8% at all bubbler tube heights (h_b). On the other hand at 5.2 mm: the discharge uniformity coefficient (Cu) was a constant value of 99.2% with h_b from 0.0 to 0.6 m, and was slightly increased to 99.4% with increase in h_b from 0.8 to 1.0 m. However, at 13.6 mm diamter: the discharge uniformity coefficient (Cu) was relatively constant value of $56 \pm 0.4\%$ with h_b of 0.0 and 0.2 m, and (Cu) was increased from 57.6 to 62.2% with increase in h_b from 0.4 to 1.0 m.

To summarize, the discharge uniformity coefficient was more sensitive to increasing bubbler height with 13.6 mm diameter than Cu at 5.2 mm. Generally, the uniformity coefficient was increased with the increase in bubbler height from 0.4 to 1.0 m, as shown in Figure 5.22b.

(iii) Initial Operating Pressure (45 kPa)

For 3.8 mm bubbler diameter: discharge uniformity coefficient (Cu) was relatively constant value $98.1 \pm 0.1\%$ at bubbler tube heights (h_b)

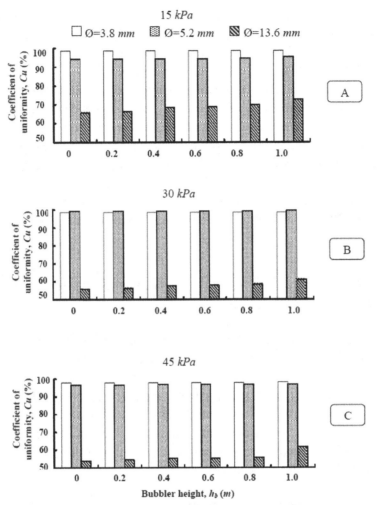

FIGURE 5.22 The relationship between bubbler height and coefficient of uniformity, for different initial operating pressures and bubbler diameters (Case 1).

from 0.0 to 0.6 m, and was slightly increased to 98.4% with h_b from 0.8 to 1.0 m. On the other hand at 5.2 mm: the discharge uniformity coefficient (Cu) was constant value of 96.8% with h_b from 0.0 to 0.8 m, and was slightly increased to 97% at h_b 1.0 m. However at 13.6 mm, the discharge uniformity coefficient (Cu) was relatively constant value of 54.3 ± 0.1% at h_b of 0.0 and 0.2 m. The Cu was increased from 55.4 to 61.8% with increase in h_b from 0.4 to 1.0 m.

In conclusion, the discharge uniformity coefficient was more sensitive to increasing bubbler height with 13.6 mm diameter than at 5.2 mm. Generally, the uniformity coefficient was increased with increase in bubbler height from 0.4 to 1.0 *m*, as shown in Figure 5.22c.

In general, there was an inverse relationship between bubbler discharge and uniformity coefficient. As a result, the discharge uniformity coefficient was increased with increase in bubbler height, due to decease in discharge (*q*), as shown in Figure 5.22. These results are in good agreement with those by El-meseery [16]. The results indicated that the highest value of discharge uniformity coefficients was obtained at initial operating pressure of 30 kPa and 5.2 mm bubbler tube diameter, for all tested ranges of bubblers tube diameters and initial operating pressures.

5.4.2.1.2 Bubbler Outlet Heights Parallel to the Hydraulic Gradient Line (Second Case)

For this experiment, the effective pressure (P_e), corresponding to the highest value of discharge uniformity coefficient obtained with bubbler outlets (h_b) at the same level, was selected to use it with h_b parallel to the hydraulic gradient line. Then (P_e) values above and below the selected one were tested to give the highest uniformity. The relationship between bubbler tube diameter (Ø), initial operating pressure (P_i), effective pressure (P_e), bubbler discharge (*q*) and coefficient of uniformity (*Cu*) are given displayed Table 5.8 and Appendix II.

a. Bubbler tube diameter (3.8 mm)

For initial operating pressure (*Pi*) of 15 kPa: the discharge (*q*) values were 0.5, 0.51 and 0.52 liters/min and the uniformity coefficient (*Cu*) values were 99.8, 99.2 and 98.6%, at mean effective pressure (P_e) of 6, 7 and 8 kPa, respectively, as shown in Figures 5.23a and 5.24a. The relative discharge and uniformity difference were calculated as –1.96 and 1.96% with equation (19), 0.6 and -0.6% with equation (20), respectively as shown in Figures 5.25 and 5.26.

TABLE 5.8 Hydraulic Properties of Bubbler for Different Locations and Internal Bubbler Diameters at the Same Effective Pressure (Case 2).

Ø ID	P_i	P_e	Mean discharge	Cu	Relative discharge difference	Relative uniformity difference
mm	kPa	kPa	Liters/min	%	–	–
3.8	15	6	0.50	99.8	–1.96	0.60
		7	0.51	99.2	0.00	0.00
		8	0.52	98.6	1.96	–0.60
	30	26	0.69	99.4	–2.85	0.20
		27	0.70	99.2	0.00	0.00
		28	0.71	98.7	1.40	–0.50
	45	38	0.74	99.2	–2.60	0.20
		39	0.75	99.0	0.00	0.00
		40	0.76	98.5	2.60	–0.40
5.2	15	10	0.91	99.2	–3.00	0.40
		11	0.95	98.8	0.00	0.00
		12	1.00	98.6	3.00	–0.20
	30	27	1.49	99.6	–1.97	0.30
		28	1.52	99.3	0.00	0.00
		29	1.55	98.7	1.97	–0.60
	45	30	1.58	99.1	–1.90	0.20
		31	1.60	98.9	0.00	0.00
		32	1.63	98.6	1.90	–0.30
13.6	15	7	6.06	95.4	–5.70	0.84
		8	6.44	94.6	0.00	0.00
		9	6.79	92.8	5.60	–1.90
	30	12	7.73	84.6	–3.40	2.17
		13	8.00	82.8	0.00	0.00
		14	8.30	80.2	3.75	–3.14
	45	22	10.20	62.4	–1.40	4.70
		23	10.35	59.6	0.00	0.00
		24	10.56	56.3	1.90	–5.50

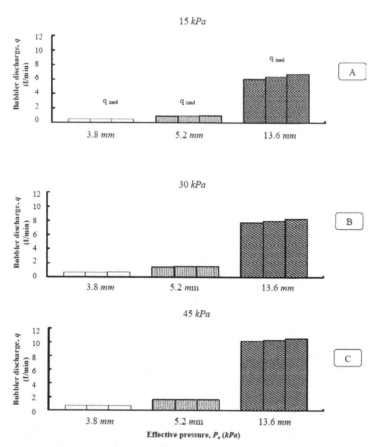

FIGURE 5.23   The relationship between effective pressure and bubbler discharge at the same initial operating pressure with different bubbler diameters (Case 2).

For (*Pi*) 30 kPa: the mean effective pressure (P_e) values were 26, 27 and 28 kPa. The discharge (*q*) values were 0.69, 0.7 and 0.71 liters/min and the discharge uniformity coefficient (*Cu*) values were 99.4, 99.2 and 98.7%, respectively as shown in Figures 5.23b and 5.24b. The relative discharge and uniformity difference values were –2.85 and 1.4% with equation (19), and 0.2 and –0.5% with Eq. (20), respectively as shown in Figures 5.25 and 5.26.

For (*Pi*) 45 kPa: the discharge (*q*) values were 0.74, 0.75 and 0.76 liters/min, and the discharge uniformity coefficient (*Cu*) values were 99.2,

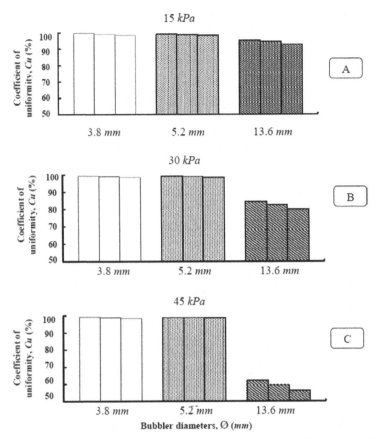

FIGURE 5.24   The relationship between bubbler diameter and coefficient of uniformity at the same initial operating pressure (Case 2).

99.0 and 98.5%, at mean effective pressure (P_e) values 38, 39 and 40 kPa, as shown in Figures 5.23c and 5.24c. The relative discharge and uniformity difference were –2.6% and 2.6% with Eq. (19), and 0.2% and –0.4% with equation (20), respectively as shown in Figures 5.25 and 5.26.

b. Bubbler tube diameter (5.2 mm)

For (*Pi*) 15 kPa: the mean effective pressure (P_e) was 10, 11 and 12 kPa. The discharge (*q*) values were 0.91, 0.95 and 1.0 liters/min and the discharge uniformity coefficient (*Cu*) values were 99.2, 98.8 and 98.6%,

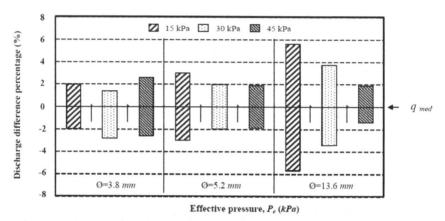

FIGURE 5.25   Relative discharge percentage based on Eq. (19): Case 2.

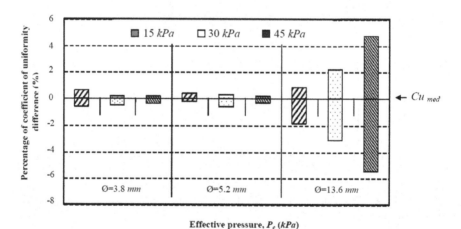

FIGURE 5.26   Relative uniformity coefficient percentage based on Eq. (20): Case 2.

respectively as shown in Figures 5.23a and 5.24a. The relative discharge and uniformity difference were -3.0% and 3.0% with equation (19), and 0.4% and -0.2% with equation (20), respectively as shown in Figures 5.25 and 5.26.

For (Pi) 30 kPa: The discharge (q) values were 1.49, 1.52 and 1.55 liters/min and the discharge uniformity coefficient (Cu) values were 99.6, 99.3 and 98.7%, with mean effective pressure (P_e) of 27, 28 and 29 kPa. respectively as shown in Figures 5.23a and 5.24a. The relative discharge and uniformity difference were −1.97% and 197% with Eq. (19), and 0.3% and −0.6% with equation (20), respectively as shown in Figures 5.25 and 5.26.

For (Pi) 45 kPa: the mean effective pressure (P_e) values were 30, 31 and 32 kPa. The discharge (q) values were 1.58, 1.6 and 1.63 liters/min and the discharge uniformity (Cu) values were 99.1, 98.9 and 98.6%, respectively as shown in Figures 5.23c and 5.24c. The relative discharge and uniformity difference was −1.9% and 1.9% with equation (19), and 0.2% and −0.3% with equation (20), respectively as shown in Figures 5.25 and 5.26.

c. Bubbler tube diameter (13.6 mm)

For (Pi) 15 kPa: the mean effective pressures (P_e) were 7, 8 and 9 kPa. The discharge (q) values were 6.06, 6.44 and 6.79 liters/min and the discharge uniformity (Cu) values were 95.4, 94.6 and 92.8%, respectively as shown in Figures 5.23a and 5.24a. The relative discharge and uniformity difference was −5.7% and 5.6% with equation (19), and 0.84% and −1.9% with Eq. (20), respectively as shown in Figures 5.25 and 5.26.

For (Pi) 30 kPa: the mean effective pressure (P_e) values were 12, 13 and 14 kPa. The discharge (q) values were 7.73, 8 and 8.3 liters/min and the discharge uniformity (Cu) values were 84.6, 82.8 and 80.2%, respectively as shown in Figures 5.23b and 5.24b. The relative discharge and uniformity difference was −3.4% and 3.75% with Eq. (19), and 2.17% and −3.14% with Eq. (20), respectively as shown in Figures 5.25 and 5.26.

For (Pi) 45 kPa: the discharge (q) values were 10.2, 10.35 and 10.56 liters/min and the discharge uniformity (Cu) values were 62.4, 59.6 and 56.3%, with mean effective pressure (P_e) values were 22, 23 and 24 kPa, respectively as shown in Figures 5.23c and 5.24c. The relative discharge and uniformity difference was -1.4% and 1.9% with equation (19), and 4.7% and -5.5% with equation (20), respectively as shown in Figures 5.25 and 5.26.

It can be noticed that bubbler discharges are proportionally the same along the lateral pipe in case of bubbler outlets, parallel to the hydraulic gradient line as shown in the Appendix II.

In general, we can conclude that there is an inverse relationship between discharge uniformity and effective pressures, for all bubbler tube diameters and initial operating pressure. It is clear that the discharge uniformity was very high, when bubbler outlets were parallel to the hydraulic gradient line compared to bubbler outlets at the same height. These results are agreement with Rawlins [37], Behoteguy and Thornton [8], Hull [24] and El-meseery [16].

In conclusion, the bubbler tube diameter 13.6 mm gave highest percentage of difference for uniformity and discharge compared to 3.8 and 5.2 mm diameters as shown in Figures 5.25 and 5.26. This study does not recommend bubbler diameter of 13.6 mm in low head bubbler irrigation systems.

In bubbler tube diameters of 3.8 and 5.2 mm, there were no significant changes in (Cu) between initial operating pressures from 15 to 45 kPa, as shown in Figure 5.27a and 5.27b. On the other hand: for bubbler diameter 13.6 mm, discharge uniformity was decreased with increase in initial operating pressure from 15 to 45 kPa as shown in Figure 5.27c. These results are in agreement with those by Ngigi [34].

5.4.2.2   Bubbler Tube Height

5.4.2.2.1   Theoretical Bubbler Tube Height

Due to importance of calculation of bubbler height to get high values of coefficient of uniformity, the elevations of each bubbler tube were calculated by equation to get the same flow of each bubbler tube along the lateral pipe.

5.4.2.2.2   Validation of Theoretical Equation for Bubbler Height

Table 5.9 and Appendix III show the relationship between experimental values of bubbler height (hbx) and estimated values of bubbler height (hbc). The results indicate that the bubbler height was decreased with the bubbler distance (l_i) downstream. Using variance analysis, T-test was used to determine *statistical* differences between the means of two groups (experimental and calculated bubbler height). Data was compared to T-values in Table 5.9.

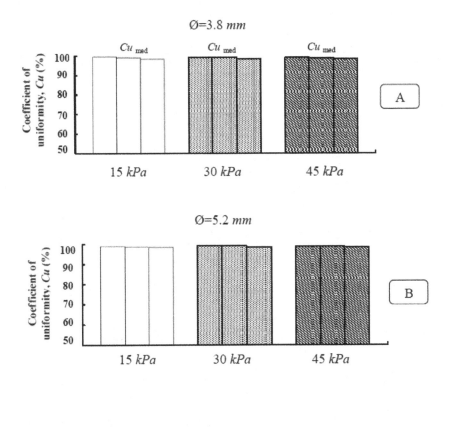

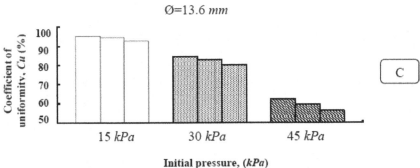

FIGURE 5.27 The relationship between initial operating pressure and coefficient of uniformity at the same bubbler diameter (Case 2).

TABLE 5.9 Mean Values of Theoretical and Experimental Bubbler Heights, for Different Bubbler Diameters and Initial Operating Pressure (Case 2)

Ø ID	P_i	Discharge q	Bubbler height, h_b			
			Experimental	Calculated	T-value	T- table
mm	kPa	liters/min		m		
3.8	15	0.50	0.38	0.39	0.20	2.3- 3.4
		0.51	0.34	0.35	0.10	2.3- 3.4
		0.52	0.29	0.30	0.10	2.3- 3.4
	30	0.69	0.99	1.00	0.68	2.3- 3.4
		0.70	0.93	0.94	0.66	2.3- 3.4
		0.71	0.90	0.91	0.66	2.3- 3.4
	45	0.74	0.87	0.88	0.86	2.3- 3.4
		0.75	0.74	0.75	0.86	2.3- 3.4
		0.76	0.67	0.68	0.85	2.3- 3.4
5.2	15	0.91	0.38	0.39	0.65	2.3- 3.4
		0.95	0.30	0.31	0.64	2.3- 3.4
		1.00	0.19	0.20	0.64	2.3- 3.4
	30	1.49	0.26	0.27	0.91	2.3- 3.4
		1.52	0.23	0.24	0.91	2.3- 3.4
		1.55	0.23	0.24	0.90	2.3- 3.4
	45	1.58	0.97	0.98	0.95	2.3- 3.4
		1.60	0.89	0.90	0.95	2.3- 3.4
		1.63	0.79	0.80	0.95	2.3- 3.4
13.6	15	6.06	0.70	0.71	0.92	2.3- 3.4
		6.44	0.67	0.68	0.91	2.3- 3.4
		6.79	0.65	0.66	0.91	2.3- 3.4
	30	7.73	1.13	1.14	0.98	2.3- 3.4
		8.00	1.10	1.11	0.97	2.3- 3.4
		8.30	1.05	1.06	0.97	2.3- 3.4
	45	10.20	0.74	0.75	0.98	2.3- 3.4
		10.35	0.70	0.71	0.98	2.3- 3.4
		10.56	0.64	0.65	0.98	2.3- 3.4

a. Bubbler tube diameter (3.8 mm)

For initial operating pressure (Pi) of 15 kPa: the mean experimental bubbler heights (hbx) were 0.38, 0.34 and 0.29 m, and the calculated bubbler heights (hbc) were 0.39, 0.35 and 0.30 m, with mean bubbler discharge (q) of 0.50, 0.51 and 0.52 liters/min, respectively as shown in Table 5.9. The calculated T-value between experimental and calculated bubbler heights was 0.2, 0.1 and 0.1, respectively.

For Pi of 30 kPa: the mean experimental bubbler height (hbx) values were 0.99, 0.93 and 0.9 m, and the calculated bubbler height (hbc) values were 1.0, 0.94 and 0.91 m, with mean bubbler discharge (q) of 0.69, 0.70 and 0.71 liters/min, respectively as shown in Table 5.9. The calculated T-value between experimental and calculated bubbler heights was 0.68, 0.66 and 0.66, respectively.

For Pi of 45 kPa: the mean experimental height (hbx) values were 0.87, 0.74 and 0.67 m, and the calculated bubbler height (hbc) values were 0.88, 0.75 and 0.68 m, with mean bubbler discharge (q) of 0.74, 0.75 and 0.76 liters/min, respectively as shown in Table 5.9. The calculated T-value between experimental and calculated bubbler heights was 0.86, 0.86 and 0.85, respectively.

b. Bubbler tube diameter (5.2 mm)

For initial operating pressure (Pi) of 15 kPa: the mean experimental bubbler height (hbx) values were 0.38, 0.30 and 0.19 m, and the calculated bubbler height (hbc) values were 0.39, 0.31 and 0.20 m, with mean bubbler discharge (q) of 0.91, 0.95 and 1.0 liters/min, respectively as shown in Table 5.9. The calculated T-value between experimental and calculated bubbler heights was 0.65, 0.64 and 0.64, respectively.

For Pi of 30 kPa: the mean experimental bubbler height (hbx) values were 0.26, 0.23 and 0.23 m, and the calculated bubbler height (hbc) values were 0.27, 0.24 and 0.24 m, with mean bubbler discharge (q) of 1.49, 1.52 and 1.55 liters/min, respectively as shown in Table 5.9. The calculated T-value between experimental and calculated bubbler heights was 0.91, 0.91 and 0.90, respectively.

For *Pi* of 45 *kPa*: the mean experimental height (*hbx*) values were 0.97, 0.89 and 0.79 m, and the calculated bubbler height (*hbc*) values were 0.98, 0.90 and 0.80 m, with mean bubbler discharge *(q)* of 1.58, 1.6 and 1.63 liters/min, respectively as shown in Table 5.9. The calculated *T*-value between experimental and calculated bubbler heights was 0.95, 0.95 and 0.95, respectively.

c. Bubbler tube diameter (13.6 mm)

For initial operating pressure (*Pi*) of 15 kPa: the mean experimental bubbler height (*hbx*) values were 0.70, 0.67 and 0.65 m, and the calculated bubbler height (*hbc*) values were 0.71, 0.68 and 0.66 m, with mean bubbler discharge *(q)* of 6.06, 6.44 and 6.79 liters/min, respectively as shown in Table 5.9. The calculated *T*-value between experimental and calculated bubbler heights was 0.92, 0.91 and 0.91, respectively.

For *Pi* of 30 kPa: the mean experimental bubbler height (*hbx*) values were 1.13, 1.10 and 1.05 m, and the calculated bubbler height (*hbc*) values were 1.14, 1.11 and 1.06 m, with mean bubbler discharge *(q)* of 7.73, 8.0 and 8.30 liters/min, respectively as shown in Table 5.9. The calculated *T*-value between experimental and calculated bubbler heights was 0.98, 0.97 and 0.97, respectively.

For *Pi* of 45 kPa: the mean experimental height (*hbx*) values were 0.74, 0.70 and 0.64 m, and the calculated bubbler height (*hbc*) values were 0.75, 0.71 and 0.65 m, with mean bubbler discharge *(q)* of 10.2, 10.35 and 10.56 liters/min, respectively as shown in Table 5.9. The calculated *T*-value between experimental and calculated bubbler heights was 0.98, 0.98 and 0.98, respectively.

Based on data in Table 5.9 and Appendix III, it was concluded that the bubbler height was decreased along the lateral line. Because values of *T*-table were greater than *T* calculated values, the hypothesis in this chapter is accepted. Therefore, Eq. (29) can be used to estimate the bubbler height along lateral line. The relationship between calculated and experimental bubbler heights is presented in Figure 5.28. It is observed that the relationship between calculated and experimental bubbler heights is linear, with a high coefficient of determination. All regression coefficients were significant at P = 0.01.

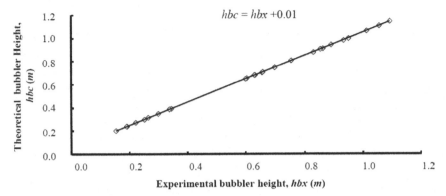

FIGURE 5.28   The relationship between theoretical and experimental bubbler heights.

The equation after adjustment for the error between calculated and experimental bubbler heights is as follows:

$$hbn = \left(H_i - \left(\frac{q}{k} \right)^{\left(\frac{1}{x} \right)} - \left(61111 \times q^{1.75} \times D^{-4.75} (s + cl) \sum_{n=1}^{N} (N - n + 1)^{1.75} \right) \right) - 1$$

(29)

5.5 CONCLUSIONS

There is a great need to apply modern irrigation systems to reduce water losses and energy consumption under Egyptian conditions. Bubbler irrigation systems are not frequently used under Egyptian conditions, because farmers lack knowledge and awareness.

Low-head bubbler system differs from other micro-irrigation systems, because it is based on gravity-flow at an operating pressure head of as low as 1 m (3.3 ft) and does not require elaborate filtration systems. The experimental work was carried out at Research Farm of Faculty of Agriculture, Suez Canal University, Rgypt. Author evaluated the effects of three bubbler tube diameters (Ø) at different initial operating pressures (P_i) on discharge uniformity coefficients (Cu) under two different cases: Case 1 – bubbler outlets at same elevation; and Case 2 – bubbler outlets parallel to the hydraulic gradient line. Also, the optimum bubbler height (h_b) of each bubbler tube diameter was determined to achieve high discharge uniformity, when bubbler outlets were parallel to the hydraulic gradient line. The variables under study were: operating pressures of 15, 30 and 45 kPa, bubblers diameters of 3.8, 5.2 and 13.6 mm with permutations and combinations of six bubbler heights (0.0, 0.2, 0.4, 0.6, 0.8 and 1.0 m), in first case; and combinations of three effective pressures (P_e) in the second case. Following conclusions were drawn based on the results in this chapter:

1. The mean effective pressure (P_e) and the mean discharge (q) were proportionally increased with increase in initial operating pressure (P_i), for all bubbler tube heights and diameters.
2. The mean effective pressure (P_e) was decreased due to increase in bubbler height. The bubbler discharge (q) was consequently decreased for all bubbler tube heights (h_b) from 0.0 to 1.0 m, at three initial operating pressures for the three-bubbler tube diameters.
3. The mean effective pressure (P_e) along the lateral pipe was decreased with increasing bubbler distance from inlet.
4. In the first case, there was an inverse relationship between discharge and uniformity. The highest values of discharge uniformity coefficients (Cu) were recorded with 5.2 and 3.8 mm diameters, while (Cu) value was considered marginal for ID of 13.6 mm.
    *   For ID 3.8 mm: the discharge uniformity (Cu) at all bubbler tube heights from 0.0 to 1.0 m was relatively constant (98.8 to 98.4%) with initial operating pressure from 15 to 45 kPa;
    *   For ID 5.2 mm: the uniformity coefficient (Cu) was fluctuated from 94.4 to 97.0% with initial operating pressure (Pi) increasing from 15 to 45 kPa;
    *   For ID 13.6 *mm*: the discharge uniformity coefficient (Cu) was decreased from 65.8 to 61.8% with initial operating pressure (Pi) increasing from 15 to 45 kPa.
5. In the second case for bubbler outlets parallel to the hydraulic gradient line: bubbler discharge was almost proportionally same along the lateral pipe. It is clear that the discharge uniformity in the second case was higher than the first case, but there were no significant changes in Cu among ID 3.8 and 5.2 mm with initial operating pressures increasing from 15 to 45 kPa compared with the ID 13.6 mm of bubbler tube diameter.

6. Due to no significant differences in (*Cu*) values between two cases of low head bubbler design, it was recommended to use a simple design in the first case than the second case with bubbler diameter 3.8 and 5.2 mm compared with 13.6 mm bubbler tube diameter.

5.6 SUMMARY

This research investigated the performance of three bubbler tube diameters of 3.8, 5.2 and 13.6 mm at three initial operating pressure of 15, 30, 45 kPa to determine optimum operating conditions to obtain high discharge uniformity. The experimental work was conducted at the farm of Agricultural Faculty, Suez Canal University, Ismailia. The coefficient of uniformity (*Cu*) was evaluated in two cases: First, when bubbler outlets heights were at the same elevation at all locations 0.0, 0.2, 0.4, 0.6, 0.8 and 1.0 m. The results show that the highest values of the coefficient of uniformity were obtained at an operating pressure of 30 kPa and for bubbler diameter of 5.2 mm, values were almost constant with an average of 99.3%. In the second case with bubbler outlets parallel to the hydraulic gradient line with three effective pressures for each initial operating pressure, the results show that all bubbler tubes were along the lateral line and gave the same discharge for 3.8 and 5.2 mm, but the discharge was different for 13.6 mm bubbler tube diameter.

The recommended bubbler diameter is 5.2 mm with 30 kPa initial operating pressure to achieve high discharge uniformity. In addition, with 5.2 mm higher length of lateral line can be used than with 3.8 mm bubbler diameter, thus minimizing initial irrigation system cost. Also, bubbler diameters 13.6 mm are not recommended for low-head bubbler systems due to poor water distribution uniformity.

KEYWORDS

- ASABE
- ASAE

- bubbler discharge
- bubbler irrigation
- bubbler irrigation performance
- Cairo University
- coefficient of uniformity
- desert land
- drip irrigation
- Egypt
- emitter
- emitter flow variation
- field evaluation
- low head irrigation
- micro irrigation
- pea
- regression coefficients

REFERENCES

1. AboZaid, M. A., M. F. Khairy., Z. Y. Abd Ellatif, and A. A. El-meseery, 1998. Bubbler irrigation system design. *Misr. J. Ag. Eng.*, 15:621–639.
2. ASABE Standards, 2006. *EP 405.1 FEB03: Design and Installation of micro irrigation systems*. St Joseph, Michigan, pages 942–945.
3. ASAE Standards, 1999. *EP 458: Field evaluation of micro irrigation systems*. St. Joseph, Michigan, pages 918–924.
4. ASAE Standards, 2000. *EP 405.1: Design and installation of micro irrigation systems*. St Joseph, Michigan, pages 889–893.
5. ASAE Standards, 2003. *EP 405.1 FEB03: Design and installation of micro irrigation systems*. St Joseph, Michigan, pages 901–905.
6. Awady, M. N. and I. M. Habib, 1992. *Irrigation methods of desert land*. Open Ed., Cairo University: pages 326–393 (In Arabic).
7. Awady, M. N., G. W. Amerhom and M. S. Zaki, 1975. Trickle irrigation trial on pea in conditions typical of Qalubia. *Egypt. Hort. Sci.,* 3(1):99–110.
8. Behoteguy, D. and J. R. Thornton, 1980. Operation and installation of a bubbler irrigation system. San Antonio Convention C., Trans. of the ASAE paper No. 80–2059.
9. Benami, A. and A. Ofen, 1984. *Irrigation Engineering*. Scientific Pub. (IESP). Technion City, Haifa, Israel. 257 pages.

10. Boswell, M. J., 1985. *Micro irrigation design manual*. El Cajon, Calif.: James Hardie Irrigation Co., USA, (6):27–30.

11. Bralts, V., 1978. *The effects of emitter flow variation on the design of single and dual chamber drip irrigation systems*. Unpublished M.SC Thesis. Hawaii University.

12. Braud, H. J. and A. M. Soon, 1981. Trickle irrigation lateral design on sloping fields. *Trans of the ASAE*, 24(4):941–944.

13. Byrnes, D., 2007. *AutoCAD 2008 for Dummies*. 2ed; Wiley Publishing. Co, Indiana.

14. Carr, M. and M. G. Kay, 1980. Bubbler irrigation. *Horticultural industry*, March 11–12.

15. El-Lithy, A. M., 1998. *A study on an appropriate design for bubbler irrigation system*. Unpublished M.Sc. Thesis, Department of Agricultural Mechanization, Faculty of Agriculture, Ain Shams University, Egypt, pages 44–45.

16. El-meseery, A. A., 1993. *A study on some factors affecting on bubbler irrigation*. Unpublished M.Sc. Thesis, Department of Agricultural Engineering, Faculty of Agriculture, Al-Azhar University, Egypt, pages 16–27 and 39–40.

17. Goyal, Megh R., 2015. Research advances in sustainable micro irrigation, volumes 1 to 10. Apple Academic Press Inc.,

18. Harrington, G. J., 1971. *The mechanics of air locks (Air binding) in small diameter pipes*. New Zealand Agriculture Engineering Institute. Internal Report No. 2 (Unpublished).

19. Hermsmeier, L. F. and L. S. Willardson, 1970. Friction factors for corrugated plastic tubing. *J. Irrig. and Drain. Div.*, ASCE 96(IR 3):265–271.

20. Howell, T. A. and E. A. Hiler, 1972. Trickle irrigation system design. *Trans of the ASAE*, 72–221, St. Joseph, Michigan 49085.

21. Howell, T. A. and E. A. Hiler, 1974. Trickle irrigation lateral design. *Trans of the ASAE*, 17(5):902–908.

22. http://weather.nmsu.edu/dripdesign/coeffofvariation.htm

23. http://www.agf.gov.bc.ca/resmgmt/publist/500Series/590304–5.pdf

24. Hull, P. J., 1981. A low pressure irrigation system for orchard tree and plantation crops. The agricultural Engineer. Summer edition.

25. Hussein, N. S., 2007. *Evaluation of trickle irrigation designs based on uniformity concept*. Unpublished M.Sc. Thesis, Department of Agricultural Mechanization, Faculty of Agriculture, Ain Shams University, Egypt, pages 56–60.

26. James, L. G., 1988. *Principles of farm irrigation system design*. John Wiley and sons, New York, pp. 260–275.

27. Jordan, T. D., 1984. *A hand-book of gravity-flow water system*. Intermediate Technology Publications. London.

28. Karmeli, D., 1977. Classification and flow regime analysis of drippers. *J. Agric. Eng. Res.*, 22:165–167.

29. Keller, J. and D. Karmeli, 1974. Trickle irrigation design parameters. *Trans of the ASAE*, 17(4):678–684.

30. Keller, J. and D. Karmeli, 1975. *Trickle irrigation design*. Rain Bird Sprinkler Manufacturing Corporation, Glendora, California, USA, pp. 24–26, and 62–66.

31. Keller, J. and R. D. Bliesner, 1990. *Sprinkle and trickle irrigation*. Van Norstrand Reinhold. New York. 652 pages.

32. Lamm, F. R., J. E. Ayars and F. S. Nakayama, 2007. *Micro irrigation for crop production: design, operation, and management.* 13th ed., Italy. Elsevier Co: pages 533–570.

33. Nakayama, F. S. and D. A. Bucks, 1986. *Trickle irrigation for crop production.* Elsevier Co: pages 11–17.

34. Ngigi, S. N., 2008. *Technical evaluation and development of low-head drip irrigation systems in Kenya. Irrig. and Drain.,* 57:450–462.

35. Perold, P. R., 1977. Design of irrigation pipe laterals with multiple outlets. *Trans. ASAE,* 103 (IR3):179–195.

36. Phocaides, A., 2000. *Technical handbook on pressurized irrigation techniques.* FAO Consultant, Rome. Italy, pages 127–132.

37. Rawlins, S. L., 1977. Uniform irrigation with a low head bubbler system. *Agriculture and Water Management,* 1:167–178.

38. Reynolds, C. A., 1993. *Design and evaluation of bubbler irrigation systems.* Unpublished M.Sc. Thesis, University of Arizona. Department of Civil Engineering and Engineering Mechanics. Faculty of Engineering.

39. Reynolds, C. A. and M. Yitayew, 1995. Low-head bubbler irrigation systems – Part II: Air lock problems. *Agriculture and Water Management,* 29:25–35.

40. Reynolds, C. A., M. Yitayew and M. S. Peterson, 1995. Low-head bubbler irrigation systems – Part I: Design. *Agriculture and Water Management,* 29:1–24.

41. Solomon, K., 1979. Manufacturing variation of emitters in trickle irrigation system. *Trans of the ASAE,* 22(5):1034–1038.

42. Solomon, K. and J. C. Bezdek, 1980. Significant features of emitter flushing mode characteristics. *Trans of the ASAE.* 23(4):903–906.

43. Steel, R. G. D., J. H. Torrie and D. A Dickey, 1996. *Principles and procedures of statistics: Biometrical approach.* 3[ed] Ed, New York: McGraw-Hill.

44. Waheed, S. I., 1990. *Design criteria for low head bubbler irrigation systems.* Unpublished M.Sc. Thesis, University of Arizona.

45. Watters, G. Z. and J. Keller, 1978. Trickle irrigation tubing hydraulics. ASAE Technical paper No. 78–2015. St. Joseph, Michigan. 17 pages.

46. WU, I. P. and H. M. Gitlin, 1973. Hydraulic and uniformity for drip Irrigation. *J. Irrig. and Drain., ASCE,* 99(IR2):157–167.

47. Wu, I. P., K. Y. Abusaki and J. M. Irudayaraj, 1985. Computer simulation of total emitter flow variation. In: *Drip/trickle irrigation in action.* Drip/Trickle Irr. Cong., ASAE Pub., MI, USA, pages 873–875.

48. Wu, I. P., T. A. Howell and E. A. Hiler, 1979. *Hydraulic design of drip irrigation systems.* Tech. Bull. No. 105, Western Reg. Res. Proj. W-128 Report, Hawaii Ag. Exp. St., Univ. of Hawaii.

49. Yitayew, M., C. A. Reynolds and A. E. Sheta, 1995. Bubbler irrigation system design and management. In: Lamm, F.R. (Ed.) *Proceedings of the fifth international Micro irrigation Congress,* April 2–6, Orlando, FL. ASAE, St. Joseph, Michigan, pages 402–413.

APPENDIX I Bubbler effective pressure (Pe), discharge (q) and uniformity (Cu) along the lateral pipe at different initial operating pressures (Pi) for internal bubbler diameters (Ø) at the same bubbler heights (h_b) versus effective pressure (Case 1)

h_b m	Ø ID mm	P_i kPa	Bubbler discharge (ℓ/min)										Cu %
			Bubbler number										
			1		2		3		4		5		
			P_e	q	P_e	q	P_e	q	P_e	q	P_e	q	
0	3.8	15	8.2	0.53	7.8	0.52	6.9	0.51	6.3	0.50	5.9	0.49	98.8
		30	27.2	0.70	25.9	0.69	24.6	0.68	23.1	0.67	21.1	0.66	98.8
		45	44.7	0.78	44.2	0.78	42.2	0.77	37.7	0.75	33.5	0.73	98.2
	5.2	15	14.8	1.11	14.1	1.08	13.3	1.05	12.8	1.03	9.8	0.90	94.4
		30	25.1	1.44	25.1	1.44	24.9	1.43	24.7	1.43	24.3	1.42	99.2
		45	37.4	1.77	36.5	1.74	35.8	1.73	34.5	1.69	33.7	1.67	96.8
	13.6	15	10.9	7.42	9.9	7.1	8.7	6.69	8.2	6.53	7.9	6.41	65.8
		30	17.8	9.22	17.3	9.1	16.5	8.89	14.5	8.39	12.5	7.90	56
		45	25.9	10.9	21.8	10.1	20.4	9.84	19.6	9.65	17.5	9.15	54.2
0.2	3.8	15	7.6	0.52	7.2	0.51	6.4	0.50	5.90	0.49	5.4	0.48	98.8
		30	26.1	0.69	25.0	0.69	23.5	0.68	22.1	0.67	20.3	0.65	98.8
		45	43.0	0.78	41.4	0.77	39.8	0.76	35.6	0.74	31.7	0.72	98
	5.2	15	13.8	1.07	12.9	1.04	12.5	1.02	12.1	0.97	9.3	0.88	94.4
		30	24.7	1.43	24.5	1.42	24.2	1.41	23.9	1.41	23.6	1.41	99.2
		45	37.0	1.75	35.7	1.72	34.6	1.70	33.6	1.67	32.4	1.64	96.8
	13.6	15	10.5	7.28	9.3	6.89	8.2	6.53	7.7	6.35	7.5	6.26	66.2
		30	16.9	9.03	16.4	8.9	15.7	8.74	13.7	8.20	12.3	7.78	56.8
		45	24.4	10.6	20.7	9.88	19.2	9.56	18.3	9.36	16.9	9.01	54.4

APPENDIX 1 Continued

h_b m	Ø ID mm	P_i kPa	Bubbler discharge (ℓ/min)										Cu %
			Bubbler number										
			1		2		3		4		5		
			P_e	q	P_e	q	P_e	q	P_e	q	P_e	q	
0.4	3.8	15	7.2	0.51	6.9	0.51	6.1	0.49	5.5	0.48	5.1	0.48	98.8
		30	25.6	0.69	24.2	0.68	22.8	0.67	21.4	0.66	19.8	0.65	98.8
		45	41.0	0.77	39.9	0.76	37.4	0.75	33.6	0.73	29.8	0.71	98.2
	5.2	15	13.6	1.06	12.8	1.03	12.3	1.01	11.8	0.99	8.7	0.85	94.6
		30	24.3	1.42	23.9	1.41	23.7	1.40	23.5	1.40	23.1	1.39	99.2
		45	36.5	1.74	34.5	1.69	33.5	1.67	32.9	1.66	31.9	1.63	96.8
	13.6	15	10.2	7.18	8.5	6.61	7.6	6.29	7.3	6.20	7.1	6.12	66.8
		30	16.4	8.89	15.8	8.76	15.2	8.63	13.0	8.06	11.7	7.69	57.6
		45	23.0	10.4	19.5	9.63	18.5	9.34	17.3	9.12	15.0	8.72	55.4
0.6	3.8	15	6.7	0.51	6.4	0.5	5.7	0.49	5.3	0.48	4.8	0.47	98.8
		30	25	0.69	23.3	0.68	22	0.67	20.7	0.66	19.3	0.65	98.8
		45	39	0.76	37.7	0.75	34.7	0.74	31	0.72	28.1	0.7	98
	5.2	15	13	1.04	12.3	1.01	11.8	0.99	11.6	0.98	8.5	0.84	94.6
		30	23.5	1.39	23.3	1.39	23.1	1.38	22.8	1.37	22.4	1.37	99.2
		45	35.5	1.72	33.5	1.67	32.9	1.65	32.2	1.63	31.2	1.61	96.8
	13.6	15	9.6	7.01	8.1	6.47	7.3	6.19	7	6.08	6.8	5.98	68.4
		30	15.6	8.76	15.6	8.68	14.8	8.5	12.7	7.92	11.6	7.62	58
		45	21.7	10.1	18.3	9.35	17.1	9.06	15.9	8.78	15	8.55	55.4

APPENDIX I Continued

h_b m	Ø ID mm	P_i kPa	Bubbler discharge (ℓ/min)										Cu %
			Bubbler number										
			1		2		3		4		5		
			P_e	q	P_e	q	P_e	q	P_e	q	P_e	q	
0.8	3.8	15	6.5	0.5	6	0.49	5.4	0.48	5.1	0.47	4.5	0.46	98.8
		30	24.3	0.68	22.6	0.67	21.1	0.66	19.8	0.65	18.7	0.64	98.8
		45	37	0.75	35.6	0.74	32.3	0.73	29.5	0.71	27.9	0.7	98.4
	5.2	15	12.6	1.02	11.8	0.99	11.3	0.97	11.1	0.96	8.3	0.83	94.8
		30	23.1	1.38	22.8	1.38	22.6	1.37	22.2	1.36	21.9	1.35	99.4
		45	35	1.7	32.7	1.65	31.7	1.63	31.2	1.61	30.4	1.59	96.8
	13.6	15	9	6.82	7.5	6.27	6.7	5.96	6.6	5.93	6.4	5.84	69.6
		30	15.3	8.61	15.1	8.56	14.2	8.37	12.4	7.8	11.4	7.5	58.6
		45	20	9.75	17.1	9.06	15.7	8.72	14.8	8.49	13.8	8.24	55.8
1.0	3.8	15	6	0.49	5.7	0.49	5.1	0.47	4.8	0.47	4.2	0.45	98.8
		30	23.4	0.68	22.1	0.67	20.6	0.66	19.4	0.65	18.2	0.64	98.8
		45	35	0.74	33.5	0.73	31.7	0.72	28.1	0.7	26.3	0.69	98.4
	5.2	15	12	1	11.3	0.97	10.8	0.95	10.6	0.94	8.1	0.82	95.6
		30	22.2	1.35	22.1	1.35	21.9	1.34	21.7	1.34	21.1	1.33	99.4
		45	34	1.68	31.7	1.62	30.8	1.6	29.9	1.58	29.5	1.57	97
	13.6	15	8.5	6.63	6.9	6.05	6.5	5.83	6.2	5.78	6.1	5.71	72.8
		30	14.7	8.44	14.2	8.37	13.7	8.19	11.9	7.68	10.9	7.41	62.2
		45	19	9.4	15.9	8.78	14.8	8.48	14.2	8.34	13.1	8.06	61.8

APPENDIX II Bubbler hydraulic properties of different locations and internal bubbler diameters (Ø) of the same effective pressure (Pe) along the lateral pipe, (second case)

| Ø ID mm | P_i kPa | P_e kPa | Bubbler discharge (ℓ/min) | | | | | Cu |
| | | | Bubbler number | | | | | |
			1	2	3	4	5	
3.8	15	6	0.50	0.50	0.50	0.50	0.49	99.8
		7	0.52	0.51	0.51	0.50	0.49	99.2
		8	0.55	0.53	0.52	0.51	0.50	98.6
	30	26	0.69	0.68	0.68	0.67	0.67	99.4
		27	0.71	0.70	0.70	0.69	0.69	99.2
		28	0.73	0.72	0.71	0.70	0.69	98.7
	45	38	0.76	0.75	0.74	0.74	0.73	99.2
		39	0.77	0.76	0.75	0.74	0.74	99
		40	0.79	0.78	0.78	0.76	0.74	98.5
5.2	15	10	0.93	0.92	0.91	0.91	0.90	99.2
		11	0.97	0.96	0.95	0.94	0.93	98.8
		12	1.02	1.02	1.00	0.99	0.98	98.6
	30	27	1.50	1.49	1.49	1.49	1.48	99.6
		28	1.53	1.52	1.52	1.51	1.50	99.3
		29	1.59	1.56	1.55	1.54	1.53	98.7
	45	30	1.59	1.59	1.58	1.57	1.56	99.1
		31	1.62	1.61	1.61	1.60	1.58	98.9
		32	1.64	1.64	1.63	1.62	1.60	98.6
13.6	15	7	6.14	6.10	6.04	6.03	6.00	95.4
		8	6.53	6.49	6.44	6.41	6.34	94.6
		9	6.95	6.88	6.80	6.77	6.72	92.8
	30	12	8.09	7.76	7.66	7.62	7.53	84.6
		13	8.37	8.05	7.90	7.90	7.76	82.8
		14	8.67	8.38	8.18	8.16	7.99	80.2
	45	22	10.9	10.35	10.12	9.90	9.50	62.4
		23	11.05	10.67	10.37	10.09	9.62	59.6
		24	11.39	10.82	10.43	10.10	10.05	56.3

APPENDIX III Theoretical (hbc) and experiment (hbx) bubbler height at different bubbler diameters (Ø) and initial operating pressures (Pi) along the lateral pipe, (second case)

ID mm	P$_i$ kPa	Discharge ℓ/min	6		12		18		24		30	
			Exp	cal	Exp	cal	Exp	cal	Exp	cal	Exp	cal
							Bubbler location "l$_i$", (m)					
							Bubbler height "h$_b$", (m)					
3.8	15	0.50	0.40	0.41	0.39	0.40	0.38	0.39	0.38	0.39	0.37	0.38
		0.51	0.35	0.36	0.34	0.35	0.34	0.35	0.33	0.34	0.33	0.34
		0.52	0.30	0.31	0.30	0.31	0.29	0.30	0.29	0.30	0.28	0.29
	30	0.69	1.04	1.05	1.01	1.02	0.99	1.00	0.96	0.97	0.95	0.96
		0.70	0.98	0.99	0.95	0.96	0.93	0.94	0.91	0.92	0.89	0.90
		0.71	0.95	0.96	0.92	0.93	0.90	0.91	0.88	0.89	0.86	0.87
	45	0.74	0.98	0.99	0.92	0.93	0.87	0.88	0.81	0.82	0.77	0.78
		0.75	0.85	0.86	0.79	0.80	0.73	0.74	0.68	0.69	0.64	0.65
		0.76	0.78	0.79	0.72	0.73	0.67	0.68	0.62	0.63	0.57	0.58
5.2	15	0.91	0.42	0.43	0.40	0.41	0.37	0.38	0.35	0.36	0.34	0.35
		0.95	0.35	0.36	0.32	0.33	0.30	0.31	0.28	0.29	0.27	0.28
		1.00	0.24	0.25	0.21	0.22	0.19	0.20	0.17	0.18	0.16	0.17
	30	1.49	0.44	0.45	0.34	0.35	0.25	0.26	0.17	0.18	0.11	0.12
		1.52	0.41	0.42	0.31	0.32	0.22	0.23	0.14	0.15	0.08	0.09
		1.55	0.40	0.41	0.30	0.31	0.22	0.23	0.14	0.15	0.08	0.09
	45	1.58	1.32	1.33	1.13	1.14	0.96	0.97	0.80	0.81	0.66	0.67
		1.60	1.24	1.25	1.05	1.06	0.88	0.89	0.72	0.73	0.58	0.59
		1.63	1.14	1.15	0.95	0.96	0.78	0.79	0.62	0.63	0.48	0.49

APPENDIX III Continued

ID mm	P_i kPa	Discharge ℓ/min	Bubbler location "l_i", (m)									
			6		12		18		24		30	
			Exp	cal	Exp	cal	Exp	cal	Exp	cal	Exp	cal
			Bubbler height "h_b", (m)									
13.6	15	6.06	0.93	0.94	0.76	0.77	0.65	0.66	0.58	0.59	0.55	0.56
		6.44	0.89	0.90	0.73	0.74	0.62	0.63	0.57	0.58	0.54	0.55
		6.79	0.86	0.87	0.70	0.71	0.60	0.61	0.55	0.56	0.53	0.54
	30	7.73	1.69	1.70	1.31	1.32	1.04	1.05	0.86	0.87	0.76	0.77
		8.00	1.64	1.65	1.27	1.28	1.00	1.01	0.83	0.84	0.74	0.75
		8.30	1.58	1.59	1.21	1.22	0.96	0.97	0.80	0.81	0.71	0.72
	45	10.2	1.69	1.70	1.05	1.06	0.59	0.60	0.28	0.29	0.10	0.11
		10.4	1.64	1.65	1.00	1.01	0.54	0.55	0.24	0.25	0.07	0.08
		10.6	1.57	1.58	0.93	0.94	0.48	0.49	0.19	0.20	0.02	0.03

CHAPTER 6

FERTIGATION TECHNOLOGY FOR DRIP IRRIGATED GARLIC UNDER ARID REGIONS

SABREEN KH. A. PIBARS

CONTENTS

In this chapter: One *feddan* = 4200 m² = one ha; One LE = Egyptian unit of currency = 0.13992 US$.

Modified and abbreviated version of: *"Sabreen Kh. A. Pibars, 2009. Fertigation technologies for improving the productivity of some vegetable crops. PhD thesis Agricultural Science (Agricultural Mechanization), Department of Agricultural Engineering, Faculty of Agriculture, Ain Shams University, Egypt".*

6.1 INTRODUCTION

The application of fertilizers (fertigation) through irrigation water is termed as fertigation and has now a common practice in micro irrigation systems. Water-soluble fertilizers, at concentrations required by plants, are conveyed with the irrigation water to the irrigated field. Some potential advantages of fertigation are: improved efficiency of fertilizer recovery, minimal fertilizer loss due to leaching, control of nutrient concentration in soil solution, control of nutrient form and ratio of the various forms particularly for N-fertilizers, and flexibility in timing of fertilizer application in relation to crop demand based on the development and physiological stage of crop. Scheduling fertilizer application on the basis of the need potentially reduces nutrient-element losses associated with conventional application methods that depend on the soil as a reservoir for nutrients. In addition, fertigation reduces fluctuations of soil solution salinity due to fertilizers, and conserves labor and energy. Fertigation through drip irrigation allows crop to be grown under conditions of precise control of water and nutrients in the root environment [74].

In Egypt, fertigation is still limited for small farmers because of unavailability of appropriate fertigation technologies. Technological obstacles are acceptance of equipment performance and suitable quality. Moreover, successful fertigation application rests on the integration among several factors like: Crop water and nutrient requirements, optimum irrigation scheduling and soil conditions especially for economical production of crops.

This study evaluates the effects of design and operating parameters on hydraulic performance of three types of fertilizer injectors (by-bass pressurized mixing tank, venturi, and positive displacement pump). The objective was to develop the relationships among these parameters.

This study, also, dealt with water and fertigation management. This chapter investigates the effects of injector types, irrigation treatments and nitrogen treatments on:

- drippers clogging as an indicator of water distribution efficiency;
- garlic yield, water use efficiency (WUE) and nitrogen use efficiency (NUE); and
- the net profit of garlic production.

6.2   REVIEW OF LITERATURE

6.2.1   CHEMIGATION AND FERTIGATION

Chemigation is the application of any chemical through irrigation water. This may include insecticides, fumigants, fungicides, nematicides, fertilizers, soil amendments, and other soluble chemical compounds. By far, the most common form of chemigation is fertigation, which refers to fertilizer application through the irrigation water [17].

The properly designed chemication system applies precise amount of chemical to the target area in a safe, efficient and uniform manner. The uniformity of the chemical application must be as uniform as the irrigation water distribution. Irrigation system location, crop type, soil type, topography, and other design factors must be considered for chemigation, if good application uniformity is to be obtained [92].

6.2.2 ADVANTAGES AND DISADVANTAGES OF FERTIGATION

6.2.2.1 Advantages

- **Uniform application:** Fertigation facilitates the uniform distribution and precision placement of fertilizers.
- **Timely application:** In most cases, fertilizers can be applied regardless of weather or field conditions.
- **Reduced application costs:** In general, fertigation cost is about one-third the cost of conventional fertilizer application methods.
- **Improved management:** Timely applications of small but precise amounts of fertilizer directly to the root zone allow growers to effectively manage fertilizer programs. This conserves fertilizer, saves money and optimizes yield and quality.
- **Reduced soil compaction:** Fertigation reduces tractor and equipment traffic in fields. This reduces soil compaction.
- **Reduced exposure to fertilizers:** Fertigation minimizes operator handling, mixing and dispensing of potentially hazardous materials.
- **Reduced environmental contamination:** When used with the recommended safety devices, properly designed and accurately-calibrated, fertigation systems preserve quality of the environment.
- **Savings:** Fertigation can save time, reduce labor requirements, and conserve energy and materials [40, 41, 42].
- **The nutrients** can be distributed more evenly throughout the entire root zone or soil profile, as the soil is wet.
- **The rate of nutrients** can be incremented gradually throughout the season to meet the actual nutritional requirements of the crop [17].

6.2.2.2 Disadvantages

- Unequal chemical distribution, when irrigation system design or operation is faulty.

- Over fertilization, in case irrigation is not based on actual water requirements.
- Leaching, if rainfall occurs at the time of fertilizer application.
- Chemical reactions in the irrigation system leading to corrosion, precipitation of chemical materials, and/or clogging of outlets [74].
- Chemicals runoff can occur as a result of high field slopes, high discharge of emitters, and low soil infiltration.
- Safety requirements like backflow prevention valves, injection devices and highly safety needs increase input costs, especially for small farms [31].
- Increased cost due to fertigation equipments. Advanced technical knowhow is needed.

6.2.3 FACTORS AFFECTING FERTILIZER SELECTION

- **Solubility of chemical:** The chemical must be readily soluble in water and stay in solution during the application process. Occasionally, fertilizers are injected as concentrated solution into the water stream and solubility may not be a factor in the selection process.
- **Desired effect of chemical applied:** The fertilizer to be applied must achieve the desired effect with fertigation.
- **Effect of chemical on soil:** Addition of some fertilizers may increase the soil acidity in the treated area, especially under a wetted region.
- **Type of irrigation system:** The uniformity, application efficiency and distribution method of the irrigation system are all important in determining if the chemical to be applied is compatible with the irrigation system.
- **Compatibility of chemical with water source:** Chemicals that react with elements in water supply after being injected should be avoided.
- **Determination of mixture of chemicals or nutrients:** Two or more chemicals that are either mixed or applied simultaneously must not react with each other to form a precipitate.
- **State of fertilizer:** Whether it is in the solid or liquid state, will determine handling, injection methods, and injection rate [92].
- **Safe to be used with irrigation system components:** The used fertilizers must not cause damage like wear, softening of plastic pipe lines or clogging any of the system components.

- **Increase crop yield** or at least don't reduce it.
- **Available in the local market** at affordable costs [31].

6.2.4   FERTIGATION SYSTEM COMPONENTS

The fertigation unit is composed of a fertigator (fertilizer injector, metering pump), a fertilizer tank for the concentrated stock solution, a non-return valve, a main filter, pressure gage and a water meter. Depending on the model of the fertigator, additional equipments (valves, pressure and flow regulators) may be required. The metal tanks may corrode and, therefore, plastic containers are preferred. To by-pass the filter, when filtering is not necessary, two injection points are recommended, one before and one after the filter. Flushing after fertigation reduces clogging, corrosion hazard and microbial growth.

6.2.5   INJECTION EQUIPMENT

The injector is the heart of the chemigation system. There are many types of injectors available, all with their own advantages and disadvantages. Some types of injection systems are not recommended due to safety hazards that are inherent with these systems. There are several methods of chemigation. These methods can be classified into four major groups and further subdivided into specific methods:

- Centrifugal pumps
- Positive displacement pumps
  - Reciprocating pumps (piston, diaphragm and piston/ diaphragm);
  - Rotary pumps (gear and lobe);
  - Miscellaneous pumps (peristaltic);
- Pressure differential methods
  - Suction line injection;
  - Discharge line injection (pressurized mixing tank and proportional mixers);
- Methods based on the venturi principle [43].

Proper selection of a chemical injector and the chemical solution tank must include the following considerations:

- type of irrigation system,
- crop grown,
- irrigation flow rate,
- irrigation operating pressure,
- injection rate,
- type of chemical to be injected,
- determination of whether a fixed volume ratio of fertilizer to water is needed,
- source of power,
- duration of operation,
- cost
- expansion requirements, and
- safety considerations.

6.2.5.1 Pressure Differential Injector

Pressure differential injector consists of a pressurized tank connected to the irrigation line through two ports. This type of injector is also known as a dilute injection system. The inlet port is connected to the mainline on the upstream side of pressure reducing valve. The outlet port is connected to the downstream side of the pressure-reducing valve. The pressure-reducing valve on the irrigation mainline is used to create a pressure difference between the inlet and outlet ports of the pressurized tank. The difference in pressure creates a flow of water through the tank. The chemical in the tank is slowly diluted over the injection time until it has all been applied [78].

6.2.5.2 Venturi System

Fertilizers can be injected into a pressurized pipe using the venturi principle. A venturi injector is a tapered constriction, which operates on the principle that a pressure drop accompanies the change in fluid velocity as it passes through the constriction. The pressure drop through a venturi must be sufficient to create a negative pressure (vacuum) relative to atmospheric pressure. Under these conditions, the fluid from the tank will flow into the injector. Most venturi injectors require at least a 20% differential pressure to initiate a vacuum. A full vacuum of 71.12 cm of mercury is attained at a differential pressure of 5% or more.

A small venturi can be used to inject small chemical flow rates into a relatively large mainline by shunting a portion of the flow through the injector. To assure that the water will flow through the shunt, a pressure drop must occur in the mainline. For this reason, the injector is used around a point of restriction such as valve, orifice, pressure regulator or other device, which creates a differential pressure.

The suction capacity depends on the fertilizer solution level in the supply tank. As the liquid level drops, the suction head increases resulting in the decreased injection rate. To avoid this problem, some manufacturers provide an additional small tank on the side of the supply tank, where the float valve maintains the level relatively constant. The fluid is injected from this additional tank [43].

6.2.5.3   Positive Displacement Pumps

Water powered injectors are considered active injectors since an external energy source is not used. The energy of the pressurized water in the irrigation system is used to drive the injector. Water powered injectors are available in turbine (impeller) or piston drive. Piston operated units use a small amount of the pressurized irrigation water supply to drive the piston. The driven water, that is expelled from the piston, is usually three times the quantity of the injected solution. The injection rate is set by controlling the amount of water entering the piston drive. Piston driven units do not reduce the irrigation system pressure [78].

The chemical injectors are compared in Table 6.1.

6.2.6   PERFORMANCE OF FERTILIZER INJECTORS

As a rule of thumb, fertilizer injection should be able to function at a rate of 0.1% of the irrigation water flow rate. A minimum pressure drop of 0.689–1.379 bars is needed for an injector, which uses a pressure differential method in a drip irrigation system [17].

The injection flow rate of by-bass pressurized mixing tank increases with increasing injection orifice size. The injection flow rate and pressure differential relationship can be represented by a power function with an exponent value of 0.5. The released concentration from a tank depends on

TABLE 6.1 Comparison of Various Chemical Injection Methods [43]

Injector	Advantages	Disadvantages
Pressure differential tanks	- Medium cost.	- Pressure differential required.
	- Easy operation.	- Variable chemical concentration.
	- Total chemical volume accurately controlled.	- Cannot be calibrated for constant injection rate.
Venturi	- Low Cost.	- Pressure drop created in the system.
	- Water powered.	
	- Simple to use.	- Calibration depends on chemical level in the tank.
	- Calibrate while operating.	
	- No moving parts.	
Positive displacement pumps	- High precision.	- High cost.
	- Linear calibration.	- May need to stop to adjust calibration.
	- Very high pressure.	
	- Calibration independent of pressure.	- Chemical flow not continuous.
		- Moving parts.

injection orifice size, tank volume, differential pressure head, and quantity of fertilizer applied; and it decreases exponentially with time. It was found that the variation pattern of the released concentration was highly controlled by the injection orifice size and tank volume [57].

A pressure drop across the venturi is required for proper operation. Minimum pressure is required for any injection to occur. The larger the pressure drop, the higher the injection rate, up to maximum drop. The injection rate of a venturi device is determined by the size of the venturi and the pressure differential between inlet and outlet ports [45]. Other parameters affecting venturi performance can be investigated like the effects of inlet and throat diameters of the venturi tube, pipe length downstream of the venturi tube, diameter of the suction pipe at the throat portion of the venturi tube, angle of the pipe downstream of the venturi tube, flow velocity at the inlet portion of the venturi tube and density and viscosity of the liquid injected into the venturi tube on liquid injection rate [70].

The injection rate of the positive displacement pump was increased with increasing pressure in the mainline. Therefore, the flow rate passed through the pump will increase and consequently will increase the number of positive displacement strokes per minute. The injection flow rate and pressure differential relationship can be represented by a power function with an exponent of 1.072. The fertilizer concentration in the irrigation water was increased at the start of fertigation; and then it became constant after this time during the injection period [1].

6.2.7   TRICKLE IRRIGATION METHOD

Drip irrigation systems generally have an emission uniformity exceeding 80% providing the system has been designed and managed correctly. Drip systems operate at efficiencies in the range of 85–90% compared to sprinkler systems that are only 65–75% efficient (depending upon system design). These systems are therefore much superior in the application of fertilizers and systemic than sprinkler systems [40, 41]. The following factors must be considered in the design of a drip system used for chemigation.

- Emitters should be spaced to effectively irrigate as much of the plants root volume as possible.
- An appropriate emitters should be selected for the terrain, crop type and water quality. Emitter flow characteristics and product durability for the conditions should be considered. An emitter with a manufacturer's coefficient of variation of less than 0.05 should be selected. Emitter flow rates at the beginning and end of the drip lines should be tested to confirm that discharge rates are within acceptable limits.
- Emitter operating pressure range should be kept within ±10% of the emitter operating pressure. If the drip irrigation system is operating on a sloping topography, pressure-compensating emitters should be used.
- The injection system must be located before the filtration system so that any precipitates that may form will have an opportunity to be filtered out before entering the irrigation system [92].

6.2.8   INJECTION OF DRY FERTILIZERS THROUGH MICRO IRRIGATION

Aboukhaled [3] reported that dry fertilizer may be placed into differential pressure tank. When the irrigation system begins to operate, water will flow into the tank through the higher-pressure inlet port, filling the tank with water and dissolving some of the dry fertilizers. Once the tank has been filled, water will flow out through the outlet port, carrying some of the dissolved fertilizers with it.

Papadopoulos [75, 76] indicated that a modification of the above pressure differential system is a tank that contains a collapsible plastic bag into which dry fertilizers can be added. Water is admitted to the area between the tank and the bag, which forces the fertilizer compound out of the bag into the irrigation system.

Bazza [10] mentioned that dry fertilizer may be completely dissolved in the holding tank and the solution is injected by an injector pump.

The fertilizers should be dissolved before adding to the irrigation water. Undissolved particles may settle out in the distribution system. The solubility of fertilizers varies according to their chemical composition and water temperature as shown in Table 6.2. Recently, several composite NPK fertilizers are being introduced, in Egypt, for use as foliar fertilizer via sprinkler irrigation and such fertilizers have potential use in drip irrigation systems.

TABLE 6.2 Solubility of Common Fertilizers (kg fertilizer/m³) in Water [13]

Fertilizer	Formula	Solubility, kg fertilizer/m³		
		[1]Water temperature, °C		
		cold	Luke-warm	Hot
Ammonium chloride	NH_4Cl	297(0)	-	758(100)
Ammonium nitrate	NH_4NO_3	1183(0)	1950(20)	3440(50)
Ammonium sulfate	$(NH_4)_2SO_4$	706(0)	760(20)	850(50)
Diammonium phosphate	$(NH_4)_2HPO_4$	429(0)	575(20)	1060(70)
Mono-ammonium phosphate	$NH_4H_2PO_4$	227(0)	282(20)	417(50)
Potassium nitrate	KNO_3	133(0)	316(20)	860(50)
Potassium chloride	KCl	280(0)	347(20)	430(50)
Potassium sulfate	K_2SO_4	69(0)	110(20)	170(50)

[1]Water temperature in brackets.

6.2.9   CHEMIGATION MANAGEMENT AND CLOGGING

6.2.9.1   Clogging

Clogging problems are associated with chemigation in addition of other agents of clogging. According to De Torch [24], the most severe is the clogging in drippers and components of system. Clogging of drippers is a major concern in drip irrigation systems. Bucks et al. [15] stated that although drip irrigation has many advantages, yet it also has some limitations. Emitter clogging is the most common. Bucks et al. [16] added that irrigation water quality affected the degree of emitter clogging. Dasberg and Bresler [23] stated that complete or partial blocking of drippers reduces the application uniformity of both water and fertilizers and negatively affects plant growth. Drip irrigation systems have low flow rates and extreme small passages for water. These passages are easily clogged by three major categories of clogging agents: physical, chemical and biological. Two or more of these clogging categories may occur at the same time [5, 14, 24].

Physical clogging is caused by inorganic and/or organic suspended particles in irrigation water, such as soil particles (clay, silt and sand) or planktonic organisms. These suspended particles can plug the narrow pathways of water within the drippers and the small openings. Pillsbury and Degan [79] stated that these suspended particles can be removed by adequate filtration. They added that the smaller is the pathway within the drippers and their openings, the finer is the screen size required. Media and cartridge filters and centrifugal separators are the main types of filters in addition to settling basins that are used in drip irrigation system to get rid of these particles [24].

Chemical precipitation is caused by the deposition of salts and/or ions inside the drippers. Hills et al. [50] mentioned that carbonate precipitation is the most common type of chemical clogging in drip irrigation. Gilbert and Ford [38] indicated that both dissolved iron and manganese (in the reduced form) may be oxidized into particulate forms that can accumulate and can block the drippers. In some cases, a combination of carbonate precipitation and some fertilizers are responsible for severe clogging of drip irrigation systems [86]. Sulfuric and hypochlorite acids are injected to reduce the pH of irrigation water to reduce the degree of chemical precipitation.

Biological clogging of drippers in drip irrigation system is due to the development of microbial slime in the lateral lines and in the drippers [4, 36]. Many drippers malfunctions are caused by a combination of the physical, chemical and biological factors. Chlorine is used to control the formation of algae and bacterial slimes. Bozkurt and Özekiei [11] carried out a study to determine the effects of different fertigation practices on clogging in in-line emitters using Samandag region well water in Turkey. Their data show that different fertilizer treatments have significant effects on emitter clogging. Fertilizers containing both Ca^{++} and $SO4^-$ caused higher clogging compared with the others. Chang [20] said that as the flow slows down and/or the chemical background of the water changes, chemical precipitates and/or microbial flocs and slimes begin to form and grow, thus emitter clogging occurs. Ravina et al. [82] stated that more clogged emitters were found at the end of the drip laterals than at the beginning (probably due to pressure head loss). This leads to non-uniformity in the discharge rate of emitters within the system. Hebbar et al. [48] found that normal fertilizers generally tend to clog the emitters causing an uneven distribution of fertilizers.

6.2.9.2   Prevention of Clogging

6.2.9.2.1   Physical Clogging

In addition to an adequate filtration system, regular flushing of the lines and emitters is desirable.

6.2.9.2.2   Chemical Clogging

Many cases of clogging can be solved by acid treatment. In sever cases, emitters must be soaked in dilute acid solution (about 1%) and even cleaned individually. For less severe cases, injection of acid to bring the water to a pH between 1 and 2 should be carried out. O'Neil et al. [69] injected orthophosphoric acid in trickle irrigation system to solve the problem of phosphorus precipitation in the irrigation lines, which causes clogging problems and eventual breakdown of the system. Hills et al. [50] injected sulfuric acid into the water source for lowering the pH. The lowering of water pH alleviated chemical clogging. Ibrahim [51] found that injecting carbon dioxide (150 ppm) through trickle irrigation system led to lowering irrigation water pH and consequently prevented emitter clogging. Padmakumari and Sivanappan [71] indicated that injecting hydrochloric acid at concentration of 2% through trickle irrigation system led to increasing the discharge rate of partially clogged emitters.

6.2.9.2.3 Biological Clogging

The standard treatment is the injection of biocide followed by thoroughly flushing to clear the system of organic matter. Chlorine gas and hydrochloride solution are the most commonly biocides. Rates of application range from 20 to 50 ppm depending on the severity of the problem and should be maintained in the lines for at least 30 minutes [4]. Sagi et al. [87] injected sodium hypochlorate (10%) every other week, at a concentration of 10 mg/L. Free chlorine for one hour prevented clogging due to sulfur bacteria.

6.2.10 FERTIGATION VERSUS CONVENTIONAL METHODS OF FERTILIZER APPLICATION

One of main benefits of fertigation is increasing efficient use of fertilizer compared to conventional method. Many studies have been conducted carried out on fertigation to prove its effectiveness in reducing fertilizer application, and production costs. Lowering fertilizer application rate increases yield and conserves environment. Rahm [81] compared fertigation with conventional methods of applying fertilizers to corn plants, and concluded that except for N, conventional methods of fertilizer application produced yields equal to those produced from fertigation. However under situations where nutrients deficiency is detected early in the growing season, the application of liquid fertilizer containing the deficient nutrient with the irrigation water may result in a substantial increase in yield.

Bravdo and Hepner [12] found that availability of N and K fertilizers was increased by fertigation and this was reflected in improved yields of grapes compared with broadcasting method. Crespo et al. [21] reported that fertigation of N fertilizer produced a higher commercial yield of sweet papers compared with banded fertilizers. Aboukhaled [3] stated that several investigations were carried out to compare conventional methods of chemical application with chemigation of soybean. He added that nitrogen fertigation received the greatest attention and has probably the largest application. Arnaout [6] performed a study to evaluate two methods of fertilizer application (fertigation and broadcasting) under three irrigation methods (surface drip, sub-surface drip, and sprinkler) on lima beans. He found that stem length, number of branches and pods/plant under fertigation were superior than those under the broadcasting method in three irrigation methods. El-Adl [30] carried out experiments to study the effects of irrigation intervals (daily, every two days, and every three days), quantities of irrigation water (100 and 120 % of ET_c) and three fertilization methods (traditional, broadcasting, and fertigation) under sprinkler irrigation system on peanut crop. The highest values of crop growth characters (plant height, number of pods/plant and weight of 100 pods) were for the fertigated treatments. These values were 51.7 cm, 42.7 and 152.2 g, respectively. However the other characters (weight of 100 seeds, and seed/pod ratio) were higher for the traditional fertilization methods. These values were 106.2 g, and 75.4%, respectively.

Bakker et al. [9] evaluated effects of fertigation and broadcast fertilization on the yield and nitrate content of lettuce; and showed that the yield was significantly higher under nitrogen fertigation. Moreover, nitrogen fertigation increased both the crop and vegetative growth compared with broadcast fertilization method. El-Gindy [34] reported that fertigation of N fertilizer increased yield of tomato by 16.1, 23.8, and 35.1% under furrow, sprinkler, and drip irrigation methods, respectively, compared with traditional method of fertilizer application. Slangen and Gals [90] compared N fertigation versus broadcast application for lettuce, spinach and Chinese cabbage crops; and concluded that the yields were comparable under both systems, but crop uniformity was better with fertigation.

Gascho [37] concluded that drip fertigation requires less P than other application methods to achieve comparable tissue P-concentration and yields, due to the placement being in the rooting zone of tomatoes. Hamdy [44] reported that fertigation resulted in 70% increase in tomato yield than that with conventional methods of N application. Imas [52] mentioned that fertigation technique was the only good way to apply fertilizers physically to the crop root zone. On high value drip irrigated crops (lettuce, tomatoes, and peppers), the level of fertigation management for achieving high yields and crop qualities exceeded compared to other fertigation methods and crops.

El-Adl [29] studied the effects of two irrigation methods (surge drip and traditional drip), two fertilization methods (traditional and fertigation), and two irrigation levels (100 and 75% of ET_c) on pea growth. His results showed best results of growth parameters (plant height, branches number/plant, pods number/plant and weight of green pods/plant) were for treatments: surge drip irrigation, fertigation, and 100% of ET_c). Average values were 83.8 cm and 766.8 g, respectively. Mishra et al. [63] found that the lower potassium uptake efficiency was obtained with banded application of 100% of recommended dose of potassium fertilizer at the time of sowing, and under drip and furrow irrigation methods. These finding were in agreement with those reported on nitrogen use efficiency for Broccoli crop by Chakraborty et al. [19], who conducted field experiments under similar set of treatments and similar soil type conditions but with different levels of Nitrogen fertilizer.

6.2.11 CROP RESPONSE TO DRIP IRRIGATION SYSTEM

Trickle irrigation is distinguished by its high efficiency in decreasing the water losses to the least limit compared to all other irrigation systems. It is important that irrigation manger should investigate the effects of drip irrigation on crop root development, soil moisture and salt distribution, and water use efficiency as well as on crop profitability. El-Gindy [33, 34] studied the optimization of water use for peppers, and found that the drip irrigation increased the pepper yield by 64.0% and water use efficiency than that under furrow irrigation.

Judah [56] found that the tomato yield was not affected by salinity under drip irrigation. He recommended that with frequent irrigation at adequate irrigation duration was able to keep salt accumulation far away from the plants. The average tomato yield was 94.5 ton/ha. Raj Kumar and Larim [82] reported that the cucumber yield was the maximum in the daily-irrigated treatment due to less salinity and better soil moisture availability near the root zone. Ebaby [26] stated that the cucumber yield was increased about 3 times under drip irrigation than that obtained under the furrow irrigation. Younis [96] mentioned that the least amount of water was required with a higher tomato yield under trickle irrigation. The percentage increase in net profit was 11.0 and 14.8% compared to furrow and sprinkler irrigation systems, respectively. Abdel-Maksoud et al. [2] studied the effects of different irrigation systems (drip, sprinkler and furrow) on tomato yield. Results indicated that yield under drip irrigation system (20 ton/fed) increased by 19.36% than that under sprinkler irrigation system (16.8 ton/fed) and by about 13.6% than that under furrow irrigation system (17.6 ton/fed). Meanwhile, WUE increased by about 19.34 and 14.14% than the sprinkler and furrow irrigation systems, respectively. Mmopharlin et al. [64] concluded that lettuce yield (plants which were trickle fertigated 6–12 times daily, and applied at 100% and 150% of the previous day's pan evaporation) showed a significant response to Nitrogen fertigation. Also, there was a significant effect on yield of irrigation x fertigation method. N in soil analysis indicated an increase in N use efficiency due to reduced nitrate leaching; a more constant nitrate concentration in the soil; a better N placement; an increased ratio of nitrate-N: ammonium-N; and a soil ammonium-N concentration below the toxic level.

Gillerman et al. [39] reported that pear yield under saline water through subsurface drip irrigation was better than under surface irrigation with tap water. Dong and Tucker [26] indicated that irrigation with saline water once a day reduced potato production by 12%. However, results obtained with irrigation frequencies of 3 and 6 times per day showed no differences in yield between fresh water and saline water. They added that plant growth, height and dry matter accumulation decreased by the saline water irrigation. Hanson et al. [46] stated that yields of lettuce were obtained under furrow and subsurface drip methods, but lower yield was obtained under the surface drip method. Water use under the drip methods was between 43 and 74% of that via furrow method. Spatial variability in plant mass and yield occurred for the drip-irrigated plots than for the furrow one.

6.2.12   EFFECTS OF FERTIGATION ON WATER AND FERTILIZER USE EFFICIENCIES

6.2.12.1   Water Use Efficiency

As a result of water resources limitation, water applications for irrigating crops must be more efficient. Fertigation increased water use efficiency (WUE) by increasing rooting depth and density, as well as the crop's ability to withstand drought stress. Where water and plant nutrients are scarce and WUE is low, increasing the nutrients availability to a level that is not limiting during plant growth leads to increasing crop production and subsequently WUE increases. Fertigation applies nutrients only to the effective root zone in a soluble form so that there is no need to excess irrigation water to dissolve fertilizers solid granules on the top layer of soil.

Morad et al. [67] studied the water distribution pattern and WUE as a function of change in emitter flow rate, distance between emitters, and fertilizer dose to achieve the optimum system management in clayey soil. The results show that the WUE was significantly affected by increasing the dose of nitrogen from 90 to 120 caused a considerable increase in WUE by 12.9 % and 27.25 %, respectively. El-Adl [30] recorded maximum WUE for peas in the surge drip irrigation with fertigation and irrigation at 75% of ET_c. WUE was 3.78 kg of green pods and 0.693 kg of dry grains per m^3 of irrigation water. On the contrary in his study (2001) on peanut crop, the maximum value of WUE was 0.42 kg of grains per m^3 of irrigation water for the treatment: irrigation every day with 100% of ET_c and traditional fertilization method.

6.2.12.2   Fertilizer Use Efficiency

The return of fertilizer unit in fertigation method was very high compared with other traditional fertilization methods. Magen [59] specified the main reasons for increasing fertilization efficiency in the introduction of fertigation, detailed studies of crop fertilizer consumption and uptake curves, the development of new fertilizers for fertigation, the intensification and calibration of plant and soil analysis and extension work. Imas [52] indicated that fertigation allows exact application of the nutrients and uniformly only to the wetted root volume, where the active roots are concentrated. This remarkably increased fertilizer use efficiency, which reduced the amount of applied fertilizer.

Arnaout [6] stated that the value of fertilizer use efficiency (FUE) under fertigation through surface drip irrigation was 16.5 compared to 12.2 kg per unit of water under broadcasting method. Also, FUE under fertigation through subsurface drip irrigation was 15.6 compared to 11.25 kg per unit of water under broadcasting method. He added that the effects of fertilization method under sprinkler irrigation was remarkably noticed, where the FUE was increased by 28.3% under fertigation compared to broadcasting method, due to high uniformity of fertilizer distribution. Morad et al. [67] revealed that FUE was significantly affected by increasing N-dose from 90 to 120. The decreases were 13.72 and 21.93%, respectively.

6.2.13   GARLIC YIELD AND QUALITY

6.2.13.1   Effects of Fertilizer Addition and Nitrogen Fertilization Rates on Yield and Some Plant Features of Garlic

Growth parameters like stem length, branches number/plant, the fresh weight per head and per bulb, pods were taken as an indicator of fertigation effectiveness. Many studies showed positive effects of fertigation on plant growth. Sadaria et al. [85] investigated the effects of N rates (25, 50 or 75 kg/ha) and P rates (11, 22 or 33 kg/ha) on garlic, grown in India. They found that the different N treatments tested had no significant effect on bulb yield, and the effects of P treatments were not clear.

Seno [89] studied the effects of four irrigation frequencies (drip irrigation at 3-, 4-, 5- or 6-day intervals) and four N rates (0, 20, 40 or 80 kg N/ha) on garlic, growing in a dark-red *podzolic, eutrophic* soil. Neither irrigation frequency nor N application had a significant effect on commercial yield. The combination effect of irrigation applied every 3 days × 20 kg N/ha resulted in the greatest bulb weight (18.40 g) and decreased the percentage of small bulbs.

Lipinski and Gaviola [58] studied the effects of N rate and N source on garlic yield, fertigated @ 0, 150, 300 or 450 kg N/ha as urea, urea-ammonium or ammonium nitrate. N was applied by fertigation. Yield was increased with increasing N rate up to 300 kg, then decreased, however there were no significant differences among sources. Marouelli et al [61] performed a study to evaluate the effects of three soil water tensions (15, 35 and 70 kPa) and three N levels (20, 100 and 500 kg per ha) on yield and quality of garlic bulbs in Brazil. Greater vegetative growth, total yield and average bulb weight were obtained under tensions between 15 and 19 kPa and N levels between 52 and 97 kg per ha. The marketable yield was maximized at 15 kPa tension, and it was linearly reduced with increasing N levels. The percentage of bulbs with secondary growth was affected by the water tension, N and the interaction of both factors, with minimum value obtained for the combination of 70 kPa tension and 20 kg-ha^{-1} of N. The percentages of dry bulbs and mass losses, between 60 and 120 days after harvest, were increased linearly with N. However, they were not affected by water tension treatments. Hence to maximize marketable yield of cultivars susceptible to secondary growth, high frequency irrigation regime is recommended.

Mohammad et al. [65] studied different N fertilization rates (0, 30, 60 and 90 ppm) fertigation under drip irrigation system and conventional fertilizer application. Results showed the yield was increased with increasing N fertigation rates. The fresh weight per head and per segment showed similar trends. However, the number of segments per head was not affected significantly by the investigated treatments. This may indicate that the zero N treatments produced heads with smaller segments compared to that produced with N application. The dry weight of shoot, segment and segment membrane responded positively to the rates of N fertigation, reaching the maximum value at 80 and 120 kg N, irrespective of N fertigation or soil application. The soil application gave a production as high as the best fertigated N rate. The fertigation treatment increased water use efficiency than the soil application at the similar N rate. Mohamad and Zuraiqi [66] evaluated the response of garlic yield, in Jordan, to nitrogen application methods (fertigation and conventional soil application) as indicated by both water and fertilizer use efficiency. The treatments were 0, 30, 60, and 90 mg of N/liter in the irrigation water. Irrigation was applied to replenish 80% of the class A pan evaporation twice each week. Yield positively responded to N regardless of the application method. Yield was increased by fertigation with 120 kg of N/ha; a higher yield response was obtained by fertigation than by soil application of N. The fresh weight per head and per bulb had similar trends. However, the number of bulbs per head was not affected by the treatments, higher NUE were enhanced due to the more efficient timing and placement of N. The overall results indicate that the yield, NUE, and WUE can be improved with fertigation.

6.2.13.2   Effects of Irrigation Water on Garlic Yield

Patel et al. [77] grew garlic under drip [trickle] irrigation @ 40, 60, 80 and 100% of cumulative pan evaporation (CPE) on alternate days. Alternatively, surface irrigation was applied each 8 to 10 day intervals at 50 mm. N fertilizer were applied at 25, 50 or 75 kg/ha. They found that highest bulb yield (75.87 100Kg/ha) was obtained under drip irrigation at 100% CPE. N uptake was highest with drip irrigation at 100% CPE. Drip irrigation used 27.20 to 62.80% less water than surface irrigation. WUE was higher with all rates of drip irrigation than with the surface one. Sankar et al. [88] studied the effects of micro irrigation systems (drip and sprinkler) and flood irrigation on garlic productivity. Among the different irrigation methods and levels tested, drip irrigation at 100% E_p recorded the highest yield of garlic (147.8 100kg/ha) followed by 75% E_p for the same system. Drip irrigation at 100% E_p recorded the tallest plant (79.3 cm). Up to 44 and 41% irrigation water were saved under drip and sprinkler systems, respectively.

Thanki and Patel [91] conducted a field experiment on garlic cv. GAUG-1 in Navsari, Gujarat, India. The treatments consisted of three irrigation regimes (0.6 and 0.8 IW/CPE ratios with 5 cm depth at each irrigation through mini sprinkler method of irrigation as well as 1.0 IW/CPE ratio with 7 cm depth at each irrigation through surface method of irrigation). The 0.8 IW/CPE ratio recorded significantly the highest value of yield attributes as well as bulb yield of garlic (5018.17 kg/ha).

6.2.13.3   Effects of Both Fertilizer and Irrigation Water on Garlic Yield

Panchal et al. [72] studied the effects of irrigation rates {IW: CPE ratios of 1.0, 1.2 or 1.4), N rates (25, 50 or 75 kg/ha) and P_2O_5 rates (25, 50 or 75 kg/ha) on garlic yield. They found that the bulb yields were highest at IW: CPE ratios of 1.2 or 1.4, at 50 or 75 kg N/ha, and at 50 or 75 kg P_2O_5 /ha. In a study using garlic (cv. G1) grown on a clay loam soil, Pandey and Singh [73] compared 5 levels of irrigation (0.5, 0.75, 1.00, 1.25 and 1.50 ID/CPE) and 3 N levels (50, 100 and 150 kg N/ha). Yields were highest (156 100kg/ha) at 1.5 ID/CPE and 150 kg N/ha (157.33 100kg/ha). These rates of irrigation and N also resulted in the greatest bulb diameter and weight of bulbs. Carvalho et al [18] planted garlic, under the conditions of 3 rates of applied N (0–120 kg/ha) and 4 rates of K (0–160 kg K2O/ha) at irrigation rates corresponding to 60, 100 and 140% of maximum evapotranspiration (401.5 -716.5 mm). Rainfall during the growing season was relatively high and no significant effects of irrigation were observed. Although, various effects of N and K on emergence and morphology were observed. Total and marketable bulb yields were affected only by N. They added that the highest yields (2400, 4440 kg/ha) were obtained using 70 and 76 kg N/ha, respectively. Sadaria et al. [85] studied the effects of irrigation water: cumulative pan evaporation ratios (IW:CPE of 1, 1.2 and 1.4). The highest bulb yield (5594 kg/ha) was obtained at IW:CPE = 1.4 and the increase in productivity obtained in this treatment was significantly higher than that at IW:CPE ratios of 1.2. It is suggested that higher water availability at 1.2 and 1.4 than at 1.0 IW:CPE increased nutrient availability, and therefore increased growth and productivity. The different N treatments had no significant effects on bulb yield, and the effects of P_2O_5 treatments were not clear.

6.2.14 CHEMIGATION ECONOMICS

The Egyptian population is a fast growing. It was 52 million in 1990 and it is expected to be 120 million in 2051. The individual share of water in Egypt decreased from 1221 m^3/year in 1990 to 1194 m^3/year in 2000. This share is expected to be 617 m^3/year in 2051 according to Mekheimer and Hegazy [62]. The individual share of 617 m^3/year is very critical compared to the standard level of water poverty (1000 m^3/year). According to the world standard, Egypt is moving very fast from abundant to stressed and scare country. Therefore, rationalization of agricultural production resources (i.e., water, fertilizers, etc.) through increasing the net income/ unit resource is a must. The first concern of the farmer is to decide the injector type, level of irrigation and fertilizer amount to increase his net income. Pillsbury and Degan]79] said that no system of irrigation should be adopted if the expected increase in farm production and income therefore would not cover the total additional cost of irrigation. They added that the choice among different feasible irrigation systems and equipment should be based either on the magnitude of the net farm income to be derived from irrigation or on cost-benefit ratios to the additional cost of irrigation. The system which is expected to yield the maximum net farm income or the highest benefit/ cost ratio should be adopted. De Torch [24] sated that an important part of irrigation system design is determining the expected annual cost of owning and operating each alternative. He added that the farm irrigation system costs include: *fixed costs* (i.e., annual depreciation, interest, and any expenditure for taxes and insurance costs), and *operating costs* (i.e., annual energy, water, maintenance, repair and labor costs). According to Threadgill [95], the costs of applying chemicals via center pivot and conventional methods were 7.84 and 9.66 LE/fed., respectively. Younis [96] reported that the maximum net profit was obtained using drip irrigation. The increases were 11 and 14.8% relative to furrow and sprinkler irrigation system, respectively. El-Gindy [34] found that chemigation method reduced the cost of production unit (LE/ ton) for both drip irrigated tomato and cucumber crops. Nagmoush [68] said that cost of hand fertilization varied from 3.5 to 4.0 and from 1.5 to 2.0 LE/ fed./application by hand and machine, respectively. EL- Gindy [34] and Abdel-Aziz [1] found that the cost of maize production was the lowest under daily drip irrigation at the rate 100% the water consumptive use (WCU) and the highest under sprinkler irrigation system (irrigation every three days at the rate of 50% of WCU). Mansour [60] and Tayel et al. [94] mentioned that both the capital and annual costs (LE/fed.) for grape crop grown in clay loam soil in Egypt, were (1000, 327), (1720, 389) and (2230, 533) under gated pipe, low head bubbler and drip irrigation systems (GPIS, LHBIS, DIS), respectively. According to the net profit (LE/ fed.), the irrigation systems were arranged in the following ascending order: GPIS>LHBIS >DIS.

6.3 MATERIALS AND METHODS

6.3.1 MATERIALS

Experiments were established at the experimental farm of Faculty of Agriculture at Ain Shams University, in Shalaquan village, Egypt. The site is located 1 km from EL-kanater EL-kjairea District (latitude 30° 13'N, latitude 31° 25'E and at 41.9 m above sea level), in Qlubia Governorate, Egypt.

6.3.1.1 Irrigation System Installation and Performance

6.3.1.1.1 Drip Irrigation System

The irrigation system consisted of: head control unit, pipe lines network (main line of 110 mm diameter PVC pipe; sub-main lines of 75 mm diameter PVC pipe; manifold of 32 mm diameter PVC pipe; laterals lines of 16 mm diameter PE hoses 22 m in length laid on the soil surface); and GR built-in emitters of 4 lph discharge at 50 cm emitter spacing.

6.3.1.1.2 Fertilizer Injectors

Two main injection techniques were tested: the ordinary closed tank (bypass) and the injector pump. Both techniques were run by pressure difference across the injector system. The injector types were mainly by-bass pressurized mixing tank, Venturi type and positive displacement pumps. Under field operating conditions, the different types of injectors were connected by a by-bass arrangement to a sub-main line with 36.5 m^3/h discharge at 2 bars of pumping pressure as shown in Figures 6.1 and 6.2. The injected fertilizer was ammonium sulfate $(NH_4)_2SO_4$ solution.

(a) By-bass pressurized mixing tank

By-bass pressurized mixing tank (Figure 6.1a) is a cylindrical, inside epoxy coated, pressurized metal tank, resistant to the system pressure, and connected by a by-bass to the main line. Tank dimensions were 40 cm in diameter and 160 cm in length, with a total volume of 200 liters. The tank

a. By-bass pressurized mixing tank.

b. Venturi.

c. Positive displacement injection pump.

FIGURE 6.1   Field installation of the three types of injectors.

was connected to submain line and controlled by 3/4" gate valves at the inlet and outlet. The flow rate of the solution was monitored by a flow meter installed on the by-bass pressurized mixing tank.

(b) Venturi injector

A venturi injector (Figure 6.1b) is a tapered constriction, which operates on the principle that a pressure drop accompanies the change in velocity of the water as it passes through the constriction. It was installed on a by-bass arrangement placed on an open container with the fertilizer solution. The injector was constructed of a PE venturi tube with 3/4" size. The venturi was devised by gate valve, which creates a differential pressure, thereby, allowing the injector to produce a vacuum.

(c) Positive displacement injection pump

This type of injector consists of a mounting bag, by-bass control knob, dosage positive displacement and suction tube fitting and hose one

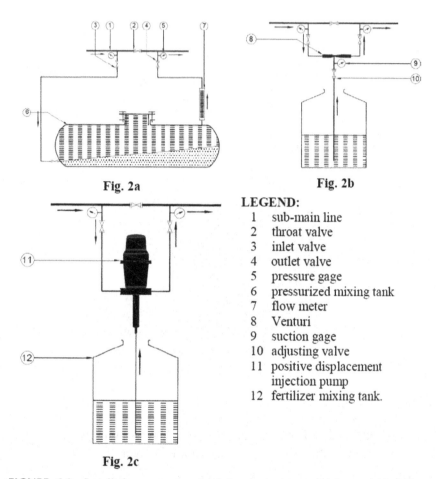

Fig. 2a

Fig. 2b

Fig. 2c

LEGEND:

1 sub-main line
2 throat valve
3 inlet valve
4 outlet valve
5 pressure gage
6 pressurized mixing tank
7 flow meter
8 Venturi
9 suction gage
10 adjusting valve
11 positive displacement injection pump
12 fertilizer mixing tank.

FIGURE 6.2 Installation arrangement of the three types of injectors: (a) by-bass pressurized mixing tank, (b) venturi, and (c) positive displacement injection pump.

inch to connect the inlet and outlet of the pump to irrigation system, and valves (Figure 6.1c). The dosage and discharge of the completely soluble fertilizer can be adjusted by the dosage positive displacement. Positive displacement injection pump was used to inject dissolved chemicals through the irrigation network with maximum rate of 150 lpm. Flow rate of this type ranges from 4.5 m³/h (maximum) at 7 bar and 19 lph (minimum) at 0.34 bar working pressure. The dosage volume spacers were exposed. The more rings on the shaft, the higher the dosage rate. Actual injection rate will vary slightly depending on water flow rate.

6.3.1.2   Soil and Irrigation Water Analysis

6.3.1.2.1   Soil Physical and Chemical Analysis

Soil physical characteristics are shown in Table 6.3. The soil type of experimental field was (Nile alluvial) silt clay loam. Table 6.4 indicates soil analysis of the site. Soil analysis was carried out according to the following standard methods of analysis:

1. Mechanical analysis was carried out according to the international pipette method as described by Piper [80].
2. Calcium Carbonate was determined as calcium carbonate using Collins calcimeter [80].
3. Bulk density (g/cm^3) was determined according to Dewis and Freitas [25].
4. Soil pH was determined in 1:2.5 soil water suspensions using a Gallen Kamp pH–meter [54].
5. The electrical conductivity was measured in 1:5 soil water extract using conductivity-meter [49].
6. Soluble ions in meq/L was determined in 1:5 soil water extract according to Hesse [49] as follows:

TABLE 6.3   Some Soil Physical Properties of the Experimental Site

Sample depth cm	Particle Size Distribution, %					θ_w %			B.D. g/cm^3
	Coarse Sand	Fine Sand	Silt	Clay	Texture class	F.C.	W.P.	A. W.	
0–15	0.81	27.8	41.44	29.95	C.L	35.45	19.2	16.25	1.25
15–30	0.7	27.5	41	30.8	C.L	35.2	19.44	15.76	1.26
30–45	0.61	27.8	38.45	33.14	C.L	34.7	19.8	14.9	1.28
45–60	0.61	28.5	37.25	33.64	C.L	34.7	20.1	14.6	1.30

F. C.: Field capacity   W. P.: Welting point   A.W. Available water   B.D.: Bulk density
C. L.: Clay loam

TABLE 6.4 Some Soil Chemical Properties at the Experimental Site

Sample depth cm	pH 1:2.5	ECe dS/m 1:5	Soluble cations, meq/L				Soluble anions, meq/L			
			Ca^{++}	Mg^{++}	Na^+	K^+	CO_3^-	HCO_3^-	SO_4^-	Cl^-
0–15	7.9	0.26	0.41	0.47	0.43	0.19	-	0.64	0.36	0.50
15–30	7.8	0.25	0.46	0.35	0.50	0.17	-	0.76	0.15	0.57
30–45	7.6	0.26	0.55	0.54	0.61	0.2	-	0.78	0.34	0.78
45–60	7.2	0.28	0.49	0.66	0.66	0.17	-	0.88	0.34	0.76

- Soluble calcium and magnesium were determined by the versinate method.
- Soluble sodium and potassium were determined using Flame photometer.
- Carbonates and bicarbonates were determined by titration with standardized H_2SO_4 solution.
- Chloride was titrated with silver nitrate.
- Sulfate was calculated by subtraction.

6.3.1.2.2 Irrigation Water Characteristics

A sample was taken from the pumped irrigation water for analyzing in the laboratory according to methods by Richards [84].

1. pH value was determined using pH meter (Table 6.5).
2. Electrical conductivity (dS/m) was measured by conductivity meter.
3. Soluble ions (meq/L), soluble Ca^{++}, Mg^{++}, Na^+, K^+, CO_3^-, HCO_3^-, SO_4^- and Cl^- were determined as previously mentioned in soil water extract analysis.
4. Adjusted SAR was calculated as described by Ayers and Westcot [8].

TABLE 6.5 Chemical Analysis of Irrigation Water at Shalaqan

pH	EC dS/m	Soluble cations, meq/L				Soluble anions, meq/L			SAR
		Ca^{++}	Mg^{++}	Na^+	K^+	HCO_3^-	SO_4^-	Cl^-	
7.37	0.85	1.72	0.85	4.78	0.85	2.18	0.14	5.88	4.22

6.3.2 METHODS AND MEASUREMENTS

6.3.2.1   Irrigation System and Fertigation Device Performance

6.3.2.1.1   Evaluation of the Irrigation System Field Performance

The drip irrigation system was evaluated to determine the actual field performance. The performance data were collected with the following procedure:

1. Four laterals had been selected. On every lateral, four emitters had been selected.
2. The emitter discharge was determined by collecting the emitted water in graduated vessels for a specific time.
3. The average emitters discharge was calculated for all emitter locations.
4. The water distribution uniformity was calculated according to El-Amoud [31] with the following equation:

$$EU = 0.5\left(\frac{Q_n}{Q_a} + \frac{Q_a}{Q_x}\right) \times 100 \qquad (1)$$

where: EU = field emission uniformity, (%); Q_n = the average of the lowest (1/4) of emitters flow rate, (lph); Q_a = the average of all emitters flow rate, (lph); Q_x = the average of the highest (1/8) of emitters flow rate, (lph).

6.3.2.1.2   Calibration of Injectors

6.3.2.1.2.1   Measurement of Injection Flow Rate Using Volume Method

The injection system must be specifically calibrated under the operating conditions that will exist when fertilizer solutions are injected. Variations in operating pressure, system flow rate, and at times even temperature can influence the calibration of the system. The calibration procedure was performed as follows [22]:

1. The system was primed to make sure that it is operating at the same pressure during injection, and that suction and discharge lines do not contain air bubbles.
2. At the same elevation point, that it will be during actual run, a known volume of N-fertilizer solution flows to the intake line of the injector.
3. The time of injection was determined at a specific pressure difference.
4. Injection flow rate was measured at a various pressure difference points.

6.3.2.1.2.2 Measurement of Fertilizer Injection Concentration Rate

Concentration rate is the rate of change of concentration by injection time. The concentration rate was estimated for the types of injectors.

1. The injection system was operated at a steady constant pressure difference and injection rate.
2. Samples of the irrigation water containing N-fertilizer were taken at 5 minutes interval.
3. Electric conductivity was measured to estimate solution concentration for every sample using conductivity-meter.

6.3.2.2 Dimensional Analysis

This is a useful technique for the investigation of problems in fluid mechanics. If it is possible to identify the factors involved in a physical situation, dimensional analysis can usually establish the form of the relationship between these variables. The qualitative equations obtained by dimensional analysis can usually be converted into quantitative results determining an unknown factor experimentally [27]. There were five variables defined for injector performance: Q, ΔP, D, ρ, μ expressed in terms of the three fundamental dimensions M, L and T. According to Buckingham π-theorem, the solution contained $5 - 3 = 2$ dimensionless groups. The dimensional units of the five variables were:

Q	ΔP	D	ρ	μ
$L^3 T^{-1}$	$M L^{-1} T^{-2}$	L	$M L^{-3}$	$M L^{-1} T^{-1}$

The two dimensionless groups were:

Coefficient of pressure difference $$C_P = \frac{\Delta P \rho D^2}{\mu^2} \qquad (2)$$

Coefficient of injection rate $$C_Q = \frac{Q\rho}{\mu D} = \frac{\pi}{4} Re \qquad (3)$$

where: Q = injection rate, (m^3/sec); ΔP = pressure difference, (N/m^2); D = injector outlet diameter, (m); ρ = liquid density, (kg/ m^3); μ = liquid viscosity, (kg/m sec).

6.3.2.3   Irrigation Requirements

Standard methods for calculating water requirements and irrigation scheduling were used as follows:

a.   Calculating of potential evapotranspiration [47] using the following equation:

$$ETP = 0.0075 \times TF \times SS \times KS \times ETR \qquad (4)$$

where: ETP = potential evapotranspiration, (mm/day); TF = mean daily temperature, (F); SS = sunshine coefficient, TF = $(100 \times n/N)^{0.5}$; N = mean daily duration of max. possible sunshine hours; n = actual mean daily duration of sunshine, hours; KS = solar radiation coefficient = 0.097 – 0.00042 × RH; RH = mean daily relative humidity, (%); ETR = extra terrestrial radiation, mm/day.

b.   Irrigation interval was estimated from the following equation:

$$I = (A.W \times A.D \times Rd/ET_a) \times Ei \qquad (5)$$

where: I = allowable intervals between two irrigation, (day); A.W. = available soil water, Aw = F.C. – P.W.P., (mm/m); F.C. = field capacity, (mm); P.W. = permanent wilting point, (mm); A.D. = allowable soil moisture depletion below field capacity; Rd = Rooting depth, (cm); ET_a = actual evapotranspiration, ETa = ETP × KC, (mm/day); KC = crop coefficient; Ei = irrigation efficiency = 90.25%.

c. Water requirements were calculated according to the following
    equation:

$$W.R. = ET_a \times I \, (1 + L.R.) \times 4.2 \hspace{2cm} (6)$$

where: W.R = water requirement, (m³/fed); L.R = leaching requirement
{$EC_{IW}/2 \, EC_{DW}$}.

6.3.2.4   Experiment Layout

Field experiments were conducted during two successive growing
seasons: 2005/2006 and 2006/2007. Figure 6.3 shows the experimental
layout. A factorial experimental design was used with three replications
combined over method of injectors. Super phosphate (15.5% P_2O_5) and
potassium sulfate (48% K_2O) were added @ 100 and 100 kg/fed., respectively, during land preparation. Chinese garlic variety (*Allium sativum*)
was planted in the second week of September of 2006. The main plots
were used for three drip irrigation rates 50 (I_1), 75 (I_2), and 100% (I_3) of
ETc (1423, 2134 and 2846 m³/fed./season) were used. On the other hand,
the subplots were used for three levels of N- fertilizer namely: 60, 90,
120 Kg/fed (N_1, N_2, N_2). N fertilizers were applied via irrigation water

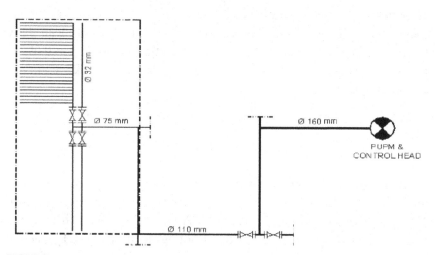

FIGURE 6.3   Layout of drip irrigation system in the experimental site.

@ N of 50, 75, 100% based on recommendations by MALR. Since three methods for fertilizer application: by-bass pressurized mixing tank (J_1), venturi (J_2), and positive displacement Injection pump (J_3) were used, the layout mentioned above was repeated three times. Garlic crop was drip irrigated every three days. The irrigation was terminated 20 days before harvesting. The crop was harvested in the last week of April (i.e., growing season lasted 165 days).

6.3.2.4.1 Evaluation of Emitter Clogging

To estimate the emitter flow rate, a collection unit and a stopwatch were used. Nine emitters for each lateral were evaluated by calculating the clogging ratio at the beginning and at the end of the growing season. Three emitters at the beginning, three at middle and three at the end of the lateral were tested for flow rate. Clogging ratio was calculated according to El-Berry et al. [32] using the following equations:

$$E = q_u/q_n \times 100 \qquad (7)$$

$$CR = (1 - E) \times 100 \qquad (8)$$

where: E = the emitter discharge efficiency, (%); q_u = emitter discharge, at the end of the growing season (Lph); q_n = emitter discharge, at the beginning of the growing season (Lph); CR = the emitter clogging ratio (%).

6.3.2.4.2 Total Yield

The total yield of each treatment was determined using a square wooden frame of 1 m × 1 m. The frame was placed randomly and the garlic plants within the frame were weighted.

6.3.2.4.3 Water Use Efficiency (WUE)

The WUE refers to garlic yield per cubic meter of irrigation water. It was calculated according to Israelson and Hanson [53] as follows:

$$WUE = Y/W \qquad (9)$$

where: WUE = water use efficiency, (Kg/m^3); Y = total crop yield, and (Kg/fed.); W = total water applied. $(m^3/fed.)$.

6.3.2.4.4 Nitrogen use efficiency (NUE)

The NUE refers to the crop yield per kilogram of Nitrogen. It was calculated according to Israelson and Hanson [53] as follows:

$$NUE = Y/C \qquad (10)$$

where: NUE = nitrogen use efficiency, (Kg/kg N); Y = total crop yield, and (Kg/fed.); C = Amount of N-fertilizer applied (kg N /fed.).

6.3.2.5 Cost Analysis

6.3.2.5.1 Irrigation Costs

Capital cost for different irrigation systems was calculated using dealer prices for 2007 for equipment and installation according to 2007 ASAE Standard [7].

Fixed Costs

The annual fixed costs of capital invested in the irrigation systems were calculating using the following equation:

$$F.C. = D + I + T \qquad (11)$$

where: F.C. = the annual fixed cost, (LE/year); D = the depreciation, (LE/year); I = the interest, and (LE/year); T = taxes and overheads ratio. (LE/year).

Depreciation was calculated using the following equation

$$D = [(I.C) - (Sv)]/(E.L) \qquad (12)$$

where: I.C. = The initial cost of irrigation system, (LE); Sv = Salvage value after depreciation and (LE); E.L = The expected life of the irrigation system components (years).

Taxes and overheads ratio was assumed as 2.0% of the initial cost. Interest on capital was calculated using the following equation

$$I = (I.C /2) \times I.R \tag{13}$$

where: I.R = the rate of interest = 14% (rate/year).

Operating Costs

The annual operating cost was calculated using the following equation:

$$R.C = L + E + (R\&M) \tag{14}$$

where: R.C. = annual operating cost, (LE/year); L = labor cost, (LE/year); E = energy cost, and (LE/year); (R&M) = repairs and maintenance costs. (LE/year).

Two farm workers for drip irrigation (60 *feddans* @ 10 LE/day for one labor) were employed. Repair and maintenance costs were taken as 2.0% of initial cost. Power cost of diesel type engine was calculated using the following equations:

$$Bp = (Q \times TDH)/K \times E \tag{15}$$

$$E.C = 1.2\, Bp \times H \times S \times F \tag{16}$$

where: Bp = break horse power, (kW); Q = discharge rate, (l/s); TDH = total dynamic head, (m); H = annual operating hours, (h); S = specific fuel consumption, (l/kW/h); F = fuel price, and (LE/l); 1.2 = factor accounting for lubrication; E.C = energy cost of diesel, (kW); K = coefficient to convert to energy unit = 102 according to James [55]; E = the overall efficiency = 45% for pump driven by internal combustion engine.

The Total Cost = Fixed Cost + Operating Cost, (LE/fed/year)

And the total costs of lifting water per m^3 were calculated as:

Total costs of lifting 1m^3 of water = {Total costs per hour (LE/h)}

$$/\{\text{Discharge (m}^3\text{/h)}\} \tag{17}$$

6.3.2.5.2 Chemical Injection Costs

There are several devices for chemical injecting through irrigation systems such as: by-pass pressurized mixing tank; Venturi and positive displacement Piston pump. The costs for different devices were calculated according to 2007 prices. The fixed and running costs for chemical injectors were calculated using the same steps as irrigation systems, described above. Following considerations were taken into account:

1. The expected life of chemical injectors was 3 years for by-pass pressurized mixing tank; and 5 years for both venturi and positive displacement piston injection pump
2. The fertigated area for different devices was: 15 fed for by-bass pressurized mixing tank; 25 fed for venturi and positive displacement pump
3. Repair and maintenance costs of injectors were 5% of initial cost.
4. Total chemigation costs = Irrigation cost + Chemical injection cost

6.4   RESULTS AND DISCUSSION

6.4.1   PERFORMANCE OF FERTILIZER INJECTORS

In this chapter, the performance of fertigation is defined as the relationship between pressure difference and the injection rate of fertilizer and variation in fertilizer concentration with injection time.

6.4.1.1   Performance of By-Bass Pressurized Mixing Tank

The relationship between pressure difference and injection rate is shown in Figure 6.4. It indicates an increment in injection rate with increasing pressure difference (ΔP) between the inlet and outlet of by-bass pressurized mixing tank. The higher-pressure difference led to higher water flow through the by-bass pressurized mixing tank. As a result, an increase in the injected fertilizer through irrigation system will be expected. The minimum injection rate was 8.3 lpm at a pressure difference of 0.1 bar; whereas, the maximum one was 40 lpm at a pressure difference of 0.7 bar.

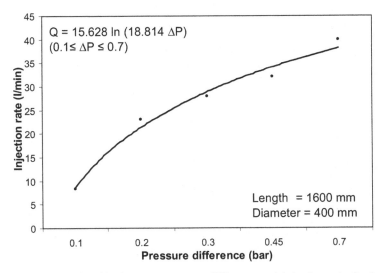

FIGURE 6.4 Relationship between pressure difference and injection rate for by-bass pressurized mixing tank.

The logarithmic relationship was observed between pressure difference and injection rate as follows:

$$Q = 15.628 \ln (18.814 \, \Delta P) \tag{18}$$

where: Q = injection rate (l/min); ΔP = pressure difference between the inlet; and outlet $(0.1 \le \Delta P \le 1)$ (bar).

At a constant pressure difference, the salt concentration (N fertilizer + salt in irrigation water) in irrigation water changed during injection time. Nonlinear regression coefficients were significant at $P = 0.05$ with a coefficient of determination of 0.98.

Figure 6.5 shows a rapid increment in fertilizer concentration at the beginning of injection time. The concentration increased from the EC_{iw} of 0.85 dS/m until it reached peak value of 1.8 dS/m at injecting time of 11 min. Then, the concentration decreased with time until it reaches nearly, the level of irrigation water after 30 minutes of injecting time. According to what had been mentioned previously, one of the main disadvantages of by-bass pressurized mixing tank is the decrease in chemical concentration with time of injection [43].

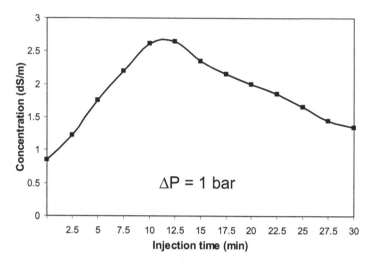

FIGURE 6.5   The change of outlet salt concentration during injection time for by-bass pressurized mixing tank.

6.4.1.2   Performance of Venturi Injector

Figure 6.6 shows the performance curve of venturi injector. It can be noticed an increase of fertilizer injection rate Q due to the increase in ΔP between the inlet and the outlet. The minimum injection rate was 0.467 L/min at ΔP value of 0.4 bar whereas the maximum was 3.01 lpm at ΔP value of 1.4 bar. The relationship between pressure difference and injection rate can be expressed in a logarithmic equation as follows, for $0.4 \leq \Delta P \leq 1.4$:

$$Q = 2.052 \ln (2.954 \ \Delta P) \qquad (19)$$

Venturi injector gave the lowest injection rate compared with the other injector types in spite of the pressure drop created in the irrigation system, due to friction losses of approximately 1.0 bar [78]. Venturi injector is more distinguished than by-bass pressurized mixing tank with its relatively constant concentration during injection time. It can be observed from Figure 6.7, concentration starts from 0.85 dS/m until it reaches to the highest concentration of 2.3 dS/m after an injection time of 7.5 minutes. Salt concentration varies in a narrow range during the rest of injection time.

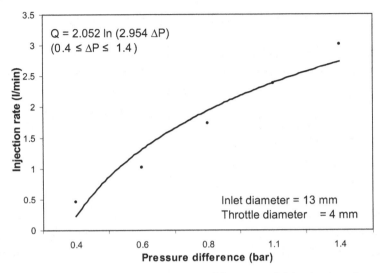

FIGURE 6.6 Relationship between pressure difference and injection rate for venturi injector.

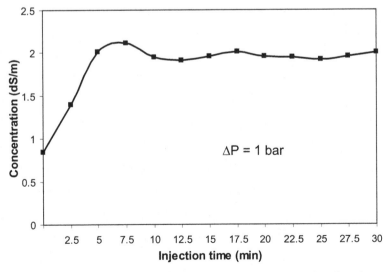

FIGURE 6.7 The change of outlet salt concentration during injection time for venturi injector.

6.4.1.3   Performance of Positive Displacement Pump

The injection rate of positive displacement pump depends on number of positive displacement strokes per minute. It was increased by increasing the pressure difference. Figure 6.8 indicates that the lowest injection rate was 7.82 lpm at 1 bar of pressure difference, whereas the highest was 240.2 lpm at 2.5 bars of pressure difference. The relationship between ΔP and Q can be represented by the following power equation, for $1 \leq \Delta P \leq 2.5$:

$$Q = 74.009 \ \Delta P^{1.183} \qquad (20)$$

Salt concentration for positive displacement pump was highest and it was a precision injector of the three tested types. It starts from 0.85 dS/m until the concentration was 2.35 dS/m after 7.5 minutes of injection time as

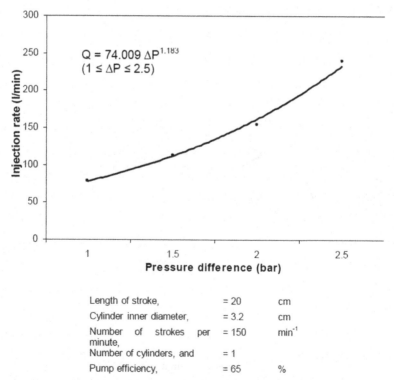

FIGURE 6.8   Relationship between pressure difference and injection rate for positive displacement injector pump.

shown in Figure 6.9. Salt concentration is constant during the rest of injection time [43].

6.4.2 DIMENSION ANALYSIS

A dimension analysis was done for the physical parameters affecting injector performance. Dimensionless qualitative equations were used to compare the three fertilizer injectors. Two dimensionless equations were obtained.

$$C_P = \frac{\Delta P \rho D^2}{\mu^2} \tag{21}$$

$$C_Q = \frac{Q\rho}{\mu D} = \frac{\pi}{4} Re \tag{22}$$

where: C_P = coefficient of pressure difference; C_Q = coefficient of injection rate; Re = Reynolds number; Q = injection rate, (m³/sec); ΔP = pressure difference between the inlet and outlet, (N/m²); D = injector outlet diameter, (m); ρ = water density, (kg/m³); and μ = water viscosity, (kg/m.sec).

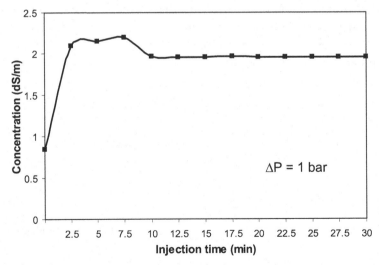

FIGURE 6.9 The change of outlet salt concentration during injection time for positive displacement injector pump.

The lowest performance was for venturi injector because the coefficient of injection rate (C_Q) was very low in spite of the relative high-consumed energy (ΔP), which is represented in the coefficient of pressure (C_p). The minimum value of C_Q was 0.36×10^3 when C_p was 1.1×10^5; and the maximum value of C_Q was 2.3×10^3 when C_p was 3.9×10^5. The next higher performance was for by-bass pressurized mixing tank. It starts injecting at a very low-pressure difference with a higher injection rate comparing with venturi injector. The minimum value of C_Q was 6.36×10^3 when C_p was 0.279×10^5; and the maximum value of C_Q was 3.06×10^4 when C_p was 1.95×10^5. The highest performance was for positive displacement pump. Although the consumed energy was high, yet the injection rate was high. The minimum value of C_Q was 6×10^4 when C_p was 2.8×10^5; and the maximum value of C_Q was 18.4×10^4 when C_p was 7×10^5.

Table 6.6 shows classification of flow characteristics based on Reynolds number (Re), under experimental operating conditions of irrigation system for 4.08 m³/h discharge at 2 bars of pressure. It can be concluded that the injection rate increases with increasing of Re for different types of injectors.

6.4.3   EMITTERS CLOGGING

Data for emitter clogging are given in Tables 6.7 and 6.8, and Figures 6.10, 6.11, and 6.12. According to clogging percentage, the injectors (J_1, J_2, and J_3) can be arranged in the ascending order: $J_3 < J_2 < J_1$, regardless of irrigation and nitrogen treatments.

TABLE 6.6   Classification of Flow Characteristics According to Re Value, for Three Injector Types

Injector type	Reynolds number, Re			
	Lower value	Flow characteristic	Higher value	Flow characteristic
Venturi	—	Laminar	< 2500	Transient
By-bass pressurized mixing tank	8000	Transient	39,100	Turbulent
Positive displacement pump	76,000	Turbulent	235,000	Turbulent

TABLE 6.7 Effects of Injector, Irrigation and Nitrogen Fertilizer Treatments on Clogging Percentage

Treatments	Clogging %	
By-bass pressurized mixing tank (J_1)	22.400	a
Venturi (J_2)	18.650	b
Positive displacement injector pump (J_3)	15.580	c
100 % of ET_c (I_1)	16.970	c
75 % of ET_c (I_2)	17.590	b
50 % of ET_c (I_3)	22.070	a
120 kg/ fed. of N (N_1)	19.530	a
90 kg/ fed. of N (N_2)	19.040	b
60 kg/ fed. of N (N_3)	18.060	c

TABLE 6.8 First and second interactions among injector, irrigation and nitrogen fertilizer treatments on clogging percentage.

Treatments	Clogging %		Treatments	Clogging %	
First order interactions			Second order interactions		
$J_1 \times I_3$	19.800	d	$J_1 \times I_3 \times N_3$	20.700	h
$J_1 \times I_2$	20.360	c	$J_1 \times I_3 \times N_2$	19.800	k
$J_1 \times I_1$	27.030	a	$J_1 \times I_3 \times N_1$	18.900	l
$J_2 \times I_3$	16.720	g	$J_1 \times I_2 \times N_3$	20.880	g
$J_2 \times I_2$	17.340	e	$J_1 \times I_2 \times N_2$	20.200	i
$J_2 \times I_1$	21.890	b	$J_1 \times I_2 \times N_1$	20.000	j
$J_3 \times I_3$	14.390	i	$J_1 \times I_1 \times N_3$	27.900	a
$J_3 \times I_2$	15.060	h	$J_1 \times I_1 \times N_2$	27.000	b
$J_3 \times I_1$	17.280	f	$J_1 \times I_1 \times N_1$	26.200	c
$J_1 \times N_3$	23.160	a	$J_2 \times I_3 \times N_3$	17.550	o
$J_1 \times N_2$	22.330	b	$J_2 \times I_3 \times N_2$	16.800	r
$J_1 \times N_1$	21.700	c	$J_2 \times I_3 \times N_1$	15.810	u
$J_2 \times N_3$	19.380	d	$J_2 \times I_2 \times N_3$	17.780	m
$J_2 \times N_2$	18.640	e	$J_2 \times I_2 \times N_2$	17.250	p
$J_2 \times N_1$	17.940	f	$J_2 \times I_2 \times N_1$	17.000	q
$J_3 \times N_3$	16.040	h	$J_2 \times I_1 \times N_3$	22.800	d
$J_3 \times N_2$	16.150	g	$J_2 \times I_1 \times N_2$	21.870	e

TABLE 6.8   Continued

Treatments	Clogging %		Treatments	Clogging %	
First order interactions			**Second order interactions**		
$J_3 \times N_1$	14.550	i	$J_2 \times I_1 \times N_1$	21.000	f
$I_3 \times N_3$	17.780	e	$J_3 \times I_3 \times N_3$	15.100	w
$I_3 \times N_2$	17.150	f	$J_3 \times I_3 \times N_2$	14.840	×
$I_3 \times N_1$	15.980	h	$J_3 \times I_3 \times N_1$	13.240	z
$I_2 \times N_3$	17.960	d	$J_3 \times I_2 \times N_3$	15.230	v
$I_2 \times N_2$	17.770	e	$J_3 \times I_2 \times N_2$	15.870	t
$I_2 \times N_1$	17.030	g	$J_3 \times I_2 \times N_1$	14.090	y
$I_1 \times N_3$	22.830	a	$J_3 \times I_1 \times N_3$	17.790	m
$I_1 \times N_2$	22.210	b	$J_3 \times I_1 \times N_2$	17.750	n
$I_1 \times N_1$	21.170	c	$J_3 \times I_1 \times N_1$	16.300	s

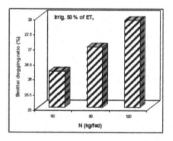
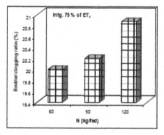
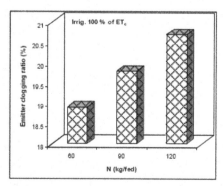

FIGURE 6.10   Effects of irrigation treatments and N-fertilization levels on emitter clogging percentage: By-bass pressurized mixing tank.

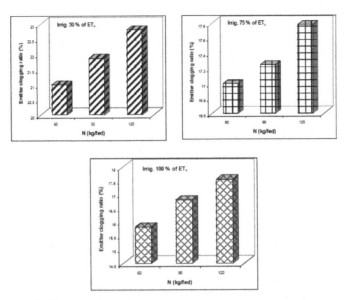

FIGURE 6.11 Effects of irrigation treatments and N-fertilization levels on emitter clogging percentage: Venturi.

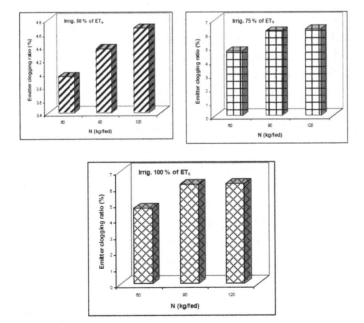

FIGURE 6.12 Effects of irrigation treatments and N-fertilization levels on emitter clogging percentage: Positive displacement injector pump.

Differences in clogging percentage were significant at the 5% level among all injector treatments (Table 6.7). It is well-known that increasing pressure within the drip irrigation system will decrease clogging potential and vice versa. The check valve used with injector J_1 decreases the pressure within the irrigation system: 0.1 bar $\leq \Delta P \leq 0.7$ bar. In the venture injector (J_2), the contracted part increases water flow velocity at the expense of the pressure, due to the increase in friction losses across the venturi. The ΔP for operating pressure in venture ranges within 0.4 bar $\leq \Delta P \leq 1.4$ bar. With respect to injector (J_3), it injects N-solution at pressure higher than that within the irrigation system. Therefore, if injector (J_3) does not increase the pressure within the irrigation system, it will not decrease it. Also, the reciprocating movement of the piston within its cylinder in the pump can cause a good agitation and good mixing of the nitrogen solution. This will lead to reduction in percentage of clogging [78]. This is clear from the range of its operating pressure: 1.0 bar $\leq \Delta P \leq 2.5$ bar. Thus, the injectors can be put in the following ascending order according to the range of operating pressures: $J_1 < J_2 < J_3$. This order is the different from the above mentioned at the beginning of this section. Data indicates that the clogging problem can be reduced with increasing the irrigation level.

According to clogging percentage, irrigation treatments can be written in the following ascending order: I_3 (100% of ET_c) $< I_2$ (75% of ET_c) $< I_1$ (50% of ET_c). Differences in clogging percentage between any two irrigation treatments is significant at the 5% level, due to that emitters are more flushed (Table 6.7). The data show that increasing the amount of the applied nitrogen fertilizer $(NH_4)_2SO_4$ from 60 (N_1) to 120 kg/fed. (N_3) increased the clogging percentage in all irrigation treatments and the three-injector treatments. According to clogging percentage, the following ascending order illustrates the role of nitrogen treatments: $N_1 < N_2 < N_3$. This can be explained on the basis that increasing nitrogen content will increase the amount of Calcium that will precipitate within the emitters and within the narrow openings after evaporation, especially in the form of CO_3^- and SO_4^-. We can conclude that the problem of emitter clogging was increased with increasing nitrogen $(NH_4)_2SO_4$ application, and decreasing the irrigation rate. The maximum and minimum values of clogging percentage were 17.79, 22.8,

27.9; and 13.24, 15.81, 18.9 under the conditions of J_3, J_2, J_1; $I_1 \times N_3$; and $I_3 \times N_1$, respectively.

According to clogging percentage, injectors can be written in the following ascending order: $J_3 < J_2 < J_1$. The results in this section are in agreement with those by Sagi [86].

6.4.3.1   First Order Interaction: Clogging Percentage (Table 6.8)

Injector × Irrigation treatments: The effects of injector types versus irrigation treatments on clogging percentage were significant at the 5% level. The maximum clogging percentage was 27.03 and the minimum was 14.39, with the interactions $J_1 \times I_1$ and $J_3 \times I_3$, respectively.

Injector × Nitrogen treatments: This interaction has a significant effect on clogging percentage at the 5% level. The maximum clogging percentage was 23.16 and the minimum was 14.55, with the interactions $J_1 \times N_3$ and $J_3 \times N_1$, respectively.

Irrigation × N treatments: This interaction has a significant effect on the problem of emitter clogging percentage at the 5% level with the exception of the cases: $I_3 \times N_3$ and $I_2 \times N_2$. The maximum clogging percentage was 22.83 and the minimum was 15.98 in the interactions $I_1 \times N_3$ and $I_3 \times N_1$, respectively.

6.4.3.2   Second Order Interactions: Clogging Percentage (Table 6.8)

J1× I × N: This interaction led to significant effects on clogging percentage at the 5% level. The highest clogging percentage was 27.9 and the lowest was 18.9 with the interactions $J_1 \times I_1 \times N_3$ and $J_1 \times I_3 \times N_1$, respectively.

J2× I × N: It caused a significant difference in clogging percentage at the 5% level. The maximum clogging percentage was 22.8 and the minimum was 15.81 with the interactions $J_2 \times I_1 \times N_3$ and $J_2 \times I_3 \times N_1$, respectively.

J3× I × N: It has a significant effect on clogging percentage at the 5% level. The highest clogging percentage was 17.79 and the lowest was 13.24, caused by the interactions $J_3 \times I_1 \times N_3$ and $J_3 \times I_3 \times N_1$, respectively.

6.4.4 GARLIC YIELD

Tables 6.9 and 6.10 and Figure 6.13 indicate the effects of injector types (by-bass pressurized mixing tank J_1, venturi J_2, and positive displacement pump J_3); irrigation treatments (50, 75, 100% of ETc: I_1, I_2, I_3), and nitrogen treatments (60, 90, 120 kg/fed: N_1, N_2, N_3) on commercial yield of garlic (ton/feddan).

6.4.4.1 Effects of Injector Types, Irrigation Treatments and Nitrogen Treatments on Garlic Yield

Table 6.9 shows that injector types, irrigation treatments and nitrogen treatments all have significant effects on garlic yield at the 5% level. These values can be put in the ascending orders according to the yield: $J_1 < J_2 < J_3$, $I_1 \Leftrightarrow I_2 < I_3$ and $N_1 < N_2 < N_3$. These results agree with those of Abdel-Aziz [1].

6.4.4.2 First Order Interactions (Table 6.10)

All the interactions ($J_1 \times I$, $J_1 \times N$, $J_2 \times I$, $J_2 \times N$, $J_3 \times I$, $J_3 \times N$ and, $I \times N$) have significant effects on garlic yield at the 5% level. The maximum yields were 5.763, 5.760, 6.103 ton/fed. and the minimum were 2.960,

TABLE 6.9 Effects of Injector Types, Irrigation Treatments and Nitrogen Treatments on Garlic Yield

Treatments		Garlic yield ton/feddan	
By-bass pressurized mixing tank	J_1	4.242	c
Venturi	J_2	5.100	b
Positive displacement injector pump	J_3	5.272	a
100 % of ET$_c$	I_1	5.399	b
75 % of ET$_c$	I_2	5.477	a
50 % of ET$_c$	I_3	3.739	c
120 kg/ fed. of N	N_1	5.436	a
90 kg/ fed. of N	N_2	5.120	b
60 kg/ fed. of N	N_3	4.059	c

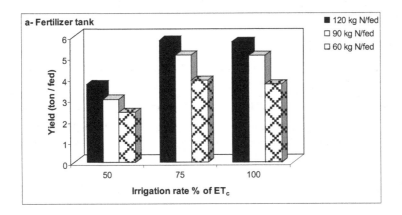

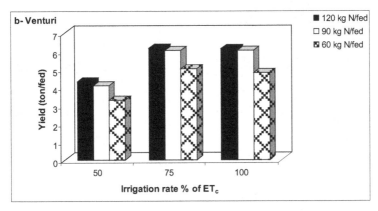

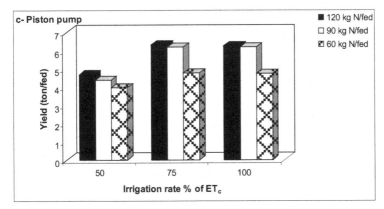

FIGURE 6.13   Effects of irrigation rates and Nitrogen fertilization levels on garlic yield under different injection methods.

3.327, 3.230 ton/fed. in the interactions $J_3 \times I_2$, $J_3 \times N_3$, $I_2 \times N_3$ and $J_1 \times I_1$, $J_1 \times N_1$, $I_1 \times N_1$, respectively.

TABLE 6.10 Effects of First and Second Order Interactions Among Injector, Irrigation and Nitrogen Fertilizer on Garlic Yield

Treatments	Yield ton/feddan		Treatments	Yield ton/feddan	
First order interactions			**Second order interactions**		
$J_1 \times I_3$	4.833	e	$J_1 \times I_3 \times N_3$	5.730	e
$J_1 \times I_2$	4.933	d	$J_1 \times I_3 \times N_2$	5.070	fg
$J_1 \times I_1$	2.960	h	$J_1 \times I_3 \times N_1$	3.700	m
$J_2 \times I_3$	5.643	c	$J_1 \times I_2 \times N_3$	5.800	e
$J_2 \times I_2$	5.734	ab	$J_1 \times I_2 \times N_2$	5.100	f
$J_2 \times I_1$	3.923	g	$J_1 \times I_2 \times N_1$	3.900	l
$J_3 \times I_3$	5.720	b	$J_1 \times I_1 \times N_3$	3.500	n
$J_3 \times I_2$	5.763	a	$J_1 \times I_1 \times N_2$	3.000	p
$J_3 \times I_1$	4.333	f	$J_1 \times I_1 \times N_1$	2.380	q
$J_1 \times N_3$	5.010	d	$J_2 \times I_3 \times N_3$	6.130	c
$J_1 \times N_2$	4.390	f	$J_2 \times I_3 \times N_2$	6.030	d
$J_1 \times N_1$	3.327	g	$J_2 \times I_3 \times N_1$	4.770	h
$J_2 \times N_3$	5.537	b	$J_2 \times I_2 \times N_3$	6.170	c
$J_2 \times N_2$	5.390	c	$J_2 \times I_2 \times N_2$	6.040	d
$J_2 \times N_1$	4.374	f	$J_2 \times I_2 \times N_1$	4.993	g
$J_3 \times N_3$	5.760	a	$J_2 \times I_1 \times N_3$	4.310	j
$J_3 \times N_2$	5.580	b	$J_2 \times I_1 \times N_2$	4.100	k
$J_3 \times N_1$	4.477	e	$J_2 \times I_1 \times N_1$	3.360	o
$I_3 \times N_3$	6.040	b	$J_3 \times I_3 \times N_3$	6.260	b
$I_3 \times N_2$	5.760	c	$J_3 \times I_3 \times N_2$	6.180	bc
$I_3 \times N_1$	4.397	e	$J_3 \times I_3 \times N_1$	4.720	hi
$I_2 \times N_3$	6.103	a	$J_3 \times I_2 \times N_3$	6.340	a
$I_2 \times N_2$	5.777	c	$J_3 \times I_2 \times N_2$	6.190	bc
$I_2 \times N_1$	4.551	d	$J_3 \times I_2 \times N_1$	4.760	hi
$I_1 \times N_3$	4.163	f	$J_3 \times I_1 \times N_3$	4.680	i
$I_1 \times N_2$	3.823	g	$J_3 \times I_1 \times N_2$	4.370	j
$I_1 \times N_1$	3.230	h	$J_3 \times I_1 \times N_1$	3.950	l

Means with different letters within each column are significant at the 5% level.

6.4.4.3   Second Interactions (Table 6.10)

The second order interactions ($J_1 \times I \times N$, $J_2 \times I \times N$ and $J_3 \times I \times N$) have significant effects on garlic yield at the 5% level. The highest yields were 5.73–5.8, 6.13–6.17, 6.340; and the lowest were 2.38, 3.36; 3.950 in the interactions: ($J_1 \times I_3 \times N_3 - J_1 \times I_2 \times N_3$), ($J_2 \times I_3 \times N_3 - J_2 \times I_2 \times N_3$), ($J_3 \times I_2 \times N_3$) and ($J_1 \times I_1 \times N_1$, $J_2 \times I_1 \times N_1$, $J_3 \times I_1 \times N_1$), respectively. Based on the results in this section, we can conclude the following:

1. J_1 and J_2 decrease pressure within the irrigation system;
2. Nitrogen concentration is not constant with time in case of injector J_2;
3. Nitrogen concentration decreased with time in case of injector J_1;
4. Injector J_3 does not decrease pressure within irrigation system and the piston movement increases N-solubility and decrease both precipitation and emitters clogging;
5. Increasing N-level increased plant growth and emitters clogging;
6. Increasing irrigation level up to 75% of ET_c increased emitter flushing, salt removal from root zone and nutrient availability, but decreased emitters clogging;
7. Yield data: The decrease in nitrogen concentration with time when J_1 is used;
8. Yield data: The fluctuation in nitrogen concentration with time when J_2 is used;
9. Yield data: Increasing N concentration increase clogging and vice versa; and
10. Yield data: Increasing irrigation rate to some extent increases yield and decreases clogging.

6.4.5   WATER USE EFFICIENCY

Tables 6.11 and 6.12 and Figure 6.14 show that effects of injectors types (J_1, J_2, J_3), irrigation treatments (I_1, I_2, I_3) and nitrogen treatments (N_1, N_2, N_3) on the water use efficiency (WUE) for garlic crop.

TABLE 6.11   Effects of Injector Types, Irrigation Levels and Nitrogen Treatments on Water Use Efficiency (WUE)

Treatments		WUE (kg/m³)	
By-bass pressurized mixing tank	J_1	2.067	c
Venturi	J_2	2.473	b
Positive displacement injector pump	J_3	2.528	a
100% of ET_c	I_1	1.935	b
75% of ET_c	I_2	2.568	a
50 % of ET_c	I_3	2.565	a
120 kg/ fed. of N	N_1	2.635	a
90 kg/ fed. of N	N_2	2.510	b
60 kg/ fed. of N	N_3	1.924	c

TABLE 6.12   Effects of First Order and Second Order Interactions Among Injector Types, Irrigation Levels and Nitrogen Levels on WUE

Treatments	WUE (kg/m³)		Treatments	WUE (kg/m³)	
First order interactions			**Second order interactions**		
$J_1 \times I_3$	1.810	i	$J_1 \times I_3 \times N_3$	2.013	r
$J_1 \times I_2$	2.311	e	$J_1 \times I_3 \times N_2$	2.118	q
$J_1 \times I_1$	2.080	f	$J_1 \times I_3 \times N_1$	1.300	v
$J_2 \times I_3$	1.985	h	$J_1 \times I_2 \times N_3$	2.717	h
$J_2 \times I_2$	2.690	d	$J_1 \times I_2 \times N_2$	2.389	j
$J_2 \times I_1$	2.743	b	$J_1 \times I_2 \times N_1$	1.827	s
$J_3 \times I_3$	2.009	g	$J_1 \times I_1 \times N_3$	2.459	i
$J_3 \times I_2$	2.704	c	$J_1 \times I_1 \times N_2$	2.108	q
$J_3 \times I_1$	2.872	a	$J_1 \times I_1 \times N_1$	1.673	t
$J_1 \times N_3$	2.396	e	$J_2 \times I_3 \times N_3$	2.154	p
$J_1 \times N_2$	2.205	f	$J_2 \times I_3 \times N_2$	2.119	q
$J_1 \times N_1$	1.600	i	$J_2 \times I_3 \times N_1$	1.683	t
$J_2 \times N_3$	2.691	c	$J_2 \times I_2 \times N_3$	2.891	ef
$J_2 \times N_2$	2.610	d	$J_2 \times I_2 \times N_2$	2.830	g
$J_2 \times N_1$	2.117	g	$J_2 \times I_2 \times N_1$	2.348	k
$J_3 \times N_3$	2.816	a	$J_2 \times I_1 \times N_3$	3.029	c
$J_3 \times N_2$	2.714	b	$J_2 \times I_1 \times N_2$	2.881	f
$J_3 \times N_1$	2.055	h	$J_2 \times I_1 \times N_1$	2.319	l

TABLE 6.12 Continued

Treatments	WUE (kg/m³)		Treatments	WUE (kg/m³)	
First order interactions			Second order interactions		
$I_3{\times}N_3$	2.122	f	$J_3{\times}I_3{\times}N_3$	2.199	n
$I_3{\times}N_2$	2.136	e	$J_3{\times}I_3{\times}N_2$	2.171	o
$I_3{\times}N_1$	1.547	h	$J_3{\times}I_3{\times}N_1$	1.658	u
$I_2{\times}N_3$	2.856	b	$J_3{\times}I_2{\times}N_3$	2.961	d
$I_2{\times}N_2$	2.706	c	$J_3{\times}I_2{\times}N_2$	2.900	e
$I_2{\times}N_1$	2.142	e	$J_3{\times}I_2{\times}N_1$	2.251	m
$I_1{\times}N_3$	2.926	a	$J_3{\times}I_1{\times}N_3$	3.289	a
$I_1{\times}N_2$	2.687	d	$J_3{\times}I_1{\times}N_2$	3.071	b
$I_1{\times}N_1$	2.083	g	$J_3{\times}I_1{\times}N_1$	2.256	m

Means with different letters within each column are significant at the 5% level.

6.4.5.1 Effects of Injector Types, Irrigation Levels and Nitrogen Treatments on WUE

Table 6.11 illustrates the main effects of injector types, irrigation levels and nitrogen treatments on WUE. The three parameters have significant effects at the 5% level on WUE. According to WUE values, parameters can be written in the ascending orders: $J_1 < J_2 < J_3$, $I_1 < I_2 < I_3$ and $N_1 < N_2 < N_3$.

6.4.5.2 First Order Interactions (Table 6.12)

The first order interactions ($J_1 \times I_1$, $J_1 \times N_1$, $J_2 \times I_2$, $J_2 \times N_2$, $J_3 \times I_3$, $J_3 \times N_3$, I × N) caused significant effects on WUE at the 5% level. The maximum WUE (kg garlic/m³ of irrigation water) were 2.87, 2.82, 2.93; and the minimum were 1.81, 1.6, 1.55 in the interactions $J_3 \times I_1$, $J_3 \times N_3$, $I_1{\times}N_3$; and $J_1 \times I_3$, $J_1{\times}N_1$, $I_3 \times N_1$), respectively.

6.4.5.3 Second Order Interactions (Table 6.12)

The second interactions ($J_1 \times I_1 \times N_1$, $J_2 \times I_2 \times N_2$ and $J_3 \times I_3 \times N_3$) led to significant effects on WUE at the 5% level. The maximum WUE values

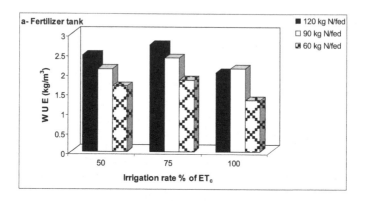

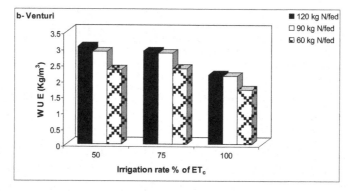

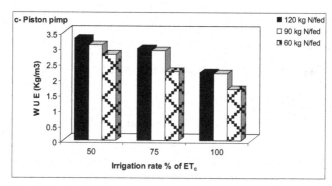

FIGURE 6.14 Effects of irrigation rates and Nitrogen fertilization levels on water use efficiency under different injection methods.

(kg garlic/m³ of irrigation water) were 2.717, 3.029, 3.289; and the minimum were 1.3, 1.683, 1.658 in the following interactions ($J_1 \times I_2 \times N_3$, $J_2 \times I_1 \times N_3$, $J_3 \times I_1 \times N_3$) and ($J_1 \times I_3 \times N_1$, $J_2 \times I_3 \times N_1$, $J_3 \times I_3 \times N_1$), respectively.

6.4.6   NITROGEN USE EFFICIENCY (NUE)

Data on nitrogen use efficiency (NUE) are given in Tables 6.13 and 6.14 and plotted in Figure 6.15.

TABLE 6.13   Effects of Injector Types, Irrigation Treatments and Nitrogen Treatment on Nitrogen Use Efficiency (NUE: kg garlic /kg N)

Treatments		NUE	
By-bass pressurized mixing tank	J_1	49.180	c
Venturi	J_2	59.750	b
Positive displacement injector pump	J_3	60.280	a
100% of ET_c	I_1	62.570	b
75% of ET_c	I_2	63.640	a
50 % of ET_c	I_3	42.990	c
120 kg/ fed. of N	N_1	45.280	c
90 kg/ fed. of N	N_2	56.170	b
60 kg/ fed. of N	N_3	67.760	a

TABLE 6.14   Effects of First Order and Second Order Interactions Between Injectors, Irrigation Levels and Nitrogen Levels on Nitrogen Use Efficiency (NUE: kg garlic/kg N)

Treatments	NUE kg garlic/kg N		Treatments	NUE kg garlic/kg N	
First order interaction			**Second order interaction**		
$J_1{\times}I_3$	55.250	f	$J_1{\times}I_3{\times}N_3$	47.750	q
$J_1{\times}I_2$	56.660	e	$J_1{\times}I_3{\times}N_2$	56.330	k
$J_1{\times}I_1$	35.610	i	$J_1{\times}I_3{\times}N_1$	61.670	i
$J_2{\times}I_3$	65.970	d	$J_1{\times}I_2{\times}N_3$	48.330	p
$J_2{\times}I_2$	67.340	a	$J_1{\times}I_2{\times}N_2$	56.660	j
$J_2{\times}I_1$	45.940	h	$J_1{\times}I_2{\times}N_1$	65.000	h
$J_3{\times}I_3$	66.500	c	$J_1{\times}I_1{\times}N_3$	29.170	x
$J_3{\times}I_2$	66.930	b	$J_1{\times}I_1{\times}N_2$	38.000	u
$J_3{\times}I_1$	47.420	g	$J_1{\times}I_1{\times}N_1$	39.670	s
$J_1{\times}N_3$	41.750	i	$J_2{\times}I_3{\times}N_3$	51.080	o
$J_1{\times}N_2$	50.330	f	$J_2{\times}I_3{\times}N_2$	67.000	f
$J_1{\times}N_1$	55.450	e	$J_2{\times}I_3{\times}N_1$	79.830	b
$J_2{\times}N_3$	46.140	h	$J_2{\times}I_2{\times}N_3$	51.420	n

TABLE 6.14   Continued

Treatments	NUE kg garlic/kg N		Treatments	NUE kg garlic/kg N	
First order interaction			Second order interaction		
$J_2 \times N_2$	59.890	c	$J_2 \times I_2 \times N_2$	67.110	f
$J_2 \times N_1$	73.220	b	$J_2 \times I_2 \times N_1$	83.500	a
$J_3 \times N_3$	47.940	g	$J_2 \times I_1 \times N_3$	35.920	w
$J_3 \times N_2$	58.290	d	$J_2 \times I_1 \times N_2$	45.560	r
$J_3 \times N_1$	74.610	a	$J_2 \times I_1 \times N_1$	56.330	k
$I_3 \times N_3$	50.330	g	$J_3 \times I_3 \times N_3$	52.160	m
$I_3 \times N_2$	64.000	d	$J_3 \times I_3 \times N_2$	68.660	e
$I_3 \times N_1$	73.390	b	$J_3 \times I_3 \times N_1$	78.670	d
$I_2 \times N_3$	50.810	f	$J_3 \times I_2 \times N_3$	52.670	l
$I_2 \times N_2$	64.180	c	$J_3 \times I_2 \times N_2$	68.780	e
$I_2 \times N_1$	75.940	a	$J_3 \times I_2 \times N_1$	79.330	c
$I_1 \times N_3$	34.700	i	$J_3 \times I_1 \times N_3$	39.000	t
$I_1 \times N_2$	40.330	h	$J_3 \times I_1 \times N_2$	37.440	v
$I_1 \times N_1$	53.940	e	$J_3 \times I_1 \times N_1$	65.830	g

Means with different letters within each column are significant at a 0.05% level.

6.4.6.1   Effects of Injector Types, Irrigation Treatments and Nitrogen Levels on NUE

Data show that the injector types, irrigation levels and Nitrogen levels had significant effects on NUE on the 5% level. According to the values of NUE, the parameters under investigation can be put in the ascending orders: $J_1 < J_2 < J_3$, $I_1 < I_3 < I_2$ and $N_1 < N_2 < N_3$.

6.4.6.2   First Order Interactions (Table 6.14)

The first order interactions ($J_1 \times I$, $J_1 \times N$, $J_2 \times I$, $J_2 \times N$, $J_3 \times I$, $J_3 \times N$, and $I \times N$) led to significant effects on NUE by garlic crop at the 5% level. The maximum values of NUE (kg of garlic yield/kg N) were 67.34, 74.61, 75.94; and the minimum were 35.61, 41.75, 34.7 in the interactions ($J_2 \times I_2$, $J_3 \times N_1$, and $I_2 \times N_1$) and ($J_1 \times I_1$, $J_1 \times N_2$, and $I_1 \times N_2$), respectively.

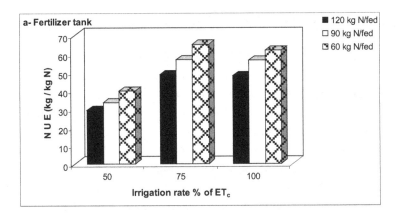

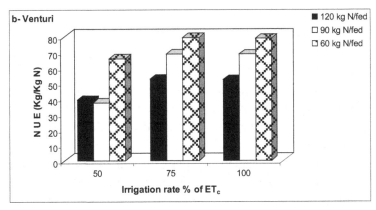

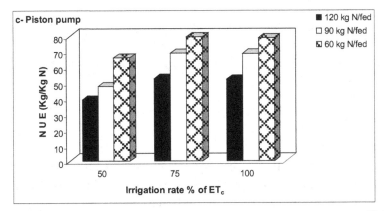

FIGURE 6.15 Effects of irrigation rates and Nitrogen levels on Nitrogen use efficiency (NUE), under different injection methods.

6.4.6.3   Second Order Interactions (Table 6.14)

In this case, the interactions ($J_1 \times I \times N$, $J_2 \times I \times N$ and $J_3 \times I \times N$) had significant effects on NUE at the 5% level. The maximum values of NUE were 65.0, 83.5, 79.33; and the minimum were 29.17, 35.92, 37.44 in interactions ($J_1 \times I_2 \times N_1$, $J_2 \times I_2 \times N_1$, $J_3 \times I_2 \times N_1$) and ($J_1 \times I_1 \times N_3$, $J_2 \times I_1 \times N_3$, $J_3 \times I_1 \times N_2$), respectively.

Results for NUE in this section reveal that the effects of the interactions – injectors types × irrigation levels, injector types × N levels and the irrigation levels × N levels – on garlic yield are significant in most cases with few exceptions. This may be due to one or more of the following reasons:

1. J_1 and J_2 decrease pressure within the irrigation system;
2. Nitrogen concentration is not constant with time in the case of injector J_2;
3. Nitrogen concentration decreased with time in the case of injector J_1;
4. Injector J_3 does not decrease pressure within irrigation system and piston movement increases N-solubility and decrease precipitation and emitter clogging;
5. Increasing N levels increases plant growth and emitter clogging;
6. Increasing irrigation levels increases emitter flushing and salt removal from root zone; and decreases soil aeration and emitter clogging;
7. According to emitter clogging %, injectors can be written the following ascending $J_3 < J_2 < J_1$ [93]; and
8. Increasing N rates and decreasing irrigation levels increased the emitter clogging [93].

6.4.7   COST ANALYSIS

Data on cost analysis for drip irrigation system and fertigation are presented in Tables 6.15 and 6.16. It is obvious that the fixed cost (depreciation + interest + taxes and overheads) of the drip irrigation system were 608.33, 184.9, 360 and 486 LE/year in J_1, J_2 and J_3 treatments, respectively. The variable costs (labor, power, and repair

TABLE 6.15 The Annual Cost for Drip Irrigation System and Fertigation with Injection Devices

Irrigation system and fertigation devices		LE/ year			
		Drip irrigation	Injector type		
			J_1	J_2	J_3
Investment cost, LE per feddan		2500	375	1000	1350
Fixed Costs	Depreciation	208.33	124.9	200	270
	Interest	350	52.5	140	189
	Taxes and Overheads	50	7.5	20	27
	Subtotal	**608.33**	**184.9**	**360**	**486**
Variable Costs	Labor	121.66	-	-	-
	Power	73.7	-	-	-
	Repair and maintenance	50	18.75	50	67.5
	Subtotal	**245.36**	**18.75**	**50**	**67.5**
Total Cost LE/Year		853.69	203.65	410	553.5
Total Cost LE/fed/Year		**853.69**	**13.58**	**16.4**	**22.14**
Total Cost LE/fed/season		426.85	6.788	8.2	11.07

J_1= by-bass pressurized mixing tank, J_2 = venturi, and J_3= positive displacement pump.

TABLE 6.16 Seasonal Total Cost of Irrigation, Fertigation Devices, Garlic Production, for Three Drip Irrigation Levels

Itemized cost	Fertigation devices		
	J_1	J_2	J_3
Irrigation Cost (I_3) LE/fed/season	426.85	426.85	426.85
Fertigation Cost (I_3) LE/fed/season	6.788	8.2	11.07
Total Cost (I_3) LE/fed/season	**433.64**	**435.05**	**437.92**
Irrigation Cost (I_2) LE/fed/season	320.14	320.14	320.14
Fertigation Cost (I_2) LE/fed/season	6.788	8.2	11.07
Total Cost (I_2) LE/fed/season	**326.93**	**328.34**	**331.21**
Irrigation Cost (I_1) LE/fed/season	213.43	213.43	213.43
Fertigation Cost (I_1) LE/fed/season	6.788	8.2	11.07
Total Cost (I_1) LE/fed/season	**220.218**	**221.63**	**224.495**

& maintenance) were 245.36, 18.75, 50 and 67.5 LE/year, J_1, J_2 and J_3, respectively. Results show the total costs (LE/fed.-garlic growing season) were 426.85, 6.788, 8.2 and 11.07 LE, J_1, J_2 and J_3, respectively. According the mentioned costs, injector devices can be arranged in the ascending order: $J_1 < J_2 < J_3$.

Tables 6.17–6.19 and Figures 6.16–6.18 show the net profit of the fertigation devices under different irrigation rates and Nitrogen levels. It can be observed that at the same irrigation and N level, there are no significant differences among the injector devices. Increasing the irrigation from I_1 to I_2 and from I_2 to I_3 increased the total cost by 48.38, 48.15; and 47.5

TABLE 6.17 Total Costs of Garlic Production Under Different Injector Types, Irrigation Levels and Nitrogen Rates

Irrigation level	Kg N/ fed.	Injector	Itemized cost, LE			
			Irrigation + Fertigation	Fertilizer NPK	Weed control	Total LE/fed.
I_1	60	J_1	220.218	452.68	160.00	1832.898
		J_2	221.63	452.68	160.00	1834.31
		J_3	224.495	452.68	160.00	1837.175
	90	J_1	220.218	599.02	160.00	1979.238
		J_2	221.63	599.02	160.00	1980.65
		J_3	224.495	599.02	160.00	1983.515
	120	J_1	220.218	745.36	160.00	2125.578
		J_2	221.63	745.36	160.00	2126.99
		J_3	224.495	745.36	160.00	2129.855
I_2	60	J_1	326.93	452.68	160.00	1939.61
		J_2	328.34	452.68	160.00	1941.02
		J_3	331.21	452.68	160.00	1943.89
	90	J_1	326.93	599.02	160.00	2085.95
		J_2	328.34	599.02	160.00	2087.36
		J_3	331.21	599.02	160.00	2090.23
	120	J_1	326.93	745.36	160.00	2232.29
		J_2	328.34	745.36	160.00	2233.7
		J_3	331.21	745.36	160.00	2236.57

TABLE 6.17 Continued

Irrigation level	Kg N/ fed.	Injector	Itemized cost, LE			
			Irrigation + Fertigation	Fertilizer NPK	Weed control	Total LE/fed.
I_3	60	J_1	433.64	452.68	160.00	2046.32
		J_2	435.05	452.68	160.00	2047.73
		J_3	437.92	452.68	160.00	2050.6
	90	J_1	433.64	599.02	160.00	2192.66
		J_2	435.05	599.02	160.00	2194.07
		J_3	437.92	599.02	160.00	2196.94
	120	J_1	433.64	745.36	160.00	2339
		J_2	435.05	745.36	160.00	2340.41
		J_3	437.92	745.36	160.00	2343.28

Add 1000 LE/fed/season as rent.

J_1 = by-bass pressurized mixing tank; J_2 = venturi; J_3 = positive displacement pump; I_1 = 100 % of ET_c; I_2 = 75 % of ET_c; I_3 = 50 % of ET_c.

TABLE 6.18 The Cost of Garlic Production and Its Sale Price Under Different Injection Methods, Irrigation Levels and Nitrogen Rates (LE/fed)

Irrigation treatment	Kg N/fed.	Injector	Total cost LE/fed. (Table 6.17)	Total yield ton/ fed.	Production sale price LE/ fed
I_1	60	J_1	1832.898	2.38	2541.84
		J_2	1834.31	3.30	3524.4
		J_3	1837.175	3.95	4218.6
	90	J_1	1979.238	3.00	3204
		J_2	1980.65	4.10	4378.8
		J_3	1983.515	4.37	4667.16
	120	J_1	2125.578	3.50	3738
		J_2	2126.99	4.31	4603.08
		J_3	2129.855	4.68	4998.24
I_2	60	J_1	2939.61	3.90	4165.2
		J_2	1941.02	5.01	5350.68
		J_3	1943.89	4.76	5083.68

TABLE 6.18 Continued

Irrigation treatment	Kg N/fed.	Injector	Total cost LE/fed. (Table 6.17)	Total yield ton/ fed.	Production sale price LE/ fed
	90	J_1	2085.95	5.10	5446.8
		J_2	2087.36	6.04	6450.72
		J_3	2090.23	6.19	6610.92
	120	J_1	2232.29	5.80	6194.4
		J_2	2233.7	6.17	6589.56
		J_3	2236.57	6.32	6749.76
I_3	60	J_1	2046.32	3.70	3951.6
		J_2	2047.73	4.79	5115.72
		J_3	2050.6	4.72	5040.96
	90	J_1	2192.66	5.07	5414.76
		J_2	2194.07	6.03	6440.04
		J_3	2196.94	6.18	6600.24
	120	J_1	2339	5.73	6119.64
		J_2	2340.41	6.13	6546.84
		J_3	2343.28	6.26	6685.68

Production sale price (LE/fed) = Total yield (ton/fed.) × 1068 (LE/ton)

TABLE 6.19 The Net Profit of Garlic Production Under Different Injector Types, Irrigation Levels and Nitrogen Rates (LE/fed)

Irrigation rates	Kg N/fed.	J_1	J_2	J_3
100% ET_c	60	708.942	1690.09	2381.425
	90	1224.762	2398.15	2683.645
	120	1612.422	2476.09	2868.385
75% ET_c	60	1225.59	3409.66	3139.79
	90	3360.85	4363.36	4520.69
	120	3962.11	4355.86	4513.19
50% ET_c	60	1905.28	3067.99	2990.36
	90	3222.1	4245.97	4403.3
	120	3780.64	4206.43	4342.4

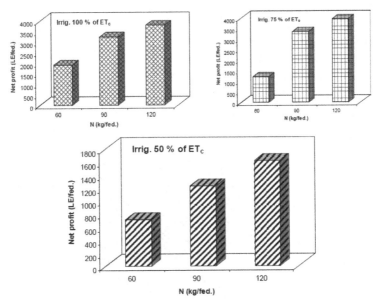

FIGURE 6.16 The net profit of garlic production at different irrigation and N-fertilization levels, for by-bass pressurized mixing tank.

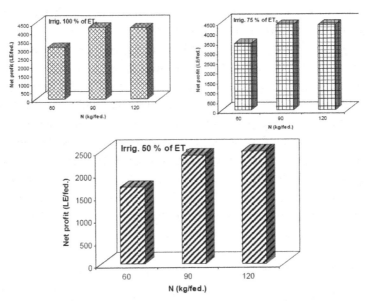

FIGURE 6.17 The net profit of garlic production at different irrigation and N-fertilization treatments, for venturi.

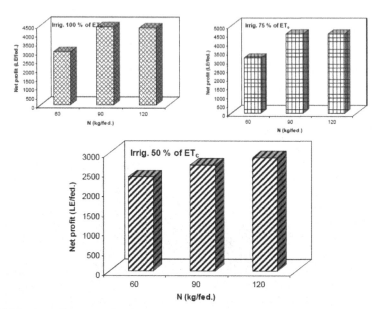

FIGURE 6.18 The net profit of garlic production at different irrigation and N-fertilization treatments, for positive displacement injection pump.

and 48% on the average, regardless of the injector device. This increase is mainly due to irrigation and fertilizer costs.

Data in Tables 6.19 and 6.20 and Figures 6.16–6.18 indicate the total production costs (LE/fed.), garlic dry yield (ton/fed.), yield price (LE/fed.) and the net profit (LE/ fed.). It can be observed that the maximum net profit and the minimum values were 4520.69 and 708.942 LE/ fed. under $J_3 I_2 N_2$ and $J_1 I_1 N_1$, respectively. This can be explained on the basis of high emitter clogging percentage [93], which was decreased under irrigation levels and N rates in the 2nd interaction ($J_1 I_1 N_1$) compared to the 1st interaction ($J_3 I_2 N_3$). The interaction ($J_3 I_2 N_2$) increased the net profit by 57.61% in addition to saving 25 and 33.33% of irrigation water and nitrogen fertilizer, respectively compared to the interaction ($J_3 I_3 N_3$).

6.5 CONCLUSIONS

The study evaluated the effects of design and engineering parameters on hydraulic performance fertilizer injectors. Three fertilizer injectors were evaluated: by-bass pressurized mixing tank (J_1), venturi (J_2), and positive displacement pump (J_3). The performance among injectors was compared. The injectors were calibrated for volume injection flow rate and the change in concentration rate. The dimensional analysis was used to compare the performance of the three-injector types. The two dimensionless groups were:

$$\text{Coefficient of pressure difference} \quad C_P = \frac{\Delta P \rho D^2}{\mu^2}$$

$$\text{Coefficient of injection rate} \quad C_Q = \frac{Q\rho}{\mu D} = \frac{\pi}{4} Re$$

The results show that the highest performance was for J_3, which gave an injection rate of 240.2 lpm at 2.5 bars of operating pressure difference. For J_1 and J_2, injection rates were 40, and 3.0 lpm at 0.7 and 1.4 bars of operating pressure differences, respectively. The J_3 and J_2 showed a relatively linear concentration rate, whereas J_1 showed changes in concentration rate during operating time.

The dimensionless coefficients showed an increase in performance degradation for J_2, J_1, and J_3. The maximum values of C_Q were 2.3×10^3, 3.06×10^4, and 18.4×10^4 at C_P values of 3.9×10^5, 1.95×10^5, and 7×10^5, respectively.

The emitter clogging percentage was measured for the drip irrigation system at the end of garlic growing season 2006–2007. The results showed that emitter clogging was increased with increasing nitrogen rates, and decreasing the irrigation rates. The maximum and minimum values of clogging percentage are 17.79, 22.8, 27.9; and 13.24, 15.81, 18.9 under (J_3, J_2, J_1), ($I_1 \times N_3$) and ($I_3 \times N_1$), respectively. According to clogging percentage, injectors can be written in the ascending order $J_3 < J_2 < J_1$.

The highest garlic yield (6.3640 ton/fed.) was obtained with treatment $J_3 \times I_2 \times N_3$ (Positive displacement pump, 75% of ET_c and 120 kg N /fed).

The maximum value of water use efficiency (WUE) was 3.289 kg garlic per m^3 of irrigation water, in the treatment $J_3 \times I_1 \times N_3$, while the minimum value was 1.3 kg garlic/m^3 of irrigation water in the treatment $J_1 \times I_3 \times N_1$ (by-bass pressurized mixing tank, 100% of ET_c and 60 kg N /fed).

The maximum values of nitrogen use efficiency (NUE) in kg garlic/ kg N was 83.50 and the minimum was 29.17 in the interactions – $J_2 \times I_2 \times N_1$ (venturi, 75 of ET_c and 60 kg N/fed.) and $J_1 \times I_1 \times N_3$ (by-bass pressurized mixing tank, 50 of ET_c and 120 kg N/fed), – respectively.

Increasing the irrigation levels from (50% of ET_c) to (75% of ET_c) and from (75 % of ET_c) to (100 % of ET_c) increased the total cost by 48.38, 48.15; and 47.5, 48% on the average, regardless of the type of the injector.

The maximum value of net profit was 4520.69 LE/fed. in the treatment $J_3 \times I_2 \times N_2$ (Positive displacement pump, 75% of ET_c and 90 kg N/fed), while the minimum value was 708.942 LE/fed. in the treatment $J_1 \times I_3 \times N_1$ (by-bass pressurized mixing tank, 100% of ET_c and 60 kg N/fed).

The interaction $J_3 \times I_2 \times N_2$ (Positive displacement pump, 75% of ET_c and 90 kg N/fed) increased the net profit by 57.61%, in addition it saved 25% of irrigation water and 33.33% of nitrogen fertilizer compared to the interaction $J_3 \times I_3 \times N_3$ (Positive displacement pump, 100% of ET_c and 120 kg N/fed).

6.6 SUMMARY

This chapter presents water and fertigation management in garlic production. This chapter investigates the effects of three injector types, three irrigation treatments and three nitrogen levels on:

- drippers clogging as an indicator of water distribution efficiency;
- garlic yield, water use efficiency (WUE) and nitrogen use efficiency (NUE); and
- the net profit of garlic production.

The results show that the highest performance was for J_3, with an injection rate of 240.2 lpm at 2.5 bars of operating pressure difference. The J_3 and J_2 showed a relatively constant concentration rate, whereas J_1 showed changes in concentration rate during operating time.

The dimensionless coefficients showed an increase in performance degradation for J_2, J_1, and J_3. The maximum values of C_Q were 2.3×10^3, 3.06×10^4, and 18.4×10^4 at C_p values of 3.9×10^5, 1.95×10^5, and 7×10^5, respectively.

The results showed that emitter clogging was increased with increasing nitrogen rates, and decreasing the irrigation rates.

The highest garlic yield (6.3640 ton/fed.) was obtained with treatment $J_3 \times I_2 \times N_3$ (Positive displacement pump, 75% of ET_c and 120 kg N/fed).

The maximum value of water use efficiency (WUE) was 3.289 kg garlic per m^3 of irrigation water, in the treatment $J_3 \times I_1 \times N_3$, while the minimum value was 1.3 kg garlic/m^3 of irrigation water in the treatment $J_1 \times I_3 \times N_1$ (by-bass pressurized mixing tank, 100% of ET_c and 60 kg N/fed).

The maximum values of nitrogen use efficiency (NUE) in kg garlic/ kg N was 83.50 and the minimum was 29.17 in the interactions – $J_2 \times I_2 \times N_1$ (venturi, 75 of ET_c and 60 kg N/fed.) and $J_1 \times I_1 \times N_3$ (by-bass pressurized mixing tank, 50 of ET_c and 120 kg N/fed), – respectively.

Increasing the irrigation levels from (50% of ET_c) to (75% of ET_c) and from (75 % of ET_c) to (100 % of ET_c) increased the total cost by 48.38, 48.15%; and 47.5, 48% on the average, regardless of the type of the injector.

The maximum value of net profit was 4520.69 LE/fed. in the treatment $J_3 \times I_2 \times N_2$ (Positive displacement pump, 75% of ET_c and 90 kg N/fed), while the minimum value was 708.942 LE/fed. in the treatment $J_1 \times I_3 \times N_1$ (by-bass pressurized mixing tank, 100% of ET_c and 60 kg N/fed).

KEYWORDS

- **by-bass pressurized mixing tank**
- **chemigation**
- **clogging**
- **clogging emitter**
- **coefficient of injection**
- **coefficient of pressure difference**
- **dimensional analysis**
- **drip irrigation**
- **Egypt**
- **feddan**

- **fertigation**
- **fertilization**
- **fertilizer**
- **garlic**
- **injectors**
- **irrigation water**
- **LE**
- **net profit**
- **Nitrogen**
- **nitrogen use efficiency**
- **operating pressure**
- **positive displacement pump**
- **rate**
- **vegetable crop**
- **venturi**
- **water use**
- **water use efficiency**

REFERENCES

1. Abdel-Aziz, A. A. 1998. *Evaluation of some modern chemigation techniques.* PhD thesis, Faculty of Agriculture, Ain Shams University, Egypt

2. Abdel-Maksoud, S. E., M. A. Hassan, M. S. El- shal and E. E. Abdel, 1992. Study on selecting the proper system of tomato irrigation, in new land. *Misr J. Agric. Eng.,* 9(3):347–358.

3. Aboukhaled, A., 1991. Fertigation and chemigation: An overview with emphasis on the Near East. Report of expert consultation on Fertigation/chemigation. FAO, Cairo, Egypt. Sept., 8–11.

4. Adin, A., 1987. Clogging in irrigation system reusing pond effluents and its prevention. *Water Sci. Technol.,* 1:19–23.

5. Alijbury, F. K., J. A., Beutel, J. W., Bigger, R. L., Branson, S., and Davis, E., 1981. *Drip irrigation management.* Division of Ag. Sci., Univ. of Calif. Leaflet 21259.

6. Arnaout, M. A., 1999. Comparative study between fertigation and conventional method of fertilizer application through different irrigation systems. *Misr. J. Agr. Eng.,* 16(2):209–217.

7. ASAE, 1997. Standards (44th Ed.). pp. 357–362. The American Society of Agricultural Engineers (ASAE), USA.
8. Ayers, R.S. and D. W. Westcot, 1976. *Water Quality for Agriculture Irrigation and Drainage*. Report No. 29, FAO Rome, Italy.
9. Bakker, M. J., H. Slangen, and W. Glas, 1984. Comparative investigation into the effect of fertigation and broadcast fertilization on the yield and Nitrate content of lettuce. *Netherlands J. of Agric. Sci.*
10. Bazza, M., 1991. Fertigation in Morocco: past, present, and perspectives. Proc. of Expert Consultation On Fertigation/Chemigation. FAO, Cairo, Egypt. Sept. 8–11.
11. Bozkurt, S. and Özekiei, B., 2006. The effects of fertigation managements on clogging of in-line emitters. *J.Appl. Sci. Res.*, 6 (15):3026–3034.
12. Bravdo, B. A., and Y. Hepner, 1987. Irrigation management and fertigation to optimize grape composition and vine performance. *Acta Horticulture*, 206:49–57.
13. Brian, B., and Obreza, T., 2002. Fertigation nutrient sources and application considerations for citrus. http://edis.ifas.ufl.edu/CH185
14. Bucks, D. A., and F. S., Nakayama, 1984. Problems to avoid with drip /trickle irrigation systems. *Proc. Amer. Soc. Civil Eng. Specially Conf. on Irri. and Drainage Div.*, pages 24–84.
15. Bucks, D. A., F. S., Nakayama and R. A. Gilberd, 1977. Clogging research on drip irrigation. *Proceedings 4th Annual international Drip irrigation Association Meeting*, pp 25–31.
16. Bucks, D. A., F. S., Nakayama and R. A. Gilberd, 1979. Trickle irrigation water quality and preventive maintenance. *Agric. Water Manage.*, 2:149–162.
17. Burt, C., O'Connor K., and Ruehr, T., 1998. *Fertigation*. Irrigation training and research center, California Polytechnic State University, San Luis Obispo, CA 93407
18. Carvalho, L. G., Silva, A. M., Souza, R. J. and Abreu, A. R., 1996. Effects of different water depths and rates of nitrogen and potassium on garlic. *Ciencia. e. Agrotecnologia*, 20(2):249–251.
19. Chakraborty, D., Singh A. K., Kumar, A., and K. S. Khanna, 1997. Effect of fertigation on nitrogen dynamics in broccoli. In Proc. Workshop on Micro Irrigation and Sprinkler system, April 28–30, Central Board of Irrigation and power, New Delhi.
20. Chang, C. A., 2008. Drip Lines and Emitters: Chlorination for disinfection and prevention of clogging. http://www.informaworld.com/smpp/content~content=a792065664~db=all~jumptype
21. Crespo, M., M. R. Goyal, C. C. Baez, and L. E. Rivera, 1988. Nutrient uptake and growth characteristics of Nitrogen fertigated sweet peppers under drip irrigation and plastic mulch. *J. of the Univ. of Puerto Rico*, 72(4):579–584.
22. Darbie M. G., Kerry, A. H., and William T. K., 2005. *Drip Chemigation: Injecting Fertilizer, Acid and Chlorine*. The University of Georgia, College of Agricultural and Environmental Sciences/Cooperative Extension Bulletin 1130.
23. Dasberg, S., and E. Bresler, 1986. *Drip Irrigation Manual*. International Irrigation Information Center, Publ. No. 9, Volcani Center, Bet Dagan, Israel, 61 pp.
24. De Torch, F., 1988. *Irrigation and Drainage*. Gent Univ., Belgium.
25. Dewis, J. and F. Freitas, 1970. *Physical and Chemical Methods of Soil and Water Analysis*. Soil Bulletin 10, FAO, Rome.

26. Dong, H. W. and D. P. Tucker, 1996. Water quality measurements for drip irrigation system, *Trans. ASAE*, 13(11):38–41

27. Douglas J. F., J. M. Gasiorek, and J. A. Swaffield, 1985. *Fluid mechanics*. (2nd Ed.). Pitman Publishing Ltd.

28. Ebaby, F. G., 1986. *Technology of irrigation for sandy soils*. Ph. D. Thesis, Fac. of Agric, Cairo Univ., Egypt.

29. El-Adl, M. A. 2000. Effect of irrigation and fertilization method on pea production. *Misr. J. Agr. Eng.*, 17(3):450–468.

30. El-Adl, M. A., 2001. Sprinkler irrigation and fertigation effects on peanut production. *Misr. J. Agr. Eng.*, 18(1):75–88.

31. El-Amoud, A. I., 1997. *Trickle irrigation systems*. Scientific Publication and Press, King Saud University, Saudi Arabia.

32. El-Berry, A. M., G. A. Baker, and A. M. Al-Weshali, 2003. The effect of water quality and aperture size on clogging of emitters. Available at: http://afeid.montpellier.cemagref.fr/Mpl2003/AtelierTechno/AtelierTechno/Papier%20Etier/N%C2%BO48%20%20EGYPTE_BM.pdf

33. El-Gindy, A. M., 1984. Optimization of water use for pepper crop. *Misr J. Agr. Eng.*, 29(1):539–556.

34. El-Gindy, A. M., 1988. Modern chemigation techniques for vegetable crops under Egyptian conditions. *Misr. J. Agr. Eng.*, 5(1):99–111.

35. El-Gindy, A. M., and A. A. Abedel-Aziz, 2003. Maximize water use efficiency of maize crop in sandy soil. *Arab Universities Journal of Agric. Sciences*, 11(1):439–452.

36. Ford, H. W., 1984. The problem of emitter clogging and methods for control. *Citrus Ind.*, 1:42–52.

37. Gascho, G. Z., 1991. Soil water nutrient interaction under fertigation system. *Proc. of Expert Consultation on Fertigation/Chemigation*. FAO, Cairo, Egypt. Sept., 8–11.

38. Gilbert, R. G., and H. W. Ford, 1986. Operational principles emitters clogging. In: *Trickle Irrigation For Crop Production, Eds. F.S. Nakayama and D.A. Bucks*, New York, pages 142–163.

39. Gillerman, l., G. Oron, Y. DeMalach and I. David, 1996, Application of brackish water through subsurface drip systems. *Int. Water and Irrig. Review*, 16(4):19–24.

40. Goyal, Megh R. (Editor), 2013. *Management of Drip/Trickle or Micro Irrigation*. Apple Academic Press Inc.,

41. Goyal, Megh R. (Editor), 2015. *Research Advances in Sustainable Micro Irrigation, volumes 1 to 10*. Apple Academic Press Inc.,

42. Granberry, D. M., K. A. Harrison, and W. T. Kelley, 2005. *Drip Chemigation-Injecting Fertilizer, Acid and Chlorine*. Cooperative Extension Service, University of Georgia, Bulletin 1130.

43. Haman D. Z., Smajstrla A. G. and Zazueta F. S., 2003. Chemical injection methods for irrigation. Available at: http://edis.ifas.ufl.edu/WI00400.pdf

44. Hamdy, A. 1991. *Fertigation: Prospects and problems*. FAO, Consultation on Fertigation/Chemigation, 8–11 Sept., Cairo, Egypt.

45. Hanson, B., 2006. *Injection Devices for Fertigation*. Available at: http://www.citrus-research.com/documents/21804bb8-f36a-4bf7–9ae6-b5777c7e7c1a.pdf

46. Hanson, B. R., L. J. Schwankl, K. F. Schulbach and G. S. Pettygroue, 1997. A comparison of furrow, surface drip and subsurface drip on lettuce yield and applied water. *Agric. Water Manag. J.*, 33:139–157

47. Hargreaves, G. H., and Z. A., Samani, 1982. Estimating potential evapotranspiration. *J. of Irri. and Drainage, ASCE*, 108(3).

48. Hebbar, S. S., B. K. Ramachandrappa, H. V. Nonjappa and M. Prabhakar, 2004. Studies on NPK drip fertigation in field grown tomato. *Eur. J. Agron.*, 21:117–127.

49. Hesse, P. R., 1971. *A Text Book of Soil Chemical Analysis*. John Murry (Publishers) Ltd., 50 Albermarle Street, 11:506–525

50. Hills, D. J., M. N. Fakher and P.M. Waller, 1989. Effect of chemical clogging on drip-tap irrigation uniformity. *Trans. ASAE*, 32(4):1202–1206.

51. Ibrahim, A. A., 1992. Response of plant to irrigation with CO_2 enriched water. *Acta Hort.*, 323:205–214.

52. Imas, P., 1999. Recent techniques in fertigation of horticultural crops in Israel. *IPI-PRII-KKV workshop on Recent trends in nutrition management in horticultural crops*, 11–12 February, Dapoli, Maharashtra, India.

53. Israelson, O. W. and Hanson, V. E., 1962. *Irrigation principles and practices*. 3rd ed. John Wiley and Sons, New York.

54. Jackson, M. L., 1967. *Soil chemical analysis*. Prentice Hall of India Pvt. Ltd., New York.

55. James, L. G., 1988. Multi-purpose and special uses. In: *Principles of Farm Irrigation System Design Handbook*. Washington State University, 405 pp.

56. Judah, M. O., 1985. Salt accumulation under various drip irrigation in the Jordan valley treatments. *Dirasat*, 12(4):39–48.

57. Li, J., Meng, Y., and Liu, Y., 2006. Hydraulic performance of differential pressure tanks for fertigation. *Transactions of the ASABE*, 49(6):1815 -1822.

58. Lipinski, V. M., and Gaviola, S., 2000. Nitrogen sources and rates used for fertigation of garlic. *Horticultura Argentina*, 18(44/45): 28–32.

59. Magen, H. 1998. *Prospects of micro irrigation and fertigation in Chinese agriculture*. Dead Sea Works Ltd., Israel. IBC conference, 1–2 Dec., Beijing, PRC.

60. Mansour, H. A. A., 2005. *The response of grape fruits to application of water and fertilizers under different localized irrigation systems*. M.Sc. Thesis, Faculty of Agriculture, Ain Shams University, Egypt.

61. Marouelli, W. A., Silva, W. L. C., Carrijo, O. A., and Silva, H . R., 2002. Production and quality of garlic crop under soil water regimes and nitrogen levels. *Horticultura Brasileira*, 20(2):191–194.

62. Mekheimer, S. and K. H. Hegazy, 1996. *Water crisis in Arabic region (facts and alternative)*. Aalam El-maarefa, Egypt. (in Arabic).

63. Mishra, P., D. K. Singh, A. K. Bhattacharya and A. K. Singh, 2002. Utilization of potassium by radish under fertigation system. Available at: http://www.mail.iari.res.in/~dksingh/publ/potassium%20symposium.Pdf

64. Mmopharlin, I. R., P. M. Aylmore and R. C. Jeffery, 1995. Nitrogen requirements for lettuce under sprinkler irrigation and trickle fertigation on a spear wood sand. *J. of Plant Nutrition*, 18(2):219–241.

65. Mohammad, M. J., Al-Omari, M., Zuraiqi, S., and Qawasmi, W., 2002. Nitrogen and water utilization by trickle fertigated garlic using the neutron gage and 15N

technologies. In: *Water balance and fertigation for crop improvement in West Asia*, IAEA-TECDOC-1266, January, pages 15–26.

66. Mohammad, M. J., and S. Zuraiqi, 2003. Enhancement of yield, nitrogen and water use efficiencies by nitrogen drip-fertigation of garlic. *J. of Plant Nutrition*, 26(9):1749–1766.

67. Morad, M. M., Arnaout, M. A., and Ramadan, T. Y., 1999. Fertilization distribution though irrigation system. *Misr. J. Agr., Eng.*, 18(1):75–88.

68. Nagmoush, S., 1991. Fertigation practice on land reclamation projects and its implication for the fertilizer industry in Egypt. *Expert Consultation of Chemigation Fertigation*, Sept. 8–11, Cairo, Egypt, 10 pp.

69. O'Neill, M. K., B. R. Gardner and R.L. Roth, 1979. Ortho-phosphoric acid as a phosphorus fertilizer in trickle irrigation. *Soil Sci. Soc. Amer. J.*, 143:283–286.

70. Ozkan, F., Mualla, O., and Ahmed, B. 2005. Experimental investigations of air and liquid injection by venturi tubes. At: http://www3.interscience.wiley.com/cgi-bin/fulltext/118593396/HTMLSTART.

71. Padmakumari, O. and R. K., Sivanappan, 1995. Study on clogging of emitters in drip irrigation and micro irrigation for a changing the environment. *ASAE*, pages 80–83.

72. Panchal, G. N., Modhwadia, M. M., Patel, J. C., Sadaria, S. G., and Patel, B. S., 1992. Response of garlic to irrigation, nitrogen and phosphorus. *Indian Journal of Agronomy*, 37(2):397–398

73. Pandey, U. B. and Singh, D. K., 1993. Response of garlic to different levels of irrigation and nitrogen. *Newsletter National Horticultural Research and Development Foundation*, 13(3/4):10–12.

74. Papadopoulos, I., 1991. Fertigation in Cyprus and some countries of the Near East Region. *The Expert Consultation of Chemigation/Fertigation*, 8–11 Sept., Cairo, Egypt, pages 67–78 .

75. Papadopoulos, I., 1993. Fertigation in Cyprus and some other countries of the wear East region: Present Situation and Future prospects. Feb 11–15, Cyprus.

76. Papadopoulos, I. and G., Eliads, 1987. A fertigation system for experimental purposes. *Plant and Soil*, 102:141–147.

77. Patel, B. G., Khanpara, V. D., Malavia, D. D, Maraviya, R. B. and Kaneria, B. B., 1995. Economic feasibility of drip irrigation in garlic. *Indian Journal of Agronomy*, 40(1):143–145.

78. Phocaides, A., 2000. *Technical handbook of pressurized irrigation techniques*. FAO, Rome, pages145–149.

79. Pillsbury, A. F and Degan, A., 1975. *Sprinkler irrigation*. FAO Agricultural Development Paper No. 88, Rome.

80. Piper, C. B., 1950. *Soil and plant analysis*. Inter Science Publisher Inc., New York.

81. Rahm, G. W., 1982. Fertigation of corn in the eastern great plains. *Proc. For the 2nd Nat. Symp. on chemigation*.

82. Raj Kumar, S. and K. Larim, 1985. Movement of salt and water under trickle irrigation and its field evaluation. *Egypt. J. Soil Sci.*, 25(2):127–132.

83. Ravina, E. P., Z. Sofer, A. Marcu, A. Shisha, G. Sagi, and Y. Lev, 1997. Control of clogging in drip irrigation with stored treated municipal sewage effluent. *Agricultural Water Management*, 33(2–3):127–137.

84. Richards, L. A., 1962. *Diagnosis and Improvement of Saline and Alkaline Soil.* USDA Handbook No. 60.

85. Sadaria, S. G., Malavia, D. D., Khanpara, V. D., Dudhatra, M. G., Vyas, M. N. and Mathukia, R. K., 1997. Irrigation and nutrient requirement of garlic under south Saurashtra region of Gujarat. *Indian Journal of Agricultural Sciences*, 67(9): 402–403.

86. Sagi, G., 1990. Water quality and clogging of irrigation systems in Israel. *Water and Irrigation Bull.*, 280:57–61 (in Hebrew).

87. Sagi, G., E. Paz, I. Ravina, A. Schischa, A. Marcu and Z. Y. Yehiely, 1995. Clogging of drip irrigation systems by Colonical Protozoa and sulfur bacteria micro-irrigation for a changing the environment. *Proc. ASAE*, pages 250–255.

88. Sankar, V., Qureshi, M. A., Tripathi, P. C., and Lawande, K. E., 2001. Micro irrigation studies in garlic. *South Indian Horticulture*, 49(Special issue):379–381.

89. Seno, S., 1997. Effects of irrigation frequency and nitrogen rates on garlic. *Cultura Agronomica*, 6(1):29–40.

90. Slangen, D. H. T., and W. Gals, 1988. The importance of fertigation for the improvement of N fertilizer use efficiency in lettuce. *Acta Hort.*, 222:13–146.

91. Thanki, J. D., and Patel, C. L., 2005. Effect of moisture regimes and herbigation through mini sprinkler on yield attributes and bulb yield of garlic. *Journal of Maharashtra Agricultural Universities*, 30(1):72–74

92. The Irrigation Association, 2000. *Chemigation.* The irrigation Association Publications, 6540 Arlington Boulevard, Fall Church, VA 22042–668, USA.

93. Tayel, M. Y., A. M. El-Gindy and A. A. Abdel- Aziz, 2008. Effect of irrigation systems on productivity and quality of grape crop. *J. Appli. Sci. Research*, 4(12): 1722–1729.

94. Tayel, M. Y., A. M. El-Gindy, K. F. El-Bagoury and Sabreen Kh. Pibars, 2009. Effect of injector types, irrigation levels and nitrogen treatments on: Emitter clogging. *Journal Earth and Environmental Sciences*, 4(3, March):131–137.

95. Threadgill, E. D., 1985. Chemigation via sprinkler irrigation: Current status and future development. *App. Eng. in Ag.*, 1(1):46–48.

96. Younis, S. M., 1986. Study on different irrigation methods in Western Nobaria to produce tomato. *Alex. J. Agric. Res.*, 31(3):11–19.

CHAPTER 7

PERFORMANCE OF DRIP IRRIGATED SOYBEAN UNDER MULCHING: EGYPT

A. RAMADAN EID, B. A. BAKRY, and M. H. TAHA

CONTENTS

In this chapter: One *feddan* = 4200 m² = one ha; One LE = Egyptian unit of currency = 0.13992 US$.

Modified from, "*Abdelraouf Ramadan Eid, Bakry Ahmed Bakry, Moamen Hamed Taha, 2013. Effect of pulse drip irrigation and mulching systems on yield, quality traits and irrigation water use efficiency of soybean under sandy soil conditions. Open source article from J. Agricultural Sciences, 4(5):249–261*".

7.1   INTRODUCTION

Pulse drip irrigation can be employed with either drip or sprinkler irrigation systems and extend conventional irrigation systems to ultra-low drip irrigation systems [7, 9]. The charge-discharge cycling will continue as long as the flow rate coming in through the inlet is less than the water flow passing out through the outlets. An irrigation flow controller at the inlet regulates the flow into the inlet.

There are several advantages of pulse drip irrigation [6], such as: No run off on heavy soils; No leaching or water loss in sandy soils; Water can be applied efficiently on shallow soils and in hilly areas; Intermittent operation of sprinklers can provide evaporative cooling; Reduce the size of growing containers in greenhouses due to very low flow rates; and Low flow inputs reduce high initial costs of supply system. Disadvantages of pulse drip irrigation are: The additional expense of buying and maintaining a pulse drip system to a pre-existing irrigation system; The system requires a minimum operating pressure of 25 psi to run efficiently; Maintaining the integrity of a pressurized water supply is of critical importance; Leaks in the piping can run up substantial costs due to the long irrigation duration; Close attention to the irrigation cycles is needed to avoid salinity build up in the soil; and the additional expense of leak prevention devices (LPDs) are recommended to be fitted at each watering distribution point along the system to ensure even distribution of water. If the discharging conduits are allowed to drain between pulses, most of the water will preferentially flow to those at the upstream of the line.

Redesign of the irrigation system is necessary if the wetted area near the emitter is too small (limiting) and pulsing is not an option [9]. Based on research studies from other states (where soil types are different), it is often believed that the size of the wetted zone can be increased if irrigation is pulsed [7]. Applying irrigation water in stages or pulses rather than all at one time can save water by giving the soil time to wet from the first pulse of water, thereby allowing it to absorb subsequent moisture more readily and reducing the total amount of water required [15]. High irrigation frequency might provide desirable conditions for water movement in soil and for uptake by roots [16].

Continuous water application is associated with increased water percolation under root zone. Intermittent irrigation strategy based on discharge pulses followed by breaks can improve water management in the field and increase irrigation efficiency [14]. Segal [16] and Steyn 19] reported that high frequency or low flow irrigation can increase water use efficiency (WUE) and yield of crops by providing desirable conditions for water movement in soil and for uptake by roots. As soil moisture becomes more uniform, plants are able to utilize water and increase production. The studies resulted in the fact that the yield of cultivated crops (clover, rye grass, timothy, fescue, onion, lettuce) was usually 1.3–2.5 times higher than the yields of crops grown under traditional regular irrigation [13], when pulse irrigation was based irrigation scheduling close to the daily crop evapotranspiration rate. Further research is necessary to evaluate these findings on other crops and to develop economically feasible methods for high frequency irrigation.

Mulch is a protective layer of either organic or inorganic material that is spread on the top soil. Mulch helps regulate soil temperature by shading it in the summer thus keeping it cooler and helps insulate it in the winter from chiling winds. This temperature regulating effect helps encourage the root growth [21]. Tropical agriculture under mulching promotes plant health and vigor. Mulching improves nutrient and water retention in the soil, encourages favorable soil microbial activity and worms, and suppresses weed growth. When properly executed, mulching can significantly improve the well-being of plants and reduce maintenance as compared to bare soil culture. Mulched plants have better vigor and, consequently have improved resistance to pests and diseases [5]. Using plastic mulches of 200 and 150 mm, two layers of cattail or rice straw mulch, and hand hoeing for controlling weeds resulted in the highest yield per tree without significant differences between these treatments [1].

Soybean (*Glycine max*) is one of the most important world crops and is grown for oil and protein. Present world production is about 176.6 million-tons of beans. The crop is mainly grown under rainfed conditions but irrigation, specifically supplemental irrigation, is increasingly used [8].

This chapter investigates the effects of pulse drip irrigation and mulching practices on soybean yield, irrigation water use efficiency and improvement of quality traits for soybean crop in arid regions of Egypt.

7.2 MATERIALS AND METHODS

7.2.1 EXPERIMENTAL SITE

During two soybean seasons from May to September of 2010–2011, field experiments were conducted at the experimental farm of National Research Center, El-Nubaria, Egypt (latitude 30.87°N, and longitude 31.17°E, and mean elevation of 21 m above sea level). The experimental area has an arid climate with cool winters and hot dry summer. Table 7.1

TABLE 7.1   The Monthly Climatic Data for the Two Soybean Growing Seasons of 2010 and 2011, at El-Nubaria city, Egypt

Date	Total rainfall		Air temperature		Relative humidity
	mm		°C		%
	sum	mean	minimum	maximum	mean
May/2010	0.0	21.4	15.4	28.8	65.8
Jun/2010	0.0	24.3	18.7	30.5	69.6
Jul/2010	0.19	26.63	20.62	33.19	73.16
Aug/2010	0.1	26.4	20.5	32.8	73.2
Sep/2010	0.19	25.06	19.05	32.22	75.13
May/2011	0.05	22.46	15.65	30.72	73.81
Jun/2011	0.0	25.3	19.6	31.9	80.2
Jul/2011	0.0	27.7	22.1	33.8	80.1
Aug/2011	0.0	27.5	21.7	34.1	79.6
Sep/2011	0.00	24.99	19.34	31.51	80.20

summarizes the monthly mean climatic data for the two growing seasons 2010 and 2011, respectively, for El- Nubaria city, which is near El-Nubaria. There was negligible rainfall during the study period that was not taken into consideration through the two seasons.

Soil moisture was determined according to the method described by [12]. The soil samples were taken by profile probe before and 2 hours after irrigation and for three soil depths (0, 15, 30 and 45 cm). In the case of 30 cm emitter spacing, the soil sample were also taken at 5, 10, and 15 cm in the x-direction (spacing between the emitters). *Contour Program Surfer version 8* was used to draw all contour maps for moisture distribution under an emitter.

Mulching treatments consisted of: 2 layers of rice straw (about 6 cm deep) = 15 tons/ha; and black plastic mulch (BPM) of 200 mm thickness. Soybean seeds were manually planted on May during both seasons. Agronomic and cultural practices were followed according to soybean cultivation recommendations by National Research Center. Characteristics of soil and irrigation water for experimental site are presented in Tables 7.2 to 7.4.

TABLE 7.2   Chemical and Mechanical Analyzes of Soil

Depth cm	Chemical analysis				*Mechanical analysis, %		
	OM	pH	EC	CaCO3	Course sand	Fine sand	Clay + Silt
	%	1:2.5	dS/m	%			
0–20	0.65	8.7	0.35	7.02	47.76	49.75	2.49
20–40	0.40	8.8	0.32	2.34	56.72	39.56	3.72
40–60	0.25	9.3	0.44	4.68	36.76	59.40	3.84

*Soil texture for all depths was sandy.

TABLE 7.3   Soil Characteristics

Depth cm	SP	F.C.	W.P.	A.W.	Hydraulic conductivity cm/h
	%	%	%	%	
0–20	21.0	10.1	4.7	5.4	22.5
20–40	19.0	13.5	5.6	7.9	19.0
40–60	22.0	12.5	4.6	7.9	21.0

TABLE 7.4   Chemical Characteristics of Irrigation Water

pH	EC dS/m	Cations and anions, meq/L								SAR
		Cations				Anions				
		Ca^{++}	Mg^{++}	Na^+	K^+	CO_3^-	HCO_3^-	Cl^-	SO_2^-	%
7.35	0.41	1	0.5	2.4	0.2	–	0.1	2.7	1.3	2.8

7.2.2   EXPERIMENTAL DESIGN

Experimental design was split plot with three replications. The main plots were for pulse drip irrigation treatments (application of daily water requirements 4 times, 8 times, and 12 times compared with irrigation application only once/day). Subplots were used for mulching treatments (covering the soil with: black plastic mulch, rice straw and soil without mulch (control)).

7.2.3   DRIP IRRIGATION SYSTEM

Irrigation system consisted of control head, pumping unit and filtration unit. It included: submersible pump with 45 m³/h discharge, screen filter, check valve, pressure regulator, pressure gauges, flow meter, and control valves. Main line was of PVC pipe with 110 mm in diameter (OD) to convey the water from the source to the main control points in the field. Sub-main line was of PVC pipes with 75 mm diameter (OD) and was connected to the main line. Lateral lines were PE tubes with 63 mm in diameter (OD) that were connected to the sub-main line. Emitters were built-in on PE tube with 16 mm diameter (OD) with a dripper spacing of 30 cm and emitter discharge of 4 lph at an operating pressure 1.0 bar operating pressure. Length of each emitter tube was 30 m.

Seasonal irrigation requirement for soybean (Figure 7.1) was 3385 m³/feddan that was estimated using the meteorological data at the Central Laboratory for Agricultural Climate (CLAC) and Penman–Monteith equation [11].

7.2.4   OTHER PERFORMANCE PARAMETERS

Irrigation water use efficiency of soybean crop (IWUE) was calculated as follows [11]:

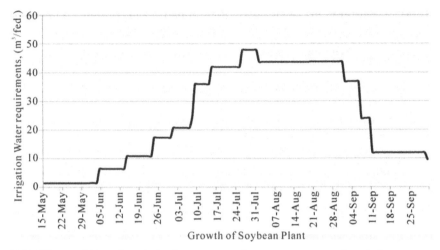

FIGURE 7.1   The soybean irrigation water requirement versus growth period.

$$\text{IWUE (kg of grain/m}^3 \text{ of water)} = \text{Total yield (kg grain/fed.)}$$
$$\text{/Total applied irrigation water}$$
$$(\text{m}^3 \text{ water/fed./season}) \qquad (1)$$

Leaf area (= leaf length × maximum leaf width × 0.75) was calculated according to Liven et al. [12] and chlorophyll content was estimated with a Span Device.

$$\text{Cost of rice straw mulch} = \text{Price of rice straw} + \text{Cost of transport of rice straw} + \text{labor}$$

$$\text{Cost of black plastic mulch} = \text{Price of black plastic roll} + \text{Cost of transport of black plastic sheet} + \text{labor}$$

$$\text{Cost with no mulch} = \text{Cost of weed control} \qquad (2)$$

Seed oil content (%) was determined by Soxhlet apparatus using petroleum ether (40°C – 60°C B.P) according to AOAC [2]. Soybean oil yield (kg/fed) was calculated as follows:

$$\text{Oil yield} = \text{Seed yield (kg/fed)} × \text{Seed oil content (\%)} \qquad (3)$$

Total N-content in seeds was determined. The protein percentage was calculated by multiplying total N-content by 6.25 according to Chapman [3]. **Soybean protein yield** (kg/fed) was calculated as follows:

$$\text{Protein yield} = \text{Protein percentage} \times \text{Seed yield (kg/fed)} \qquad (4)$$

7.2.5 STATISTICAL ANALYSIS

The standard analysis of variance procedure was used for split-split plot design with three replications, described by Snedecor [18]. All data were average of values for the two growing seasons 2010 and 2011. The treatments were compared according to L.S.D. test at 5% level of significance.

7.3 RESULTS AND DISCUSSION

7.3.1 SOIL MOISTURE DISTRIBUTION IN ROOT ZONE

The Figures 7.2–7.4 indicated the effects of number of irrigation pulses and mulching systems on soil moisture distribution in the root zone (SMDRZ). SMDRZ was improved after irrigation and before the next irrigation by increasing number of irrigation pulses per day than the SMDRZ for irrigation requirements continuously (AIRC) and for irrigation requirements (AIR) with 12 pulses.

SMDRZ was also improved with black plastic mulch (BPM) and rice straw mulch (RSM), respectively compared with control treatment (without mulch, WM). Improvement of SMDRZ by increasing number of irrigation pulses may be due to the increase in water movement in horizontal direction than in vertical direction. These results are agreement with those obtained by different investigators [4, 7, 9, 14, 17]. Not only SMDRZ is improved by increasing number of pulses but wetted soil volume (WSV ≥ FC) was also improved. Increasing of WSV ≥ FC in the root zone will increase the volume of available water in root zone.

WSV ≥ FC decreased after AIR on 12 pulses. Improving of SMDRZ with BPM and RSM compared with without mulch was due to increase

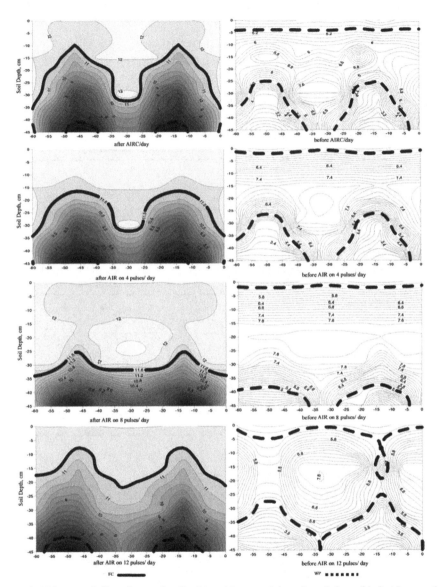

FIGURE 7.2   Soil moisture distribution without mulch and applying of irrigation needs continuously (AIRC).

of initial moisture content before irrigation. Increasing of initial moisture content in the root zone with BPM and RSM was due to decrease in evaporation process from soil surface.

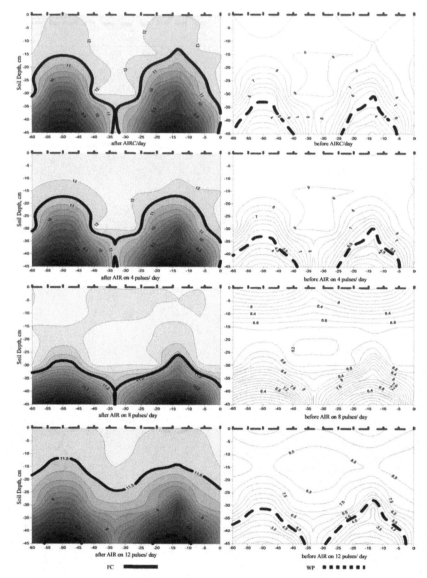

FIGURE 7.3 Soil moisture distribution under rice straw mulch and AIRC (applying of irrigation requirements continuously).

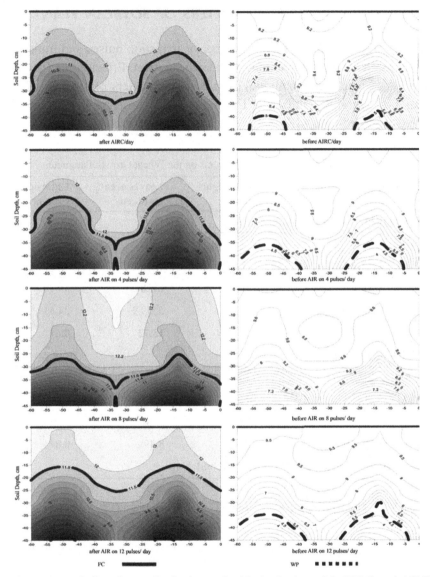

FIGURE 7.4 Soil moisture distribution under black plastic mulch (BPM) and AIRC (Applying of irrigation requirements continuously).

7.3.2   PERFORMANCE PARAMETERS OF SOYBEAN PLANT

Table 7.5 shows the effects of number of irrigation pulses and mulching systems on the growth parameters of soybean plant. Effect of the first factor (pulse irrigation technique) on growth parameters of soybean plants was positive and significant. The highest values of growth parameters were observed with AIR on 4 and 8 pulses/day, respectively. Effect of the

TABLE 7.5   Effect of Experimental Treatments on the Wheat Growth Parameters

Treatment		Dry weight/ plant, g	Leaves area/ plant, cm^2	Chlorophyll content, %
Pulsed drip irrigation technique				
AIRC/day		103.89	5876.3	30.61
AIR on 4 pulses/day		112.67	6328	37.80
AIR on 8 pulses/day		111.22	6178	36.16
AIR on 12 pulses/day		94.11	5539.2	24.51
L.S.D. at 5% level		**2.35**	**13.72**	**0.61**
Mulching systems				
WM		97.42	5558.3	25.98
RSM		107.75	5921.8	33.02
BPM		111.25	6461	37.81
L.S.D. at 5% level		**2.47**	**8.49**	**0.81**
Interaction between pulse irrigation and mulching				
AIRC/day	WM	100.33	5825.3	27.20
	RSM	105.00	5887.3	31.67
	BPM	106.33	5916.3	32.97
AIR on 4 pulses/day	WM	110.33	5954.3	34.33
	RSM	112.00	6007.7	36.33
	BPM	115.67	7022.0	43.03
AIR on 8 pulses/day	WM	99.00	5434.0	23.03
	RSM	114.33	6045.0	40.03
	BPM	120.33	7055.0	45.40

TABLE 7.5   Continued.

Treatment		Dry weight/ plant, g	Leaves area/ plant, cm^2	Chlorophyll content, %
AIR on 12 pulses/day	WM	80.00	5020.0	19.67
	RSM	99.67	5747.0	24.03
	BPM	102.67	5850.7	29.83
L.S.D. at 5% level		**4.95**	**16.98**	**0.61**

RSM: rice straw mulch, BPM: black plastic mulch, WM: without mulch, AIRC: applying of irrigation needs continuously, AIR: applying of irrigation needs.

second factor (mulching systems) on growth parameters of soybean plant was also positive and significant. The highest values of growth parameters were observed under BPM and RSM, respectively compared with control treatment (without mulch).

Table 7.5 shows that the effect of interaction between pulse drip irrigation and mulching systems on growth parameters of soybean plants was positive and significant. The highest values of growth parameters were observed under AIR on 8 pulses per day and BPM. These increases this may be attributed to improvement in SMDRZ and increasing of WSV ≥ FC in the root zone. BPM will decrease evaporation from soil surface and salt accumulation. These results are agreement with those at GardenWeb [22].

7.3.3   GRAIN YIELD OF SOYBEAN

Table 7.6 shows the effect of number of irrigation pulses and mulching systems on soybean yield. Effect of the first factor (pulse irrigation technique) on soybean yield was positive and significant. The highest values of soybean yield were 1.67 (ton/fed.) and 1.61 (ton/fed.) under AIR on 4 and 8 pulses/day, respectively and the differences were not significant between the yields under AIR on 4 and 8 pulses/day. Effect of the second factor (mulching systems) on soybean yield was also positive and significant. The highest value of soybean yield was 1.71 (ton/fed.) under BPM and the differences were significant among BPM and other treatments.

Table 7.6 and Figure 7.5 show the effect of interaction between pulse drip irrigation and mulching systems on soybean yield. The highest value

TABLE 7.6 Effect of Experimental Treatments on the Soybean Yield and Irrigation Water Use Efficiency

Treatment		Seed yield, ton/fed.	IWUE, kg of seed per m^3 of water
Pulsed drip irrigation technique			
AIRC/day		1.36	0.389
AIR on 4 Pulses/ day		1.67	0.478
AIR on 8 Pulses/ day		1.61	0.478
AIR on 12 Pulses/ day		1.02	0.300
L.S.D. at 5% level		**0.06**	**0.051**
Mulching systems			
WM		1.08	0.317
RSM		1.46	0.425
BPM		1.71	0.492
L.S.D. at 5% level		**0.04**	**0.027**
Interaction between pulse irrigation and mulching			
AIRC/day	WM	1.13	0.333
	RSM	1.40	0.400
	BPM	1.53	0.433
AIR on 4 pulses/ day	WM	1.53	0.433
	RSM	1.67	0.500
	BPM	1.80	0.500
AIR on 8 pulses/ day	WM	0.90	0.300
	RSM	1.73	0.500
	BPM	2.20	0.633
AIR on 12 pulses/ day	WM	0.73	0.200
	RSM	1.03	0.300
	BPM	1.30	0.400
L.S.D. at 5% level		**0.08**	**0.054**

RSM: rice straw mulch, BPM: black plastic mulch, WM: without mulch, AIRC: applying of irrigation requirements continuously, AIR: applying of irrigation requirements.

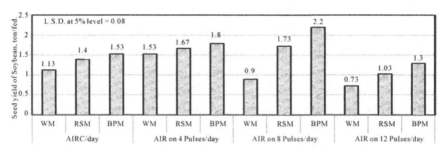

FIGURE 7.5   Effects of interactions between pulse drip irrigation and mulching systems on soybean yield.

of soybean yield was 2.2 (ton/fed.) under AIR on 8 pulses per day and under BPM and the differences were significant among these conditions and other treatments. Increasing the value of soybean yield under AIR on 8 pulses per day and with BPM may be due to the increase in the available nutrients in the root zone and these nutrients were more available due to improvement in SMDRZ and increasing of WSV $\geq$ FC in the root zone. Also BPM will decrease evaporation from soil surface and salt accumulation.

In general, soybean yield was increased by increasing number of irrigation pulses under BPM and RSM until AIR on 8 pulses/day and then values of soybean yield were decreased at AIR on 12 pulses/day, due to low irrigation water with every pulse at AIR on 12 pulses/day, and the increase in time-off events. This implies that there were not sufficient irrigation applications to remove water stress in the root zone. Under no-mulch and after AIR on 4 pulses/day, the values of soybean yield were decreased by increasing number of irrigation pulses. This may be due to the increase of evaporation rate from soil surface by increasing number of irrigation pulses and increasing of time-off events. This implies that salt concentration around the plant increased the osmotic potential, thus increasing the plasmolysis (loss of water through osmosis is accompanied by shrinkage of protoplasm away from the cell wall), and decreasing the yield.

7.3.4   IRRIGATION WATER USE EFFICIENCY

Table 7.6 shows the effect of number of irrigation pulses and mulching systems on irrigation water use efficiency (IWUE) for soybean. Effect of the first factor (pulse irrigation technique) on IWUE was positive and significant. The highest IWUE was 0.478 kg of seed/m^3 of water with AIR on 4 and 8 pulses/day and the differences were not significant between AIR on 4 and 8 pulses/day. Effect of the second factor (mulching systems) on IWUE was also positive and significant. The highest IWUE was 0.492 kg of seed/m^3 of water under BPM and the differences were significant among BPM and other treatments.

Table 7.6 and Figure 7.6 show the effects of interactions between pulse drip irrigation and mulching systems on IWUE. The highest IWUE was 0.633 (kg of seed/m^3 of water) with AIR on 8 pulses per day and under BPM and the differences were not significant between these treatments and other treatments. Increasing the value of IWUE with AIR on 8 pulses per day and under BPM may be due to the increase in soybean yield. This implies that the highest soybean yield was 2.2 ton/fed. with AIR on 8 pulses per day and under BPM compared with 1.13 ton/fed. under AIRC/day and without mulch (control treatment) under the same amount of irrigation water. Thus there was a 50% increase in yield.

7.3.5 OIL CONTENT AND OIL YIELD

Table 7.7 shows the effects of number of irrigation pulses and mulching systems on oil content and oil yield. Effect of the first factor (pulse irrigation technique) on the oil content was positive and significant. The highest values of oil content were 19.77% and 19.37% under AIR on 4 and 8 pulses/day, respectively and there were no significant differences between AIR on 4 and 8 pulses/day. However, there were significant

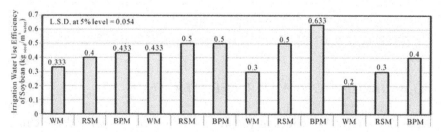

FIGURE 7.6   The effects of interactions between pulse drip irrigation and mulching systems on irrigation water use efficiency of soybean.

TABLE 7.7   Effect of Experimental Treatments on the Oil Content, Oil Yield, Protein Content and Protein Yield

Treatments		Seed yield, ton/fed.	Oil content, %	Oil yield, kg/fed.	Protein content, %	Protein yield, kg/fed.
Pulsed drip irrigation technique						
AIRC/day		1.36	17.69	240.58	29.38	399.57
AIR on 4 pulses/day		1.67	19.77	330.16	34.53	576.65
AIR on 8 pulses/day		1.61	19.37	311.86	34.73	559.15
AIR on 12 pulses/day		1.02	16.04	163.61	25.09	255.92
L.S.D. at 5% level		**0.06**	**0.5**	**14.36**	**1**	**26.02**
Mulching systems						
WM		1.08	16.94	182.95	28.38	306.50
RSM		1.46	18.26	266.60	31.02	452.89
BPM		1.71	19.45	332.60	33.4	571.14
L.S.D. at 5% level		**0.04**	**0.57**	**8.1**	**1.14**	**15.09**
Interactions between pulse irrigation and mulching						
AIRC/day	WM	1.13	16.97	191.76	27.93	315.61
	RSM	1.40	17.7	247.80	29.4	411.60
	BPM	1.53	18.4	281.52	30.8	471.24
AIR on 4 pulses/day	WM	1.53	18.6	284.58	32.2	492.66
	RSM	1.67	20.07	335.17	35.13	586.67
	BPM	1.80	20.63	371.34	36.27	652.86
AIR on 8 pulses/day	WM	0.90	16.63	149.67	29.27	263.43
	RSM	1.73	19.87	343.75	35.73	618.13
	BPM	2.20	21.6	475.20	39.2	862.40
AIR on 12 pulses/day	WM	0.73	15.57	113.66	24.13	176.15
	RSM	1.03	15.4	158.62	23.8	245.14
	BPM	1.30	17.17	223.21	27.33	355.29
L.S.D. at 5% level		**0.08**	**1.31**	**18.07**	**2.63**	**34.84**

RSM: rice straw mulch, BPM: black plastic mulch, WM: without mulch, AIRC: applying of irrigation requirements continuously, AIR: applying of irrigation requirements.

differences between the highest values of oil yield. Effects of the second factor (mulching systems) on oil content and oil yield were positive and significant. The highest value of oil content and oil yield was 19.45% and 332.60 kg/fed. under BPM and there were significant difference between using BPM and other treatments.

Table 7.7 and Figure 7.7 show effects of interaction between pulse irrigation and mulching systems on oil content and oil yield. The highest value of oil content and oil yield was 21.6% and 475.20 under AIR on 8 pulses per day and BPM, respectively and there were significant differences between these conditions and other treatments. Increasing the value of oil content and oil yield with AIR on 8 pulses per day and BPM may be due to the increase in the available nutrients in the root zone and these nutrients were more available by improvement of SMDRZ and increasing of WSV ≥ FC in the root zone. Also BPM will decrease the salt accumulation.

7.3.6 PROTEIN CONTENT AND PROTEIN YIELD

Table 7.7 shows the effects of number of irrigation pulses and mulching systems on protein content and protein yield. Effect of the first factor (pulse irrigation technique) on the protein content was positive and significant. The highest values of protein content were 34.73% and 34.53% under AIR on 8 and 4 pulses/day, respectively and differences were not significant between AIR on 8 and 4 pulses/day. The highest values of protein yield were 576.65 and 559.15 kg/fed. under AIR on 4 and 8 pulses/day, respectively and there were no significant differences between AIR on 4 and 8 pulses/day. Effect of the second factor (mulching systems) on

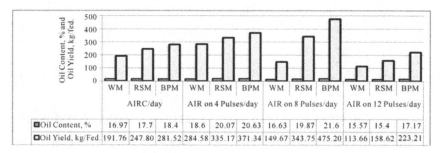

FIGURE 7.7 Effects of interactions between pulse irrigation and mulching systems on oil content and oil yield of soybean.

protein content and protein yield was positive and significant. The highest value of protein content and protein yield was 33.4% and 571.14 kg/fed. under BPM and there were significant difference between BPM and other treatments.

Table 7.7 and Figure 7.8 show the effects of interactions between pulse irrigation and mulching systems on protein content and protein yield. The highest value of protein content and protein yield was 39.2% and 862.40 kg/fed., respectively with AIR on 8 pulses per day and BPM and there were significant differences between these conditions and other treatments. Increasing the value of oil content and oil yield under AIR on 8 pulses per day and BPM may be due to the increase in the available nutrients in the root zone and these nutrients were more available for plant by improvement of SMDRZ and increasing of WSV≥FC. Also BPM will decrease the salt accumulation.

7.3.7 ECONOMICAL ANALYSIS

Table 7.8 and Figure 7.9 show the effects of interaction between pulse irrigation and mulching systems on costs of mulching and total income – CMP (costs of mulching process).

The highest value of total income – CMP was 8660 LE/fed under AIR on 4 pulses per day with WM. Although, mulching with black plastic is very expensive, yet there were no significant differences between 8660 and

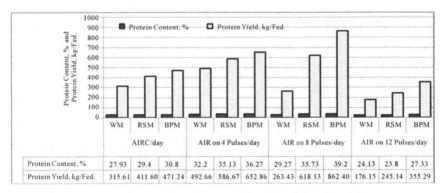

	WM	RSM	BPM	WM	RSM	BPM	WM	RSM	BPM	WM	RSM	BPM
	AIRC/day			AIR on 4 Pulses/day			AIR on 8 Pulses/day			AIR on 12 Pulses/day		
Protein Content, %	27.93	29.4	30.8	32.2	35.13	36.27	29.27	35.73	39.2	24.13	23.8	27.33
Protein Yield, kg/Fed.	315.61	411.60	471.24	492.66	586.67	652.86	263.43	618.13	862.40	176.15	245.14	355.29

FIGURE 7.8 Effects of interactions between pulse irrigation and mulching systems on protein content and protein yield of soybean.

TABLE 7.8 Effect of Experimental Treatments on Costs of Mulching Process and Total Income–CMP

Treatment		Costs of mulching, LE/fed.	Total income CMP, LE/fed.
Pulsed drip irrigation technique			
AIRC/day		2533	5586
AIR on 4 pulses/day		2533	7480
AIR on 8 pulses/day		2533	7153
AIR on 12 pulses/day		2533	3573
L.S.D. at 5% level		**N.S.**	**371**
Mulching systems			
WM		600	5840
RSM		2000	6775
BPM		5000	5230
L.S.D. at 5% level		**0.05**	**241.9**
Interaction between pulse irrigation and mulching			
AIRC/day	WM	600	6180
	RSM	2000	6440
	BPM	5000	4140
AIR on 4 pulses/day	WM	600	8660
	RSM	2000	8080
	BPM	5000	5700
AIR on 8 pulses/day	WM	600	4800
	RSM	2000	8360
	BPM	5000	8200
AIR on 12 pulses/day	WM	600	3620
	RSM	2000	4220
	BPM	5000	2880
L.S.D. at 5% level		**0.06**	**483.9**

RSM: rice straw mulch, BPM: black plastic mulch, WM: without mulch, AIRC: applying of irrigation requirements continuously, AIR: applying of irrigation requirements.

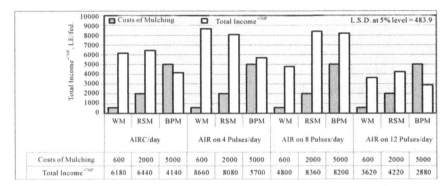

FIGURE 7.9 Effects of interactions between pulse drip irrigation and mulching systems on costs of mulching and total income – CMP.

8360 and 8200 LE/fed with AIR on 4 and 8 pulses/day + RSM and AIR on 8 pulses/day + BPM, respectively. Therefore, according to economical view, AIR on 8 pulses/day + BPM was the best combination resulting in highest soybean yield under the same amount of irrigation water.

7.4 SUMMARY

Two field experiments were conducted to study the effects of pulse drip irrigation and mulching practices on water saving, and performance parameters yield of soybean. The treatments were pulse drip irrigation technology (irrigating 4 times, 8 times, 12 times daily compared with once daily) and mulching systems (black plastic mulch, BPM, rice straw mulch, RSM and soil surface without mulch, control WM). The following parameters were studied to evaluate the effects of pulse drip irrigation and mulching systems on: Soil moisture distribution under an emitter, growth parameters and yield of soybean, irrigation water use efficiency, oil content and oil yield, protein content and protein yield, and cost economics. It was concluded that irrigating @ 8 pulses daily and BPM provided best conditions to give highest yield and quality, and IWUE. The differences were significant among treatments. Pulse irrigation improved moisture distribution and wetted volume in root zone. The BPM decreased evaporation rate from soil surface and weed growth around emitter.

KEYWORDS

- arid region
- black plastic mulch
- drip irrigation
- Egypt
- irrigation water use efficiency
- mulching systems
- oil yield
- protein yield
- pulse drip irrigation
- pulse irrigation
- rice straw
- soybean
- water use efficiency

REFERENCES

1. Abouziena, H. F., Hafez, O. M., El-Metwally, I. M. and Sharma S. D., 2008. Comparison of weed suppression and mandarin fruit yield and quality obtained with organic mulches, synthetic mulches, cultivation, and glyphosfate. *HortScience,* 43:795–799.
2. Association of Official Analytical Chemists (AOAC), 1980. *Official methods of the association of official analytical chemists.* 11th Edition, Association of Official Analytical Chemists, Washington DC.
3. Chapman, H. D. and Pratt, R. F., 1978. *Methods of analysis for soil, plant and water.* University of Georgia, Athens – GA: USA.
4. El-Adi, A. M., 2000. Effect of irrigation and fertilization methods on pea production. *Misr. Journal of Agriculture Engineering,* 17:450–468.
5. Elevitch, C. R. and Wikinson, K. M., 1998. Greater plant and soil health for less work. *Permanent Agriculture Resources,* Holualoa. http://www.agroforestry.net/
6. El-Gindy, A. M. and Abdel Aziz, A. A., 2001. Maximizing water use efficiency of maize crop in sandy soils. *Arab University Journal of Agriculture Science,* 11:439–452.
7. Eric, S., David, S. and Robert, H., 2004. To pulse or not to pulse drip irrigation that is the question UF/IFAS. Horticultural Sciences Department, University of Florida.
8. FAO, 2001. *Soybean water management.* FAOSTAT http://www.fao.org/-landandwater/aglw/cropwater/soybean.stm#links.

9. Helen, R., 2007. *Citrus irrigation*. Department of Agriculture and Food, Waroona – Australia. www. Agric .wa .gov.au

10. Helmy, M. A., Gomaa, S. M., Khalifa, E. M. and Helal, A. M., 2000. Production of corn and sunflower under conditions of drip and furrow irrigation with reuse of agricultural drainage water. *Misr Journal of Agriculture Engineering*, 17:125–147.

11. James, L., 1988. *Principles of farm irrigation system design*. John Willey & Sons Inc., Washington DC.

12. Liven, P. C. and Van, F. C., 1979. Effect of discharge rate and intermittent water application by point-source irrigation on the soil moisture distribution pattern. *Soil Science Society of America Journal*, 43:5–8.

13. Nosenko, V. F., Balabanand, E. I. and Landes, G. A., 1991. Aspects of agro-biological and environmental assessment of irrigation technologies with different volumes of water delivery. *Land Reclamation and Water Management*. Central Bureau of Scientific and Technical Information.

14. Oron, G., 1981. Simulation of water flow in the soil under sub-surface trickle irrigation with water uptake by roots. *Agricultural Water Management*, 3:179–193.

15. Scott, C., 2000. *Pulse irrigation*. Water savings report by Indiana Flower Growers Association. In Cooperation with Department of Horticulture and Landscape Architecture Cooperative Extension Service, Purdue University, West Lafayette, pages 907–1165.

16. Segal, E., Ben-Gal, A. and Shani, U., 2000. Water availability and yield response to high-frequency micro-irrigation in sunflowers. *The 6th International Micro Irrigation Congress*.

17. Shock, C. C., Flock, R. J., Eldredge, E. P., Pereira, A. B. and Jensen, L. B., 2006. *Successful potato irrigation scheduling*. Oregon State University Extension Service publication.

18. Snedecor, G. W. and Cochran, W. G., 1982. *Statistical methods*. 7th Edition, Iowa State University Press, Iowa.

19. Steyn, J. M., Duplessis, H. F., Fourie, P. and Roos, T., 2005. *Irrigation scheduling of drip irrigation potatoes*. Northern Province Department of Agriculture and Environment, P/Bag X9487, Pietersburg 0699, Za.

20. Stickler, F. C. and Pauli, A. W., 1961. Leaf area determination in grain sorghum. *Agronomy Journal*, 53:187–188.

21. USDA, FAQ by Elevitch, C. R. and Wikinson, K. M. http://www.nrcs.usda.gov/feature/backyard/Mulching.html.

22. *Weeds, FAQ: Organic weed control methods and mulching*. GardenWeb. http://faq.gardenweb.com/faq/lists/weeds/2005083919029708.html.

CHAPTER 8

SMART IRRIGATION SCHEDULING OF DRIP IRRIGATED TOMATO: SAUDI ARABIA

FAWZI SAID MOHAMMAD, HUSSEIN MOHAMMED AL-GHOBARI, and MOHAMED SAID ABDALLA EL MARAZKY

CONTENTS

In this chapter: One U.S. dollar = 3.75 riyals (KSA currency).

Modified from, "*Fawzi Said Mohammad, Hussein Mohammed Al-Ghobari, and Mohamed Said Abdalla El Marazky, 2013. Adoption of an intelligent irrigation scheduling technique and its effect on water use efficiency for tomato crops in arid regions. Open Access Article in Australian J. of Crop Science (AJCS), 7(3):305–313*".

8.1   INTRODUCTION

Tomatoes (*Lycopersicon esculentum Mill*) are an important global vegetable crop [4], and require a high water potential for optimal vegetative and reproductive development [34]. Production areas are typically intensively managed with high inputs of fertilizer and irrigation. Planting tomatoes in Saudi Arabia accounted for 13% of the total land planted with vegetables in 2008 [24]. Tomato is one of the most important vegetables because of its special nutritive value, and is the world's largest vegetable crop after potato and sweet potato. Considerable quantities of irrigation water are required, depending on the soil and weather conditions. To reduce the total amount of irrigation water needed by a tomato crop without affecting the yield and fruit quality, the grower must develop management strategies. To achieve better control and management of water in tomato production, irrigation schedules should be based on crop water requirements according to FAO guidelines [1, 8]. Another approach is the development of a daily water balance to calculate ETc and to schedule irrigation events according to effective soil water storage capacity and estimated crop water removal [13, 14].

These methods for irrigation scheduling can be very efficient, but this is difficult and expensive to implement at a farm level. In most of the world, irrigated agriculture has been faced with increased limitations in the water supply over the last few decades. Major efforts have been made by researchers and irrigators to increase and to conserve this vital source by many means. One of these means is the application of irrigation scheduling using sensors and electronic control devices [13, 14, 31]. Irrigation scheduling is a technique designed to accurately give water to a crop in a timely fashion [10]. Irrigation scheduling methods are based on two approaches: soil measurements and crop monitoring [13–15]. However, the use of more efficient technologies often increases, rather than decreases, water consumption [11, 38]. Improved irrigation scheduling can reduce irrigation costs and increase crop quality. Irrigation scheduling based upon crop water status is more advantageous since crops respond to both the soil and aerial environments [39]. Drip irrigation has been practiced for many years due to its effectiveness in reducing soil surface evaporation. It has been used widely for crops in both greenhouses and the field [9]. Uniform water application in drip irrigation is affected by field topography as well as the hydraulic design parameters of the drip system such as energy losses in laterals and emitter characteristics [25, 40, 42]. An intelligent irrigation system (IIS) is integrated with intelligent controllers and uses microclimate data to schedule water irrigation. Intelligent irrigation technologies are regarded as a promising tool to achieve landscape water savings and reduce non-point source pollution [27]. This technique is under evaluation at the trial farm in Dookie, Egypt, and the initial results indicate up to 43% (average 38%) water savings over conventional irrigation control methodologies [7]. In the past 10 years, intelligent irrigation controllers have been developed by a number of manufacturers and have been promoted by water purveyors in an attempt to reduce over-irrigation [23]. There are many intelligent irrigation systems that compute the amount of water applied and ET based on climate conditions [19, 20, 21]. These systems differ in their accuracy and reliability. Intelligent irrigation systems usually depend on modern electronic sensors, which are capable of collecting data, analyzing and decision making to start/stop irrigation. These devices generally transmit the decisions to electronic controller devices, which control the sprinkler or drip irrigation system. Several moisture sensors are commercially available, such as tensiometers and watermarks. They can generally be used for manual readings to guide irrigation scheduling, while some of them can also be interfaced directly with the irrigation controller in a closed loop control system [41] to automate irrigation. Some researchers have used tensiometer sensors in irrigation scheduling for tomato under drip irrigation systems [21, 32, 33]. Water use efficiency

(WUE) has been reported to decrease with increased irrigation times and the amount of irrigation water per growing season [31]. Several studies have found that drip irrigation increases yields and WUE by large amounts compared with those with sprinkler or surface irrigation [17, 28].

The automation of irrigation systems based on soil moisture sensing (SMS) has the potential to provide maximum WUE. Such systems maintain soil moisture within a desired range, which is optimal or adequate for plant growth and/or quality [23, 26]. Therefore, based on prevailing conditions and water shortages, the optimum irrigation schedules for the tomato crop in a region should be determined.

This chapter investigates the effects of different aspects of smart irrigation system on ET, yield, WUE and irrigation water use efficiency (IWUE) for drip-irrigated tomato under arid climatic conditions.

8.2 MATERIALS AND METHODS

8.2.1 EXPERIMENTAL SITE

Field experiments were performed at the King Saud University Experimental Farm of the College of Food and Agriculture Sciences, Riyadh (24°43' N latitude, 46°43' E longitude and 635 m altitude) during the spring seasons of 2010 and 2011. Generally, the climate in this region is classified as arid, and the climatological data at the experimental site are shown in Table 8.1. The weather station measured the climate parameters that were used to compute evapotranspiration (ETo). These values were then compared with the values obtained from the IIS in the tomato crop.

TABLE 8.1 Metrological Data at the Experimental Site

Month	Tmax	Tmin	MRH	Total rainfall	SR 104W-2SR	WS	ETo
	°C	°C	%	mm		m/s	mm/day
2010 season							
February	26.28	13.40	26.96	0.00	41.29	5.76	4.62
March	30.03	16.39	19.02	0.01	51.51	5.53	5.97
April	32.86	21.41	28.53	0.27	46.01	6.94	6.20
May	37.64	25.25	25.06	0.18	48.22	5.93	6.90

TABLE 8.1   Continued

Month	Tmax	Tmin	MRH	Total rainfall	SR 104W-2SR	WS	ETo
	°C	°C	%	mm		m/s	mm/day
2011 season							
February	23.44	12.41	36.23	0.00	38.71	1.53	4.29
March	25.39	14.77	31.69	0.54	40.34	1.94	5.28
April	30.83	19.86	24.18	0.04	39.59	1.92	6.02
May	35.40	23.29	20.97	0.09	51.63	1.59	6.96

MRH = Maximum relative humidity, SR = Solar radiation.

The IIS was programed in situ for the crop type and environmental conditions. This smart device was then calibrated and configured to implement the next phase of the study prior to collecting real data.

8.2.2 FIELD LAYOUT AND EVALUATION OF IRRIGATION PRACTICES

The study site was divided into two equal plots with a 5 m buffer zone in the middle (Figure 8.1). Each tomato plot size was 7.2 m × 12 m (86.4 m²), and the plots were irrigated via nine drip lines that were 16 mm in diameter at distances of 0.8 m and mounted with 30 drippers. The distance between drippers on the drip line was 0.4 m. The soil type in the plot area was sandy clay loam based on physical properties that are presented in Table 8.2. One of the two fields was irrigated automatically with the IIS, while the other was irrigated manually based on ETc values and using climatological data from the weather station installed at the site.

The drip irrigation system was installed in both plots and equipped with controllers to regulate the pressure and a flow meter to quantify the amount of water added during each irrigation event. The drip system was evaluated in the field according to the methodology by ASABE Standard, S346.1 [3]. The intelligent irrigation system required a complete database for each station (or "zone") to be controlled. It was easy to set up this database with little effort, and the operator was completely responsible for the accuracy of both input information and output results from the database. Each system was carefully observed and monitored after initial installation for the best results. Generally, most systems require adjustment,

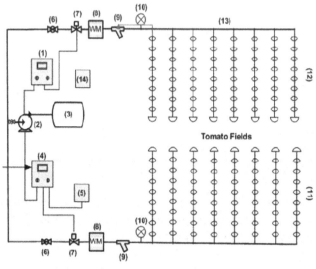

1	ICS Control panel	6	Ball Valve	11	Intelligent irrigation Treatment (IIS)
2	Pump	7	Solenoid Valve	12	Control Treatment (ICS)
3	Water Tank	8	Water Meter	13	Main Lines
4	IIS Control panel	9	Filter	14	Weather Station
5	ET sensor (Fig 2)	10	Pressure Gauge		

FIGURE 8.1 Schematic diagram of drip irrigated tomato field for both intelligent irrigation (IIS) and control irrigation systems (ICS).

TABLE 8.2 Physical Properties of Different Soil Layers in the Studied Field

Layer depth	Particle size distribution, %			Soil texture class	FC %	PWP %	BD
cm	Sand	Silt	Clay		$m^3\ m^{-3}$		$g.cm^{-3}$
0–20	68.5	12.0	19.5	Sandy clay loam	13.65	6.83	1.48
20–30	68.7	11.0	20.3	Sandy clay loam	14.34	7.17	1.46
30–60	58.7	15.0	26.3	Sandy clay loam	16.67	8.33	1.40
Average	65.3	12.7	22.1	Sandy clay loam	14.89	7.44	1.45

BD = bulk density, PWP = permanent welting point, FC = field capacity.

at the station level, for some time after installation to provide ideal results. Evaluation tests were conducted by checking the performance index values under the operating field conditions. All evaluation index values were within acceptable limits with good water distribution uniformity (over 90%). The control experiment was used for comparison purposes.

8.2.3 COMPONENTS AND INSTALLATION OF THE SMART IRRIGATION SYSTEM

The intelligent irrigation system chosen for this study was the Hunter ET-System. The smart controllers integrate many disciplines to produce a significant improvement in crop production and resource management [29]. This system is not considered the best system, but it was inexpensive and available in the local market. The IIS was installed according to the manufacturer's instructions in the field. It can be customized by station (or "zone") for specific plant, soil and drip types. The system uses digital electronic controllers and modules, and its platform can be wired to an ET module that can sense the local climatic conditions via different sensors that measure wind speed, rainfall, solar radiation, air temperature and relative humidity (Figure 8.2). The ET module then receives data from the ET sensor and applies it to the individual fields (zones) of irrigation. The IIS automatically calculates crop evapotranspiration (ETc) for local microclimates based on a modified

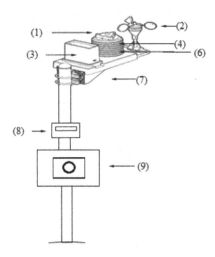

1	Solar Radiation	6	Relative Humidity
2	Wind Speed	7	Sensor Base
3	Rain Gauge	8	ET Module
4	Air Temperature	9	Irrigation Controller (Pro C)

FIGURE 8.2   The components of smart irrigation system.

Penman equation [1, 13, 14] and creates a scientific program that it downloads to the controller. Here, the ET module was plugged into the irrigation controller Pro-C, which was called the *Controller Intelligent Port*, and adjusted the irrigation run duration to only replace the amount of water lost from the plant canopy, at a rate at which it can be effectively absorbed by the soil. Hence, the IIS relayed data acquisition of climatic and system parameters (pressure, flow, etc.). The state of the system was compared with a specified desired state, and a decision as to whether or not to initiate an action is based on this comparison. In the case of a decision taken by the ET sensor (Figure 8.2) to initiate irrigation, a signal was transmitted to open the solenoid valve and pump to supply the required depth of irrigation water.

8.2.4 PERFORMANCE PARAMETERS FOR DRIP IRRIGATED TOMATO

The processor-interfaced IIS was used as an electronic controller to monitor, record ETo based on measured weather parameters and automatically adjust the depth of irrigation water. The daily and weekly averages of the ETc rates for tomato crops under IIS and ICS treatments were calculated using the daily records during the two growing seasons (See Table 8.3).

TABLE 8.3 Daily and Weekly Averages of Tomato ETc for Both Irrigation Systems

Growth period	ET_c for IIS	ET_0	K_c	ET_c for ICS
Week[-a]		mm/day	—	mm/day
1	2.34	4.22	0.70	2.95
2	3.15	4.65	0.70	3.25
3	3.94	4.98	0.93	4.54
4	3.95	5.56	1.15	6.39
5	4.36	5.61	1.15	6.46
6	4.58	5.78	1.15	6.64
7	4.87	5.28	1.15	6.08
8	4.56	5.92	1.03	6.30
9	5.26	6.71	1.03	6.84
10	5.10	6.67	0.90	6.00

TABLE 8.3 Continued

Growth period	ET$_c$ for IIS	ET$_0$	K$_c$	ET$_c$ for ICS
Week^{-a}		mm/day	—	mm/day
11	4.93	6.54	0.90	5.89
12	5.00	6.87	0.90	6.18
13	4.85	6.56	0.83	5.53
14	4.60	6.64	0.83	5.53
15	5.81	7.49	0.90	6.74
16	4.83	6.96	0.75	5.22
17	5.07	7.17	0.75	5.38
Avg.	4.54			5.64
Sum.	540.42			671.57

^{a}Weeks after transplanting of tomato seedlings.

The values of ETc for ICS treatment were derived by the product of the reference evapotranspiration (ETo) and the crop coefficient (Kc) for different stages of tomato crop development.

In IIS treatment, irrigation was scheduled and initiated automatically based on ETo prediction. This system is equipped with special options, such as addition of more or less water depending on the plant need. The water quantities and timings were monitored and recorded. The ETo values for ICS were determined using the modified Penman method, FAO version [1] and used efficiently to schedule irrigation at different growth stages. Based on research studies for KSA, the growth stages were approximately 30, 70, 110, and 135 days after transplanting, respectively, and were considered to estimate Kc. These stages were: initial, crop development, mid-season and late season.

8.2.5 ESTIMATION OF OPERATION TIME

To calculate the ETc and the irrigation water requirements of tomatoes, daily ETo values were estimated using the meteorological data and were then multiplied by crop coefficients and the water application efficiency. Based on the area of the field (86.4 m^2) and the discharge rate from the drippers (1.220 lph), the required water quantity per event and actual operation time were determined as follows:

$$T, \text{min.} = [V, \text{liters}/Q_s, \text{in lpm}]$$
$$= [K_c \times ET_o \text{ in mm} \times A \text{ in m}^2 \times P_w]$$
$$/[E_a \times (1 - LR) \times Q_s \text{ in lpm}] \qquad (1)$$

$$= [K_c \times ET_o \text{ in mm} \times 86.4 \times 0.40]$$
$$/[0.90 \times (1 - 0.10) \times (1.220/60)]$$
$$= [K_c \times ET_o \text{ in mm}] \times 2.31 \qquad (2)$$

where: T is the actual operation time required in minutes, V (liter) refers to the water volume to be added, Q_s (lpm) is the discharge from the irrigation system, K_c represents crop coefficient, A (m²) refers to the area of the field, ETo (mm) is the reference evapotranspiration, LR refers to the leaching requirement = 0.1 on the least water area [18], Pw (40%) refers to the wetted area percentage and E_a (90%) refers to the water application efficiency given by Eq. (3).

$$E_a = K_s \times Eu \qquad (3)$$

where: E_a = irrigation efficiency coefficient (<1) and expressed as a fraction and = 0.90 for drip irrigation = crop root zone to be used by the crop/applied water; K_s is a coefficient (< 1) which expresses the water storage efficiency soil (0.9 in sandy soils, 1.0 in clay or loam soils); Eu is a uniformity of water application (<1: a properly designed and well-managed drip system should reach Eu values of 0.85–0.95). This coefficient should be measured for each system regularly [35]. The net irrigation requirement Dg must replenish the crop evapotranspiration (ETc), as rainfall and other components of the water balance are normally unimportant in the irrigated area. The gross irrigation requirements (Dg)t must increase the Dg in order to compensate the irrigation efficiency and to leach salts.

$$(Dg)_t = Dg/[E_a (1 - LR)] \qquad (4)$$

The irrigation system was turned on and off manually in the control experiments in the ICS plots. The net depth of the irrigation water (Dg) for IIS under the drip irrigation system was calculated based on the difference in the flow meter readings before and after irrigation.

Irrigation water used efficiency (IWUE) was calculated as the ratio between the total fresh fruit yield (FY) and the seasonal applied irrigation water $(Dg)_t$ [22]. Water use efficiency (WUE) was the relationship between the yield and the ETc [36]. Thus, WUE was calculated as the fresh tomato fruit mass (kg) per unit land area (Y, kg.m^{-2}) and divided by the units of water consumed by the crop per unit land area (ETc, m^3.m^{-2}, usually reported in mm) to produce that yield. In this case, WUE is presented in kg.m^{-3}, and crop evapotranspiration ETc can be expressed as the water depth (mm). Another key parameter for evaluating system efficiency is the irrigation water use efficiency (IWUE, kg.m^{-3}). The WUE and IWUE were calculated using Eqs. (3)–(6), respectively.

$$WUE = [Y/ETc] \tag{5}$$

$$IWUE = [Y/(Dg)_t] \tag{6}$$

In these equations, Y is the economical yield (kg.m^{-3}), ETc is evapotranspiration (mm) and $(Dg)_t$ is the amount of seasonally irrigation water applied (mm).

8.2.6 AGRONOMIC PRACTICES

Tomato plants (*Lycopersicon esculentum Mill, GS-12*) were transplanted into the fields on February 14, 2010 and February 7, 2011. The seedlings were planted in a single row in each bed, with a row spacing of 0.8 m and an interplant space of 0.4 m per row. Other cultivation practices were performed according to a scheduled tomato crop program. Daily and weekly ETc rates for tomatoes during the growth period were determined for the IIS and ICS treatments. The irrigation water depths (Dg) and accumulative depths added to the tomato crop under the two treatments were monitored by flow meters and were recorded through the growing season. The last irrigation was on 31 May of 2010 and 27 May of 2011 in the first and second seasons, respectively. Fruit yield and its components were evaluated from eight plants from the central plot rows during the harvest period. The growth characteristics of tomato plants grown during the two seasons (2010 and 2011) were recorded to study effects of IIS scheduling on tomato growth and productivity parameters. The mature fruits were harvested once or twice a week, to measure: the plant height (cm), branch number, fruit length (cm), fruit diameter (cm), fruit shape index (length/diameter), average fruit weight (g), and total yields (kg per m^2 and ton/ha) were calculated and recorded.

The data obtained from the two growing seasons were tabulated and subjected to variance analysis and least significant difference analysis (LSD) using CoHort Software [6]. Treatment mean values were compared using the least significant difference (LSD) at P = 5%. Water consumption was considered in this analysis. Statistical analysis was conducted using CoHort Software [6] program version 6.311. A t-test was used to compare the average of the two irrigation methods following a normal distribution. This t-test was also used to find significant differences between IIS and ICS irrigation treatments.

8.3   RESULTS AND DISCUSSION

8.3.1   TOMATO EVAPOTRANSPIRATION (etc)

The accumulated rainfall for the 2010 and 2011 growing seasons were 14 mm and 16.6 mm, respectively, which are considered to be not significant for irrigation. The Table 8.3 shows daily and weekly averages of the ETc rates for tomato crops under IIS and ICS treatments. From the data in Table 8.3, it can be observed that the total ETc values for tomato crops were 540.42 and 671.57 mm under the IIS and ICS treatments, respectively; with significant water saving equal to 20% with IIS treatment compared to ICS. Values of ETc during the first four weeks of crop growth were lower under IIS treatment, then increased during plant booming and development, peaking approximately 55 days (8 weeks) after transplantation. After this point, values of ETc began to retreat gradually with leaf senescence, most significantly during weeks 9 to 15, and a similar was observed in ICS management.

In this study, marked variation in the ETc of the tomato crop was observed between the two irrigation treatments during the two seasons. The results indicate the importance of adopting IIS due to its effectiveness in providing irrigation water, which requires extraordinary effort to obtain especially in arid regions, which suffer from water scarcity, such as Saudi Arabia. Also, this system will improve irrigation practices and ultimately minimize labor inputs. In general, this superiority in saving water may be due to the fact that the IIS has the feature of increasing or reducing irrigation water according to the needs of the plants. Despite this, to initiate the process of irrigation at 80% of ETc, the analysis showed that the ETc value of the control treatment was higher than that of the IIS in both seasons. Therefore, a comprehensive understanding of the relationship between the effect of the IIS technique and moisture distribution in the root zone is imperative. This may be due to the increased accuracy of the irrigation scheduling which leads to even distribution of water with sufficient quantities in the root zone. Moreover, the differences might have occurred due to application of the incompatible Kc values, which were selected from Allen et al. [1] and used for the prediction of ETc. Insignificant differences were found in the ETc values between the treatments only in the initial development stage, while marked differences were observed in other growth stages, with higher values under ICS treatment (Table 8.3). Simultaneously, the steepness of ETc for the control treatment may have resulted from an erroneous prediction of ETo, especially when selecting the crop coefficients, Kc, and the length of the crop growth stages. Additionally, the intelligent irrigation system was designed especially for irrigation scheduling of landscape, although it gave satisfactory results in drip irrigated tomato crop. Moreover, the soil distribution could also be responsible for the ICS results, since the field consisted of entirely moved soil. The results of the second season were found to be consistent with the findings of the first season within each treatment, but significant differences were found among treatments. The consistency was a result of nonsignificant differences in microclimatic parameters at the experimental site and due to minor variations in available soil moisture depletion levels.

8.3.2 IRRIGATION MANAGEMENT

ETc for the ICS treatment was higher than that for the IIS, with a similar trend during the entire growth season. The averages of weekly irrigation water (Dg) added in both treatments were calculated and are given in Table 8.4. The total applied irrigation water, Dg for IIS and ICS, was 614.26 and 825.47 mm, respectively. This indicates that there was a 26% savings in irrigation water in the case of IIS compared to the control treatment. Therefore, the results of this study show that IIS significantly conserved water compared to ICS. Moreover, the data revealed that ETc values were close in the initial developmental stages, but their values gradually diverged during the rest of the season. Hence, a change in irrigation frequency and application stage could significantly affect the available soil water during the tomato-growing season. In any case, these amounts are

TABLE 8.4   Averages of Irrigation Water Depth (Dg) and Accumulative Depths (Dg)t for the Tomato Crop Under the Intelligent and Control Irrigation Systems

Growth period	Avg. Dg for Tomato, IIS			Avg. (Dg) for Tomato, ICS		
	Water added	Irrigation depth Dg	Acc. depth (Dg)t	Water added	Irrigation depth Dg	Acc. depth (Dg)t
week	m³	mm	mm	m³	mm	mm
1	0.65	18.83	18.83	0.89	25.81	25.81
2	0.90	25.94	44.77	0.97	28.09	53.90
3	1.07	30.99	75.76	1.32	38.29	92.19
4	1.12	32.53	108.29	1.93	55.91	148.10
5	1.21	35.08	143.37	1.91	55.15	203.25
6	1.26	36.43	179.80	1.91	55.33	258.58
7	1.35	39.18	218.98	1.82	52.54	311.12
8	1.24	35.87	254.85	1.86	53.78	364.90
9	1.41	40.91	295.76	2.07	59.85	424.75
10	1.42	41.03	336.79	1.84	53.16	477.92
11	1.34	38.78	375.57	1.74	50.28	528.20
12	1.34	38.89	414.46	1.85	53.41	581.61
13	1.35	39.02	453.48	1.61	46.64	628.24
14	1.24	35.78	489.26	1.67	48.39	676.63
15	1.60	46.22	535.47	1.92	55.51	732.14
16	1.31	37.99	573.46	1.60	46.25	778.39
17	1.41	40.79	614.26	1.63	47.08	825.47
Sum	21.23	614.26	—	28.53	825.47	—

Dg = irrigation water depth, (Dg)t = accumulative depth

greater than the amount of irrigation water usually delivered by the farmers in the region.

8.3.3 AGRONOMICAL CHARACTERISTICS

The effects of IIS scheduling on tomato growth and productivity parameters were investigated. The growth characteristics of tomato plants grown during the two seasons (2010 and 2011) are shown in Table 8.5. This study reveals that the IIS had a clear impact on agronomical plant characteristics. The average plant heights were 45.3 and 38.8 cm for the IIS and ICS treatments, respectively. The average branch numbers were 6.31 and 5.05 per plant for the same treatments, and the average tomato yield for the two seasons were 39.55 and 37.05 ton/ha for the IIS and ICS, respectively. The IIS was superior to the ICS in terms of plant height, number of branches, fruit length, average fruit weight, fruit yield, WUE and IWUE by 16%, 26%, 11%, 6%, 8%, 38% and 43%, respectively. The variation in the fruit yield between IIS and ICS treatments was 5–9%. The higher yield under

TABLE 8.5 Performance of Tomato Growth and Water Use Efficiencies (WUE and IWUE) for Irrigation System (IIS and ICS) During the 2012 and 2011 Winter Seasons

Parameter	Units	2010 Season			2011 Season		
		IIS	ICS	t-value	IIS	ICS	t-value
Plant height	cm	44.0	39.0	*	46.7	38.7	**
Number of branches		6.0	5.0	*	6.63	5.10	**
Fruit length	cm	6.3	5.7	*	7.1	6.3	**
Fruit diameter	cm	4.6	4.8	*	5.8	5.1	**
Fruit shape index	—	1.23	1.31	*	1.22	1.30	**
Avg. fruit weight	grams	95.0	93.0	*	93	84	**
Early yield	tons/ha	23.60	24.0	*	26.52	22.60	**
Total yield	tons/ha	39.00	37.4	*	40.08	36.71	**
WUE	kg/m^3	7.50	5.72	*	7.15	5.33	**
IWUE	kg/m^3	6.56	4.70	*	6.32	4.30	**

*, ** = t-value is significant at P = 0.05 and P = 0.01, respectively.

IIS = Smart irrigation system, ICS = Control irrigation system, ETc = Crop evapotranspiration, WUE = Water use efficiency, IWUE = Irrigation water use efficiency.

IIS than the ICS can be attributed to differences in the amount of water applied in the two treatments.

An increased moisture level in the root zone is vital for increasing the agronomical factors, especially when more irrigation water was added (Dg) in the ICS treatment. The low amount of irrigation water added in the IIS treatment affected all the agronomical parameters compared to the control treatment. The results indicate that each 1 mm of water depth applied in both treatments yielded 65.57 and 63.24 kg/mm during 2010 and 2011 seasons for IIS, while these values were 46.97 and 42.94 kg/mm for ICS. The average values of both seasons for the IIS and ICS systems were 64.41 and 44.95 kg/mm, respectively. Conserving water is very important in areas experiencing severe drought such as Saudi Arabia.

The lower amounts of water used correspond inversely with higher water use efficiency. This agrees with the results noted by Faberio et al. [12] and Almarshadi and Ismail [2]. Similar findings were also obtained by Oktem et al. [30] and Wan and Kang [36], who found that a low irrigation frequency resulted in higher water use efficiency values when compared to a high irrigation frequency.

8.3.4   WATER USE EFFICIENCY

Table 8.6 illustrates the effects of the IIS and ICS treatments on tomato water use efficiency during the two growing seasons. It was observed that the values of WUE and IWUE were higher in the IIS treatment. The WUE values were 7.50 and 7.15 kg.m^{-3} in the IIS treatment during 2010 and 2011, respectively. The IWUE values were 6.56 and 6.32 kg.m^{-3} in the IIS treatment during 2010 and 2011, respectively.

The tomato yield, in the case of IIS treatment, was 39 and 40.08 ton/ha for both seasons, respectively; and a similar trend was observed for WUE and IWUE. Moreover, the amounts of applied irrigation water were 5947.6 and 6337.6 m^3/ha for consecutive seasons, respectively (Table 8.6). Consequently, the maximum and minimum values of WUE were 7.50 and 5.33 kg/m^3, respectively. However, the results indicate that irrigation water was used more effectively through IIS treatment. The Table 8.6 also shows that the highest and lowest values of IWUE in both seasons were 6.56 and 4.30 kg.m^{-3} in IIS and ICS, respectively. The comparison of the

TABLE 8.6   Effects of the IIS and ICS Treatments on Tomato Water Use Efficiency During the Two Growing Seasons

Irrigation treatment	ETc		AIW		WUE	IWUE
	mm	m³/ha	mm	m³/ha	kg/m³	kg/m³
2010 growing season						
IIS	520.3	5203	594.76	5947.6	7.50	6.56
ICS	653.7	6537	796.15	7961.5	5.72	4.70
2011 growing season						
IIS	560.5	5605	633.76	6337.6	7.15	6.32
ICS	689.2	6892	854.79	8547.9	5.33	4.30

IIS = Smart irrigation system, ICS = Control irrigation system, ETc = Crop evapotranspiration, WUE = Water use efficiency, IWUE = Irrigation water use efficiency, AIW = Applied irrigation water.

IIS with the ICS shows that the increases in IWUE were 39% and 47% in the 2010 and 2011 seasons, respectively. In contrast, the smallest amount of irrigation water was 594.76 mm in case of IIS, while the largest amount applied was 854.79 mm in the control treatment.

Generally, IWUE can be increased by reducing irrigation losses [30]. Irrigation water use efficiency can also be affected by soil type, cultural and management practices [35]. Generally, in IIS, increased yields are obtained while minimal water is applied, which eventually results in higher IWUE. This finding is consistent with a study by Sammis and Wu [32], who reported that IWUE was increased under soil moisture stress, and is also consistent with the observations of Camp et al. [5], Howell et al. [16], Oktem et al. [30] and Wan and Kang [36], who reported that low irrigation frequencies resulted in higher WUE values than do high irrigation frequencies. For both seasons, the IIS resulted in higher WUE and IWUE values compared to the ICS. In general terms, this study suggests that IIS should be implemented to supply irrigation water to crops. The decreased WUE and IWUE observed under the ICS treatment can be attributed to the increasing level of applied irrigation water. Hence, it can be concluded that the effects on IWUE accuracy were significant for the IIS, amounting to a 26% decrease in the amount of seasonal irrigation water required (Table 8.6). The same trend was observed for WUE and IWUE, in which higher values were obtained with the IIS in both seasons (Tables 8.5 and 8.6). Therefore, the IIS resulted in higher WUE and IWUE values than the ICS. In general, the results in Table 8.5 show that all agronomical characteristics of IIS treatments were significantly superior compared to those of ICS. The fact that the yield of 2011 was lower with the ICS treatment could be due to the excess of irrigation water which was applied.

8.3.5   STATISTICAL ANALYSIS OF AGRONOMICAL FACTORS

The results of this study clearly show a large influence of the IIS technology on tomato agronomical factors in both years. For instance, the highest amount of irrigation water applied was detected with the ICS in both seasons, while less water was applied with the ICS. The data suggest that the IIS technique had a highly significant effect on the average fruit weight. However, there were no such effects on either fruit diameter (cm) or fruit shape. Meanwhile, the agronomical data for the IIS treatment revealed a significant difference in plant height (cm), number of branches, fruit length (cm), average fruit weight (g), total yield (kg.m^{-2}), total yield (ton/ha), WUE, and IWUE (kg.m^{-3}) compared those in the control.

WUE and IWUE were significantly affected by the IIS (P = 0.05) in both growing seasons, as shown in Table 8.5. Their averages were different, depending on the schedule of the IIS. However, WUE and IWUE ranged from 5.53 to 7.33 kg.m^{-3} and from 4.50 to 6.44 kg.m^{-3}, during the two seasons, respectively. Furthermore, the results presented in Tables 8.5 and 8.6 show that both efficiencies under the IIS were higher than those under the ICS. Maximum values of WUE (7.50 and 7.15 kg m-3) were obtained in the IIS, whereas minimum values (5.72 and 5.33) were obtained in the ICS treatment. These results indicate that water was used more effectively in the IIS treatment.

The results also indicate that the IWUE for IIS was higher than that for ICS treatment. The maximum values of IWUE (6.56 and 6.32 kg.m^{-3}) were obtained in the IIS for both years, whereas the minimum values (4.70 and 4.30) were obtained in the ICS. IWUE was higher with the IIS compared to ICS by 29% and 32% during the 2010 and 2011 seasons, respectively. Thus, the WUE and IWUE values decreased with increased amounts of applied irrigation water (Table 8.6). Furthermore, the higher respective values (7.50 and 4.75 kg.m^{-3}) in the first season were achieved with the

IIS treatment, while the corresponding values for the second season were 7.15 and 4.30 kg m^{-3}.

8.4   CONCLUSIONS

The highest actual yield was observed for the IIS (40.08 ton/ha for the second season), which shows the relevance of this system to field crops, although it was only intended for scheduling water in landscaping as instructed by the manufacturer's manual. As a result of this two-year field study and using the IIS for irrigation water scheduling, it was found that the IIS offered a significant advantage in managing the irrigation of tomato crop under arid conditions. In comparison with the control treatment, the IIS significantly reduced irrigation water by 26% due to improved moisture distribution in the root zone. The lowest amount of water supplied was recorded with the IIS (614.26 mm), while the highest value was obtained with the ICS (825.47 mm) treatment during the two seasons. Also, the results indicate that the values of WUE (7.50 kg. m^{-3}) and IWUE (6.56 kg.m^{-3}) were higher in the IIS than the ICS. The results indicate that water was used most effectively in the IIS treatment. All tomato growth parameters in the IIS were significantly superior compared to those under the ICS. The IIS had significant effect on WUE and IWUE. The IIS technique conserved irrigation water by 26% compared to the amount provided by the control system. This study has demonstrated possible modifications and developments to the proposed system for improved and more efficient irrigation scheduling. It can be concluded that an economic benefit can be achieved with saving of irrigation water when applying advance irrigation scheduling techniques such as smart irrigation under arid conditions.

8.5   SUMMARY

The intelligent irrigation technique is a valuable tool for scheduling irrigation and quantifying water required by plants. This study was carried out during two successive seasons of 2010 and 2011. The main objectives were to investigate the effectiveness of the smart irrigation system (IIS) on water use efficiency (WUE), irrigation water use efficiency (IWUE) and to assess its potential for monitoring the water status and irrigation scheduling of drip irrigated tomato under arid climate conditions. The intelligent irrigation system was implemented and tested for drip irrigated tomato crop (*Lycopersicon esculentum Mill, GS-12*). The results obtained with this system were consequently compared with the control system (ICS). The results reveal that plant growth parameters and amount of water were significantly affected by IIS irrigation. The water use efficiency in IIS was generally higher (7.33 kg.m^{-3}) compared to that under ICS (5.33 kg/m^3), resulting in maximal water use efficiency for both growing seasons (average 6.44 kg.m^{-3}). The application of IIS technology therefore provides significant advantages in terms of both crop yield and WUE. In addition, IIS conserves 26% of the total irrigation water compared to the control treatment, and simultaneously generates higher total yield. These results show that this technique can be flexible and practical tool for improving irrigation scheduling. This technology can therefore be recommended for efficient automated drip irrigation systems. The intelligent irrigation technique may be extendable for use in other similar agricultural crops.

ACKNOWLEDGMENTS

This project was supported by King Saud University, Deanship of Scientific Research, College of Food and Agriculture Sciences, and Research Center. The use of the trade name does not imply promotion of this product; it is mentioned for research purposes only.

KEYWORDS

- applied irrigation water
- arid region
- ASABE
- crop coefficient
- crop water requirement
- depth of irrigation water

- drip irrigation
- evapotranspiration
- FAO
- intelligent irrigation system
- irrigation control system
- irrigation discharge
- irrigation method
- irrigation scheduling
- irrigation water use efficiency
- King Saud University
- Kingdom of Saudi Arabia
- leaching requirement
- least significant difference
- plant growth parameters
- precision irrigation
- smart irrigation
- soil moisture sensing
- tensiometer
- tomato
- tomato yield
- total depth of irrigation water
- water application efficiency
- water use efficiency

REFERENCES

1. Allen, R. G., Pereia, L. S., Raes, D., and Smith, M., 1998. *Crop evapotranspiration guidelines for computing crop water requirements*. FAO Irrigation and Drainage Paper No 56, pages 301.
2. Almarshadi, M. H., and Ismail, S. M., 2011. Effects of precision irrigation on productivity and water use efficiency of alfalfa under different irrigation methods in arid climates. *J. Appl. Sci. Res.,* 7(3):299–308.

3. ASABE Standard S436.1, 2007. Test procedure for determining the uniformity of water distribution of center pivot and lateral move irrigation machines equipped with spray or sprinkler nozzles. American Society of Agricultural and Biological Engineers, St. Joseph.

4. Berova, M. and Zlatev, Z., 2000. Physiological response and yield of paclobutrazol treated tomato plants (*Lycopersicon esculentum Mill*). *Plant Growth Reg.*, 30(2):117–123.

5. Camp, C. R., Sadler, E. J., and Busscher, W. J., 1989. Subsurface and alternate middle micro irrigation for the Southeastern Coastal Plain. *Trans ASAE*, 32:451–456.

6. CoHort Software, 2005. *CoSat Statistical package* (version 6.311). P.O. Box 1149, Berkeley, CA, 94701, USA.

7. Dassanayake, D. K., Dassanayake, H., Malano, G. M., Dunn Douglas, P., and Langford, J., 2009. Water saving through intelligent irrigation in Australian dairy farming: use of intelligent irrigation controller and wireless sensor network. 18th World IMACS/MODSIM Congress, Cairns, Australia, pp. 4409–4417.

8. Doorenbos, J., and Pruitt, W. O., 1977. *Guidelines for predicting crop water requirements*. FAO Irrigation and Drainage Paper 24. FAO, Rome, Italy.

9. Du, T., Kang, S., Zhang, J., and Li, F., 2008. Water use and yield responses of cotton to alternate partial root-zone drip irrigation in the arid area of north-west China. *Irr. Sci.*, 26:147–159.

10. El-Tantawy, M. M., Ouda, S. A., and Khalil, F. A., 2007. Irrigation scheduling for maize grown under Middle Egypt conditions. *Res J of Agric and Bio Sci.*, 3(5):456–462.

11. English, M., Solomon. K., and Hoffman, G., 2002. A paradigm shift in irrigation management. *J. Irrig. Drain. E-ASCE*, pp. 267–277.

12. Faberio, C., Martin de Santa Olalla, F., De Juan, J. A., 2001. Yield and size of deficit irrigated potatoes. *Agric. Water Manage.*, 48:255–266.

13. Goyal, Megh R., 2012. *Management of Drip/Trickle or Micro Irrigation*. Apple Academic Press Inc.

14. Goyal, Megh R., 2015. *Research Advances in Sustainable Micro Irrigation*. Apple Academic Press Inc.,

15. Hoffman, G. J., Howell, T. A., and Solomon, K. H., 1990. *Management of farm irrigation system*. ASAE, St. Joseph, MI (USA). ISBN 0–929355–11, pages 1094.

16. Howell, T. A., Schneider, A. D., and Evett, S. R., 1997. Subsurface and surface micro irrigation of corn – Southern High Plains. *Trans ASAE*, 40:451–456.

17. Kamilov, B., Ibragimov, N., Esanbekov, Y., Evett, S. R., and Heng, L. K., 2003. Drip irrigated cotton: irrigation scheduling study by use of soil moisture neutron probe. *Int. Water Irrig.* 1:38–41.

18. Stegman, E. C., Fargo, N. D., Musick, J. T., Stewart, J. I., 1980. Irrigation water management. In: *Design and Operation of Farm Irrigation System* (ed. M. E. Jensen). ASAE, pages 763–816.

19. Lozano, D., and Mateos, L., 2007. Usefulness and limitations of decision support systems for improving irrigation scheme management. *J. Agric. Water Manage.*, 95(4):409–418.

20. McCready, M. S., Dukes, M. D., and Miller, G. L., 2009. Water conservation potential of intelligent irrigation controllers on St Augustine grass. *J Agric. Water Manage.*, 96:1623–1632.

21. Mendez-Barroso, L. A., Payan, J. G., Vivoni, E. R., 2008. Quantifying water stress on wheat using remote sensing in the Yaqui Valley, Sonora, Mexico. *Agric. Water Manage.*, 95(6):725–736.

22. Michael, A., 1978. *Irrigation and theory practice.* Vikas Pub. House Pvt. Ltd., New Delhi, India.

23. Michael, D., and Dukes, M. D., 2008. Water conservation potential of smart irrigation controllers. Paper No. 10–9520 at 5th National decennial irrigation conference proceedings, Phoenix convention center, Phoenix, AZ, USA. ASABE, 2950 Niles Road, St. Joseph, MI 49085.

24. Ministry of Agriculture (MOA), 2008. *Agriculture Statistical Year Book.* Agricultural Research and Development Affaires, Department of Studies Planning and Statistics.

25. Mofoke, A. L. E., Adewumi, J. K., Mudiare, O. J., and Ramalan, A. A., 2004. Design, construction and evaluation of an affordable continuous flow drip irrigation system. *J Appl. Irrig. Sci.*, 39(2):253–269.

26. Munoz-Carpena, R., and Dukes, M. D., 2005. Automatic irrigation based on soil moisture for vegetable crops. IFAS Extension Bulletin. University of Florida.

27. Nautiyal, M., Grabow, G., Miller, G., and Huffman, R. L., 2010. Evaluation of two smart irrigation technologies in Cary, North Carolina. ASABE Annual International Meeting. David L. Lawrence Convention Center, Pittsburgh, PA, June 20–23.

28. Nazirbay, I., Evett, S. R., Esanbeko, V. Y., Kamilov, B. S., Mirzaev, L., and Lamers, J. P. A., 2007. Water use efficiency of irrigated cotton in Uzbekistan under drip and furrow irrigation. *Agric. Water Manage.*, 90(1):112–120.

29. Norum, M. N., and Adhikari, D., 2009. Smart irrigation system controllers. At 7th World Congress on Computers in Agriculture Conference Proceedings, Reno, Nevada. ASABE, St. Joseph, Michigan.

30. Oktem, A., Simsek, M., and Oktem, A. G., 2003. Deficit irrigation effects on sweet corn (*Zea mays saccharata Sturt*) with drip irrigation system in a semi-arid region, I: water-yield relationship. *Agric. Water Manage.*, 61:63–74.

31. Qui, G. Y., Wang, L., He, X., Zhang, X., Chen, S., Chen, J., and Yang, Y., 2008. Water use efficiency and evapotranspiration of wheat and its response to irrigation regime in the North China Plain. *Agric. and Forest Meteo.*, 148:1848–1859.

32. Sammis, T. W., and Wu, I. P., 1986. Fresh market tomato yields as affected by deficit irrigation using a micro irrigation system. *Agric Water Manage.*, 12:117–126.

33. Smajstrla, A. G., and Locascio, S. J., 1997. Tensiometer controlled, drip scheduling of tomato. *Appl. Eng. Agric.*, 12(3):315–319.

34. Vermeiren, I., and Jobling, G. A., 1980. *Localized irrigation: design, installation, operation, evaluation.* FAO Irrigation and Drainage paper 36 FAO, Rome. 203 pp.

35. Waister, P. D., and Hudson, J. P., 1970. Effect of soil moisture regimes on leaf deficit, transpiration and yield of tomatoes. *J Hort Sci.*, 45:359–370.

36. Wan, S., and Kang, Y., 2006. Effect of drip irrigation frequency on radish (*Raphanus sativus L.*) growth and water use. *Irrig. Sci.*, 24:161–174.

37. Wanga, D., Kang, Y., and Wana, S., 2007. Effect of soil matric potential on tomato yield and water use under drip irrigation condition. *Agric. Water Manage.* 87:180–186.

38. Whittlesey, N., 2003. Improving irrigation efficiency though technological adoption: when will it conserve water? In: *Alsharhan, A.S. and Wood, W.W. (eds.) Water resources perspectives: evaluation, management and policy.* Elsevier Science, Amsterdam, pp 53–62.

39. Yazar, A., Howell, T. A., Dusek, D. A., and Copeland, K. S., 1999. Evaluation of crop water stress index for LEPA irrigated corn. *Irrig. Sci.* 18:171–180.

40. Yildirim, G., 2007. An assessment of hydraulic design of trickle laterals considering effect of minor losses. *Irrig. and Drain.*, 56(4):399–421.

41. Zazueta, F. S., Smajstrla, A. G., and Clark, G. A., 1994. *Irrigation system controllers.* Sheet SS-AGE-22. Department of Agricultural and Biological Engineering, Florida Cooperative Extension Service, Institute of Food and Agriculture sciences, University of Florida.

42. Zhu, D. L., Wu, P. T., Merkley, G. P., and Jin, D. J., 2009. Drip irrigation lateral design procedure based on emission uniformity and field microtopography. *Irrig. and Drain.*, Published online in Wiley Inter Science. doi: 10.1002/ird.518.

CHAPTER 9

PERFORMANCE EVALUATION OF DRIP IRRIGATION SYSTEMS IN ARID REGIONS

H. M. AL-GHOBARI and M. S. A. EL MARAZKY

CONTENTS

In this chapter: One U.S. dollar = 3.75 riyals (KSA currency).

9.1 INTRODUCTION

Water scarcity is an increasingly important issue in many parts of the world. All water users must share responsibilities to conserve water quantity and quality. Agricultural producers are facing decreasing water supplies and are becoming increasingly aware of the need to conserve limited water resources. One way to address these concerns is to utilize irrigation technologies with high application efficiency, such as drip irrigation systems DI. In the arid and semi-arid areas, DI is frequently used to reach the maximum water use efficiency [12]. DI has the potential to use scarce water resources most efficiently to produce vegetable and field crops [19]. However, DI is an irrigation system to supply water under low pressure directly near the plant roots [13, 14, 22].

Uniformity of water application is an important parameter in the design, maintenance and management of DI [13, 14]. The modeling of crop response to water application indicated that there was a dependent relationship between crop yield and irrigation uniformity and a more uniform application of water that leads to a better distribution of nutrients in the soil and consequently to a higher crop yield [16]. Several design and evaluation standards for drip irrigation uniformity have been developed in different countries [3]. ASAE Standards [2, 3] recommend a design emission uniformity (EU) of 70% to 95% depending on the point or line source, crop, emitter spacing, and field slope. Wu and Barragan [29] estimated optimal emitter flow uniformity and provided design criteria for DI based on the availability of water resources while considering environmental pollution and groundwater contamination.

The design of a DI can have a major impact on the initial cost because the cost increases with the level of uniformity [28]. The costs of DI installation and operation may be reduced if the systems are designed using a uniformity that is lower than the values recommended by the current standards. However, it remains unclear whether lower system uniformity will result in a decreased yield and quality. Therefore, the effects of irrigation uniformities on plant growth, yield and quality have been a crucial topic for several decades [17, 26, 31].

Crop production is a measure of irrigation uniformity and efficiency [26]. Therefore, crop yield should be more uniform than the water applied.

Li and Rao [18] demonstrated that the uniformities of winter wheat (*Triticum aestivum L.*) yield were higher than those for water application, and the yields seemed to be insensitive to spatial variation of applied water even though sprinkler uniformities varied from 57% to 89% during the irrigation season. Bordovsky and Porter [4] did not observe significant differences in cotton yield among subsurface drip irrigation treatments at flow variations (qvar) of 5%, 15% and 27%. To obtain the gross depth of application for satisfying crop water requirements, soil water uniformity should be considered rather than water application uniformity of individual irrigation events. A good approximation of soil water uniformity is the value corresponding to the set of irrigation events, at least when the irrigation interval is less than 3 or 4 days [20, 23]. In a semi-humid region, the effects of system uniformity on the mean yield and quality indexes of drip irrigated cabbage and their uniformities were insignificant [31].

The distribution uniformity of an irrigation system depends both on the system characteristics and managerial decisions [24]. Uniformity of water application in micro-irrigation system depends on system application uniformity and spatial uniformity in the field [29]. System uniformity is affected by design factors such as parallel diameter, emitter spacing [30], manufacturing variation [7] and emitter clogging [8]. The parameters used to evaluate micro-irrigation system application uniformity are coefficient of uniformity (Cu), emitter flow variation (qvar), and coefficient of variation (CV) of emitter flow [6]. The uniformity is also affected by the field topography as well as the hydraulic design parameters of DI such as energy losses in lateral lines and emitter characteristics [32]. These factors of an irrigation system must be correctly managed to ensure that the distribution uniformity is at an acceptable level. This will ensure the optimal use of water resources.

The causes of non-uniformity include unequal drainage and unequal application rates [9]. Overall, minimizing non-uniformity of the DI requires: a design, which considers the topography of the field [29], and irrigation scheduling (volume and frequency) [10]. Higher irrigation uniformity can be achieved by using pressure-compensating emitters in drip surface [25]. Some researchers found that the non-uniformity may reduce irrigation efficiency and crop yield, and thus adversely affects the economics of crop production [21].

In order to achieve this, the uniformity with which the irrigation system applies water will have to be high. The distribution uniformity of a system affects system application efficiency and crop yield [16, 17, 26]. The distribution uniformity of an irrigation system depends both on the system characteristics and on managerial decisions [24].

This chapter evaluates effects of field performance of drip irrigation system for three successive seasons on crop yield. This can contribute to achieve higher irrigation performance, when combined with improvements in irrigation methods and irrigation scheduling.

9.2 MATERIALS AND METHODS

9.2.1 SITE DESCRIPTION

This study was performed at the experimental farm of the College of Food and Agriculture Sciences of King Saud University, Riyadh (24°43' N latitude, 46°43' E longitude and 635 m elevation above msl) during three successive spring seasons from 2011 to 2013. The experimental site was irrigated by a drip irrigation system. The field was further divided into three plots. Before the start of the experiment, intact soil cores were collected from different locations in the field to determine soil physical properties. Locations were selected to represent the dominant soil conditions in the field. Three soil samples were taken from each plot at three different depths (0–20, 20–30 and 30–60 cm) to determine soil texture. The soil type in the plot area was sandy loam (70.82% sand, 13.94% silt and 15.24% clay). The site was provided with automatic weather station and installed at the experimental location to measure the climate parameters to compute reference evapotranspiration (ET_o). Generally, the climate in this region is classified as arid, and the climatological data measured at the experimental site during this study period are given in Table 9.1.

9.2.2 EXPERIMENTAL LAYOUT, IRRIGATION SETUP AND MANAGEMENT

The study area was 1200 m² (40×30 m), which was divided into three fields separated with buffer zone of 5 m wide. Drip irrigation system

TABLE 9.1   Average Metrological Data of the Experimental Site During Three Seasons

Month	Average three seasons from 2011–2013				
	Tmax (°C)	Tmin (°C)	RH (%)	Rainfall (mm)	SR (MJm^{-2}d^{-1})
January	28.9	9.7	39.61	0.00	13.00
February	33	7.3	27.18	0.00	17.4
March	37	9.8	24.18	0.01	19.3
April	38.6	17	35.6	0.04	20.13
May	43.5	18.5	24.14	1.15	23.00
June	45.4	25.9	9.58	0.00	25.00
July	46.5	27.3	10.36	0.00	30.68
August	47.2	26.6	11.1	0.00	23.00
September	46.2	21.4	13.8	0.00	20.74
October	40.7	17.3	17.44	0.00	18.23
November	35.6	10.8	51.81	0.08	14.60
December	28.9	9.7	39.61	0.00	13.00

Tmax = Maximum temperature, Tmin = Minimum temperature, RH = Relative humidity, SR = Solar radiation.

was installed in each field. Three fields were managed by an automatic weather station, which is based on reference evapotranspiration, (ET$_o$), utilizing the climatic data.

Controllers were used to control the pressure and flow to quantify the water added during each irrigation event. Each plot size was 7×10 (70 m^2), consisting of 7 drip lines per plot apart running from west to east. The 7 drip lines in each plot were connected to a common submain at the inlet side of each plot. A common flush line and flushing valve were provided at the distal end of the plot. The drip line consisted of 16 mm inside diameter (I.D.) thin-wall PE tubes with welded-on emitters (NETFIM USA Standards, 50 cm dripper spacing) with 20 drippers with a nominal emitter discharge of 3.5 L/h at a design pressure of 80 kPa. Irrigation amounts were metered separately in each plot using commercial municipal-grade flow meters. The experimental sequences steps of preparation completed the installation of the rest of irrigation network, such as valves, flow meter and pressure meters. The hydraulic aspects of the design for drip system were aimed at obtaining uniform application of irrigation water.

Tomato seedlings (*Nema tomato cv.*) were transplanted at plant spacing of 0.5 m on February 5, 7, 10, for each of three seasons from 2011–2013, respectively. Other cultivation practices were performed according to a scheduled tomato crop program.

Net irrigation requirements were computed using estimated long-term reference ET_0 and effective precipitation. Daily and weekly crop evapotranspiration (ET_c) rates for tomatoes during the growth period were determined based on the control treatments. The irrigation water depths (D_g) and accumulative depths for the tomato crop under the three treatments were monitored by flow meters and were recorded throughout the growing season.

Ripe tomatoes were manually picked up and weighed twice a week. Fruit yield and its components were evaluated for eight plants from the central rows in each plot. Other performance parameters, total fruit yield, were also calculated for each plot to obtain the gross yield (t.ha^{-1}).

Flow rates and manifold operating pressures were monitored during the three-year growing period. Drip system was evaluated in the field according to the methodology by ASABE Standards S346.1 [3]. Evaluation tests were conducted for drip irrigation system by checking values of the performance indexes under operating field conditions. All indices values were found to be within acceptable limits and with good water distribution uniformity.

9.2.3 SYSTEM MEASUREMENTS

New DI was purchased at the start of season 2011, and was evaluated in the field using the pressure regulating valves. Emitter discharge rate was determined by collecting the water from every emitter on a single, 10-m length of lateral (total of 20 emitters) for 15 minutes. This evaluation was repeated for two other laterals of the same length. Water volume was determined using a graduated cylinder. Water flow rate, pressure, and temperature were measured for each test. Each test was replicated three times for each of the three laterals of each field.

The field evaluation of the same drip irrigation system was carried out at the beginning of each season for three successive seasons from 2011 to 2013.

9.2.4   DETERMINATION OF UNIFORMITY OF DRIP IRRIGATION SYSTEM

Coefficient of Variation, C_V

$$C_V = \frac{S_d}{q_a} \tag{1}$$

where, s_d = standard deviation of emitters discharge, and q_a = average emitter discharge.

Emitter Flow Rate Variation, q_{var}

$$q_{var} = \left(1 - \frac{q_n}{q_m}\right) \times 100 \tag{2}$$

where, q_{var} = variation of the average flow rate, (%), q_n = average quarter for the discharge of emitters, and q_m = average maximum eighthly for discharge of emitters.

Uniformity Coefficient, Cu

Uniformity coefficient was defined by Christiansen [11] and is modified to give the value in percentage. The following equation by Keller and Karmeli [15] was used to compute the uniformity coefficient of the drip irrigation system:

$$Cu = \left(1 - \frac{\sum_{i=1}^{i=n} |q_i - q_a|}{n \cdot q_a}\right) \times 100 \tag{3}$$

where, n = represents number of emitters under evaluation, q_a = average discharge of emitters, and q_i = discharge for the i-th emitter.

Field Uniformity Coefficient, Eu_f

$$Eu_f = \frac{q_n}{q_a} \times 100 \tag{4}$$

where, q_n = Average quarter, at least for the discharge of emitters, q_a = Average discharge of emitter.

Design Uniformity Coefficient, Eu_d

$$Eu_d = \left(1 - \frac{1.27 \times C_v}{\sqrt{Np}}\right) \times \frac{q_n}{q_a} \times 100 \qquad (5)$$

where, Np = Number of drippers per tree, C_v = manufacturer's coefficient of variation, q_n = Average quarter for the discharge of emitters, and q_a = Average discharge of emitters.

Coefficient of Absolutely Uniformity, Eu_a

$$Eu_a = 0.5 \left(\frac{q_n}{q_a} + \frac{q_a}{q_m}\right) \times 100 \qquad (6)$$

where, q_n = average quarter for the discharge of emitters, q_a = average discharge of emitters, and q_m = average maximum eighthly for discharge of emitters.

9.3 RESULTS AND DISCUSSION

9.3.1 FIELD EVALUATION

Newly installed DI system at the start of 2011 season was evaluated in the field using pressure-regulating and flow metering valves. Performance parameters were calculated to develop the relationships between the operating pressure and the coefficient of variation (C_v), emitter flow rate variation (q_{var}), uniformity coefficient (Cu), field coefficient uniformity (E_{uf}), design coefficient uniformity (Eu_d) and coefficient of absolute uniformity (Eu_a).

Results indicate that performance parameters measured at 75 kPa for all emitters were almost close to the design flow rate given by the manufacturer. The hydraulic characteristics of an irrigation system eventually translate to observable system performance parameters, which are of more direct implication to crop growth and yield. The performance of an irrigation system is represented by the measured level of achievement in terms of one or several parameters chosen as indicators of the system. The ultimate goal of the system is to provide relatively high water application uniformities and thus provide the potential for high irrigation efficiency.

The uniformity data show that the operating appropriate pressure for drip irrigation system is one bar with a uniformity coefficient of 90.33%. This implies that the systems are designed adequately. The emitter discharge was increased linearly by increasing the operating pressure. The characteristics of all tested emitters by the manufacturer were influenced by the operating pressure. Also, results indicated that measured C_V values of emitters at the recommended operating pressure (50 to 200 kPa) were close to the design C_V as claimed by the manufacturer.

The manufacturer's coefficient of variation (C_V) is a function of the emitter type and the quality control exercised during the manufacturing process. The Figure 9.1 shows the coefficient of variation (C_V), during the tomato growing from 2011 to 2013. The emitters with q = 3.5 L/h were for the three laterals combined (n = 60 emitters). The C_V values ranged from 0.09, 015 and 0.22 with a mean value of 0.15 and R^2 = 0.998, in each season. This indicates the desirability of larger sample sizes and the existence of high variation among laterals than among emitters within a single

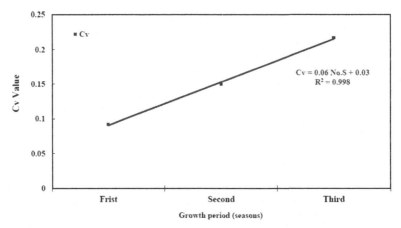

FIGURE 9.1   The coefficient of variation, *CV*, during the tomato growing seasons of 2011–2013.

lateral. Finally, the emitter performance coefficient of variation, C_V during three seasons, shows that the differences were caused by factors other than hydraulic design characteristics (e.g., emitter plugging). The coefficients of variation of flow rates were based on the ASAE [3] classification. The low C_V indicated a good performance of the system throughout the cali-brated period (Figure 9.1). This implied that the results were in agreement with those by Bralts [7, 8]. The ASAE [2] has established C_V range for line-source drip lines: A C_V of less than 10% is considered good, from 10 to 20%, average, and greater than 20%, marginal to unacceptable. The C_V of an emitter should be obtained from the manufacturer to aid in decision regarding suitability of the product for a particular installation.

The Figure 9.2 shows the emitter flow rate variation at an operating pressure of 1.00 bars. The q_{var} values were 22.54%, 32.01% and 42.53% with a mean value of 32.36% and $R^2 = 0.999$, in each of the growing season. Finally, the performance coefficient of emitter flow rate variation value indicates that the drip irrigation system had a higher variation com-pared to that by the manufacturer. It can be seen that emitter flow rate variation was different in each of three seasons.

System operating parameters, including system flow rate and manifold operating pressure, were monitored during the three-year operation period. Figure 9.3 shows performance evaluation of parameters q_a (average emitter

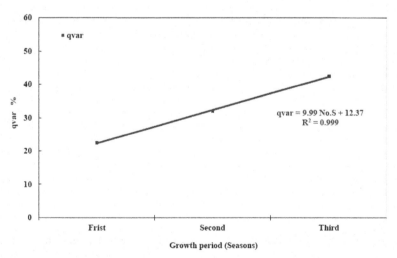

FIGURE 9.2   The coefficient of variation, q_{var}, for three seasons, 2011 to 2013.

discharge), q_n (average quarter for the emitter discharge) and q_m (average maximum eighthly for emitter discharge) for three growing seasons from 2011–2013. It can be observed that the field variation of emitter flow rate, q_{var}, is influenced by different factors q_a, q_n and q_m, which were probably introduced into the system during construction and/or repair operations.

Values of emitter flow rate variations (q_{var}) are generally higher at the beginning of the first season and the end of the experiment until the third season. Discharge variations in drip system arrangement can be attributed to differential clogging of emitters, problems in punching holes that are of the same size, and pressure head differences among others. Higher values of discharge variations (qvar) signify that clogging and variability in punching uniform orifice outlets are the strongest factors affecting the performance.

It is obvious from the Figure 9.3 that the average discharge emitters (q_a) fluctuated at least gradually from the first season to the third season and this value was decreased slightly. Also, average quarter for the discharge of emitters (qn) varied between 3.12 and 2.60 for the emitters in this study. This value represents a 17% decrease between the first and the third seasons. Average maximum eighthly for discharge of emitters (q_m) were 4.03, 4.20 and 4.53 for each of three seasons. There is an increase of 12% in the third season compared to the first season. These emitter

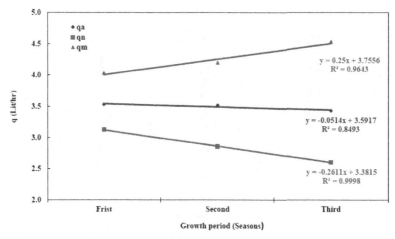

FIGURE 9.3   The different factors, q_a, q_n and q_m, that affect the variation in emitter flow rate throughout the growing period for each of three seasons.

discharge coefficients of variation (q_a, q_n and q_m) are due to hydraulics of the surface drip system indicating fewer differences among the three seasons for this parameter.

The general variation in emitter discharge can also be attributed to major and minor losses, which was suggested by Zhu et al. [32] due to water temperature and pressure variation in joints and fittings, from the tank to the emitters.

9.3.2 EFFECTS OF UNIFORMITY PARAMETERS ON TOMATO YIELD

The variation in uniformity parameters (C_u, E_{uf}, E_{ud} and E_{ua}) can be observed in Figure 9.4. Prior to installation of laterals, the emitter flow rates of the uniformity coefficient (C_u) were measured by collection of water in cans at the start of 2011 and end of 2013)] to confirm that the C_u values of the assembled laterals were comparable with the design values. From Figure 9.4, it is clear that the uniformity coefficients were 92.41%, 86.26% and 81.20% for drip irrigation in each of three different years of study. The distributions were classified as "excellent", "good" or almost

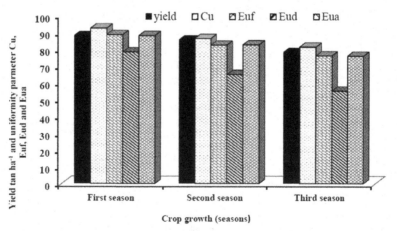

FIGURE 9.4 Variation in mean yield of tomato and uniformity parameters (Cu, E_{uf}, E_{ud} and E_{ua}) for each of three seasons.

excellent and "poor" according to the classification by Bralts [5]. These values of uniformity coefficient were classified as good according to the ASAE Standard EP458 [2], which indicated that the field received an appreciable amount of water distributed throughout the irrigated area suggesting a uniform plant development.

It can be inferred that the evaluation of uniformity of water distribution of drip irrigation is crucial for proper planning and operation of the system [4].

Similarly, coefficient of field uniformity (Eu_f) was 88.55, 82.33 and 75.95% for each of three seasons, respectively. These values of coefficient uniformity field, Eu_f were fair and good according to ASAE [2]; and acceptable and good according to Keller and Karmeli [15]. The results showed a relatively uniform water application, which would have a positive impact on the crop production. Ortega [23] calculated the average E_U as 84%, which is almost similar to results in this chapter (82.28%). Also, the coefficient field uniformity (Eu_f) was decreased over time from year to year.

The coefficient of design uniformity Eu_d was 78.22%, 64.83% and 55.04% for each of three seasons, respectively. It is obvious from Figure 9.4 that the coefficient of design uniformity Eu_d was decreased at the end of the experimental period of three years. These results may be due to a poor design and management of drip irrigation system with considerable pressure variation, and clogged emitters mainly affecting uniform water distribution in study areas. Therefore, after carefully designing and installing drip systems, service and maintenance and repair operations should be carefully performed to obtain uniform water distribution throughout the field.

Correspondingly, coefficient of absolute uniformity E_{ua} was 88.01%, 82.49% and 75.84% for three years, respectively. Coefficient of absolute uniformity varied from high value to lower value with an average of 82.11%. It is clear that the coefficient of absolute uniformity Eu_a was decreased at the end of three years. The variations were due to different factors such as topography and length of laterals. It is, therefore, difficult to precisely characterize the pattern of uniformity without a large number of measurements in drip systems. According to minimum and maximum values, Eu_a was classified as poor and excellent, respectively [27].

Similar studies were conducted in semi- arid regions of Turkey [1]. It was found that Eu_a was 75–85% and 31–74% [1]. The results in this chapter are in line with these two studies.

Figure 9.4 shows tomato yield as affected by uniformity coefficients during 2011 to 2013. Tomato yield for first year was better than second and third years. The field uniformity coefficient was decreased at the end of three years. Yields were significant differences at 5% level among the three years. Total yield values of 88.23, 85.71 and 78.36 tons ha^{-1} were obtained during 2011, 2012 and 2013, respectively.

Eventually, uniformity of water application plays an important role in crop growth and can have a significant economic impact. The yield differences among the three years might be ascribed to seasonal variations in performance parameters and due to less water with non-uniformity (Figure 9.4). The hydraulics finally translates to observable system performance parameters, which are of more direct implication to crop growth and yield. The performance of an irrigation system is represented by the measured level of achievement in terms of one or several parameters chosen as indicators. The goal is to provide relatively high water application uniformities, and thus provide the potential for high irrigation efficiency. This result is in close agreement with previous studies, which indicate that the non-uniformity may reduce irrigation efficiency and crop yield, and thus adversely affect the economics of crop production [21].

9.4 CONCLUSIONS

Field experiments were conducted in the arid environments of KSA to evaluate the effects of uniformity parameters of drip irrigation on tomato growth. The variations in parameters can be attributed to spatial and environmental conditions. However, the values of the indices are higher than 90%, which indicate a good performance and high efficiency of the system under local conditions. Generally, it can be concluded that the uniformity parameters of water application for the drip irrigation system (C_v, q_{var}, Cu, Eu_p, Eu_d and Eu_a) in each of three years were higher at the beginning of the study compared to the values at the end of 2013 season.

The spatial variations of total yield under non-uniform water application were small and insensitive to the spatial variation of water application. This conclusion is significant for the design of drip surface irrigation system. The surface system uniformity and measured emitter uniformity values were generally slightly different for those of unused tubing of the same age. The uniformity parameters evaluated classified the irrigation set as good for drip irrigation surface. Finally, drip irrigation uniformities might be used if sufficient surface drip irrigation methods events are conducted to approach a uniform distribution of water in arid region. Hence, in order to conserve water resources, close attention must be paid to the performance of surface drip irrigation systems. The results of the study show that more attention needs to be given towards improving managerial skills for a drip irrigation system.

9.5 SUMMARY

There were significant differences in tomato yield among the three years. Total yield values were 88.23, 85.71 and 78.36 tons.ha^{-1} for each of three years, respectively. Generally, it can be concluded that the uniformity parameters of water application for the drip irrigation system (C_V, q_{var}, Cu, Eu_p, Eu_d and Eu_a) for all three years were higher at the beginning of the study compared to those at the end of third season. This discharge variations (q_{var}) value represents a 17% decrease between the first and the third season. Uniformity coefficients Cu were 92.41%, 86.26% and 81.20% for DI in three different years studies, respectively. Increasing uniformity of water (Cu) corresponds to 12% increase in yield for the common irrigation strategy in the three years. The results showed relatively uniform water application, which had a positive impact on crop production. Eventually, uniformity plays an important role in crop growth and can have a significant economic impact.

ACKNOWLEDGMENT

This project was supported by NSTIP Strategic Technologies Programs (No. 11-AGR1476–02) in the Kingdom of Saudi Arabia.

KEYWORDS

- *ASAE*
- coefficient of variation
- crop yield
- discharge variation
- drip irrigation
- emitter
- emitter flow
- field evaluation
- field uniformity
- flow rate
- Kingdom of Saudi Arabia
- manufacturing variation
- operating pressure
- pressurized irrigation
- subsurface drip irrigation
- tomato
- Turkey
- uniformity coefficient
- uniformity parameters
- water application

REFERENCES

1. Acar, B., Topak, R., Direk, M., 2010. Impacts of pressurized irrigation technologies on efficient water resources uses in semi-arid climate of Konya Basin of Turkey. *Int. J. of Sustainable Water & Environmental Systems*, 1(1):1–4.
2. ASAE, 2001. Field evaluation of micro irrigation Systems. In: *ASAE Standard engineering practices data: EP 458*. Pages 792–779. American Society of Agricultural Engineers. St. Joseph, MI.
3. ASAE, 2003. EP405.1. *Design and Installation of Micro irrigation Systems*. 50th ed. ASAE, St. Joseph, MI.

4. Bordovsky, J. P., Porter, D. O., 2008. Effect of subsurface drip irrigation system uniformity on cotton production in the Texas high plains. *Applied Engineering in Agriculture*, 24:465–472.
5. Bralts, V. F., Edwards, D. M., and Kesner, C. D., 1985. Field evaluation of drip/ trickle irrigation submain units. *Third International Drip/Trickle Irrigation Congress*, Fresno-California, U. S. A., pp. 274–280.
6. Bralts, V. F., and Kesner, C. D., 1983. Drip irrigation field uniformity estimation. *Transactions of the ASAE*, 26(5):1369–1374.
7. Bralts, V. F., Wu, I. P., and Gitlin, H. M., 1981. Manufacturing variation and drip irrigation uniformity. *Transactions of the ASAE*, 24(1):113–119.
8. Bralts, V. F., Wu, I .P., and Gitlin, H. M., 1981. Drip irrigation uniformity considering emitter plugging. *Transactions of the ASAE*, 24(5):1234–1240.
9. Burt, C. M., 2004. Rapid field evaluation of drip and microspray distribution uniformity. *Irrigation and Drainage Eng.,* 18:275–297.
10. Burt, C. M., Clemmens, A. J., Strelkoff, T. S., Solomon, K. H., Bliesner, R. D., Hardy, L. A., Howell, T. A., and Eisenhauer, D. E., 1997. Irrigation performance measures: efficiency and uniformity. *J. of Irri. and Drain. Eng.* 123(6):423–442.
11. Christiansen, J. E., 1942. *Irrigation by Sprinkling*. California Agriculture Experiment Station Bulletin No. 670. University of California, Berkeley, USA.
12. Fabeiro, C., de Santa Olalla, F., and Martin de Juan, J. A., 2002. Production of muskmelon (*Cucumis mela L.*) under controlled deficit irrigation in a semi-arid climate, *Agricultural Water Management*, 54:93–105.
13. Goyal, Megh R., 2012. Management of Drip/Trickle or Micro Irrigation. Apple Academic Press Inc.,
14. Goyal, Megh R., 2015. Research Advances in Sustainable Micro Irrigation. Apple Academic Press Inc.,
15. Keller, J., and Karmeli, D., 1974. Trickle irrigation design parameters. *Transaction of ASAE*, 7:687–684.
16. Letey, J., 1985. Irrigation uniformity as related to optimum crop production – additional research is needed. *Irrig. Sci.,* 6:253–263.
17. Letey, J., Vaux, H. J., and Feinerman, E., 1984. Optimum crop water application as affected uniformity of water infiltration. *Agron. J.,* 76:435–441.
18. Li, J., and Rao, M., 2000. Effects of sprinkler uniformity on spatial variability of soil moisture and winter wheat yield. *Journal of Hydraulic Engineering,* 1:9–14.
19. Locascio, J. S., 2005. Management of irrigation for vegetables: past, present, future. *HortTechnology,* 15(3):482–485.
20. López-Mata, E., J. M. Tarjuelo, J. A. de Juan, R. Ballesteros, and A. Domínguez, 2010. Effect of irrigation uniformity on the profitability of crops. *Agricultural Water Management,* 98:190–198.
21. Mahmoud, Y., McCarty, T. R., and Ewiny, L. K., 1994. Optimum center pivot irrigation system design with tillage effects. *J. Irrig. And Drain. Eng., ASCE*, 118(2):291–305.
22. Nautiyal, M., Grabow, G., Miller, G., and Huffman, R. L., 2010. Evaluation of two smart irrigation technologies in Cary, North Carolina. ASABE Annual International Meeting. David L. Lawrence Convention Center. Pittsburgh, Pennsylvania. June 20–23.

23. Ortega, J. F., Tarjuelo, J. M., de Juan, J. A., and Carrión, P. 2004. Distribution uniformity and its economics effects in the irrigation management of semi-arid zones. *J. Irrig. Drain. Eng.*, 130:257–268.

24. Pereira, L. S., 1999. Higher performance through combined improvements in irrigation methods and scheduling: a discussion. *Agric. Water Manage.*, 40(2):153–169.

25. Schwankl, L. J., and Hanson, B. R. 2007. Surface drip irrigation. In: *Microirrigation for crop production*, Eds. Lamm, F. R., Ayars, J. E., Nakayama, F. S. Elsevier, pages 431–472.

26. Solomon, K. H., 1984. Yield related interpretations of irrigation uniformity and efficiency measures. *Irrigation Science*, 5:161–172.

27. Tarjuelo, J. M., Ortega J. F., and Montero Juan, J. A., 2000. Modeling evaporation and drift losses in irrigation with medium size impact sprinklers under semi-arid conditions. *Agric. Water Manag.*, 43:263–284.

28. Wilde, C., Johnson, J., and Bordovsky, J. P., 2009. Economic analysis of subsurface drip irrigation system uniformity. *Applied Engineering in Agriculture*, 25:357–361.

29. Wu, I. P., Barragan, J., and Bralts, V. F., 2007. Field performance and evaluation. In: *Microirrigation for crop production design, operation and management*, Eds. Lamm, F. R., et al., F.S. Elsevier, pages 357–387.

30. Wu, I. P., and Barragan, J., 2000. Design criteria for microirrigation systems. *Transactions of the ASAE*, 43:1145–1154.

31. Zhao, W., Li, J., Li, Y., and Yin, J., 2012. Effects of drip system uniformity on yield and quality of Chinese cabbage heads. *Agricultural Water Management*, 110:118–128.

32. Zhu, D. L., Wu, P. T., Merkley, G. P., and Jin, D. J., 2009. Drip irrigation lateral design procedure based on emission uniformity and field microtopography. *Irrig. and Drain.*, Published online in Wiley Inter Science.

CHAPTER 10

ENHANCING IRRIGATION EFFICIENCY IN SAUDI ARABIA

KHODRAN H. ALZAHRANI, SIDDIG E. MUNEER, ALLA S. TAHA, and MIRZA B. BAIG

CONTENTS

In this chapter: One U.S. dollar = 3.75 riyals (KSA currency).

Modified from: "*Khodran H. Alzahrani, Siddig E. Muneer, Alla S. Taha and Mirza B. Baig, 2012. Appropriate cropping pattern as an approach to enhancing irrigation water efficiency in the KSA. The Journal of Animal & Plant Sciences, 22(1):224–232*".

10.1   INTRODUCTION

Water is becoming scarce not only in arid areas but also throughout the world. Scarcity of fresh water resources represents one of the major challenges facing the world in general and Kingdom of Saudi Arabia (KSA) in particular. Population growth, high living standards as well as development plans in KSA will cause ever-increasing demands for good quality water in the municipal and industrial sectors. At the same time, more irrigation water will also be needed to meet the increasing food and fiber needs of the growing population; and for environmental use such as aquatic life, wildlife, recreation, and scenic values. Thus, increased competition for water can be expected in future, requiring efficient water demand management (WDM). Under these circumstances, improved management of water resources is the key for future sustainable development in the KSA. Where sustainable development and improvement of standards of living require urgent reduction of dependence on non-renewable water resources, leaving these as a strategic reserve for drinking and household uses in the first place [17, 18].

Unlike water resource rich countries pursuing a policy of water supply management, countries with scarce water resources need to pursue a water demand management policy. Water demand management (WDM) includes any measures/initiatives that will result in reduction of water usage/demand based on taking action and necessary incentives to achieve efficient water use and increase awareness about water scarcity and the limited nature of water resources [21, 27]. This can help to reduce water demand up to 30–50% without any deterioration in the standard of living [27].

The agricultural sector in the KSA uses about 85% of freshwater total consumption. The increase in demand for freshwater will increase the competition between municipal, industrial and agricultural sector. It often ends up in reduction of share in the agricultural sector. This phenomenon is expected to continue with lesser freshwater available for agricultural use, intensifying in less developed arid region countries that already suffer from water, food, and health problems. Consequently, irrigated agriculture despite being the largest water – consuming sector faces challenge to produce more food with less water.

Poor water management in the agricultural sector is the most frequent cause of inefficient water use in irrigation projects [8, 11, 22]. This necessitates that water management in the agriculture sector must be coordinated with, and integrated into, the overall water management in the water – starved countries. Irrigation efficiency is affected by several factors including water availability, accuracy of the design of the irrigation system, method of irrigation, soil type and properties, and good field management [8, 9]. There are wide ranges of options available for improving irrigation water use efficiency and productivity at the farm level [10, 19].

Water demand management (WDM) in the agricultural sector under conditions of water scarcity includes practices and management decisions of multiple natures such as: agronomic, economic, and technical. Its objectives are reduction of irrigation requirements, the adoption of practices leading to water conservation and savings in irrigation, reducing the demand for water at the farm, and increasing yields and income per unit of water used [12]. Virtual water, i.e. importing commodities having high water requirements and focusing production on other commodities that require less water, is considered a promising option for WDM in the agricultural sector [13].

During the past four decades water consumption in KSA has increased about five folds with the agricultural sector being the main consumer (86.5%) of the total water consumption. Non-renewable water is the principal source of water supply, which is a strategic stock and will be depleted if not handled with great care and used optimally. By 1986, Saudi Arabia's irrigation projects had begun to produce sufficient wheat for most of its needs, and it started to export wheat to the world market; this is argued to be use of non-renewable fossil water, as production of cereals and wheat requires large amounts of water [7]. This is an example of the lack of allocative efficiency for high-yielding crops, low technical efficiency as well as not exploiting the trade in virtual water.

In view of the foregoing, this chapter identifies the most important aspects of inefficiency in water consumption at the national level; measures for improving water use efficiency at the farm level; and visualize a model for WDM in the agricultural sector in Kingdom of Saudi Arabia.

10.1.1 WATER USE EFFICIENCY

Water use efficiency (WUE) can be improved through two approaches: (i) technical or productive efficiency, and (ii) allocative or economic efficiency [4]. The term WUE captures what farmers, industries, services, communities, national water departments and national governments have to consider for achieving improvements in water use. The technical efficiency can be achieved by using more efficient technologies such as drip and sprinkler irrigation instead of traditional flood irrigation and by more efficient irrigation scheduling [9]. Allocative efficiency is based on the principal that which activity brings the best return to water?, and is relevant at the farm level when the return to a high value crop such as fruit and out-of-season vegetables for an international or local market would be much more than to a crop such as wheat or rice. In services and industry economic returns per cubic meter of water can be thousands times more than that of agriculture.

Another WDM tool is via trade in '*Virtual Water*' [6]. Water deficit areas can minimize the water use by importing commodities that take a lot of water to produce like food and electric power from other areas or countries that have more water. The receiving areas then are not only getting the commodities, but also the water that is necessary to produce them. Since this water is 'virtually' embedded in the commodity, it is called virtual water [5]. It is probably the easiest way to achieve peaceful solutions to water conflicts.

10.1.1.1 Creating an Enabling Environment for Improved WUE

The continued depletion of water resources in many parts of the world is a clear indication that options for improving WUE are not being adopted. This is largely because enabling policies are missing. Most, if not all, of the policy measures used to support agriculture currently act as powerful dis-incentives against sustainability [20]. For example, Saudi Arabia during the last three decades produced sufficient wheat for most of its needs, and started to export wheat to the world market. This was a result of a policy of direct support to producers of wheat.

10.1.1.2   Improved WUE at Farm Level

Increasing water productivity is an obvious way to reduce stresses that come from overuse of water in agriculture [15]. Technical efficiency includes improving efficiency by using less water. In developing countries, up to 30% of fresh water supplies are lost due to leakage, and in some major cities losses can run as high as 40% to 70%. In KSA, leakage losses are about 35% [25]. Improving irrigation efficiency means how to take advantage of available irrigation water effectively.

Estimates revealed that about 30% of water is wasted in storage and conveyance, about 44% of the total water available at the source is lost as runoff and drainage, and 13–18% of the initial water resource is lost through transpiration by crops in irrigated agriculture [23]. Therefore, improving the efficiency of water use through water use efficiency, water conveyance efficiency, and water application efficiency is the key solution to reduce water demand in the agricultural sector.

There are wide ranges of options available for improving irrigation WUE and productivity at the farm level (Table 10.1). Poor management is cited as the most frequent cause of inefficient water use on irrigation projects [22]. Indeed, it is clear that few of the options listed in Table 10.1 will result in a significant increase in efficiency if the overall management is poor.

Many of the water demand management options (Table 10.1) are available to diffusion and adoption by farmers. For example, demand-based on irrigation scheduling, deficit irrigation and use of treated municipal wastewater for irrigation can be used to reduce water demand for irrigation.

10.1.2   IRRIGATION SCHEDULING AND METHODS

Irrigation scheduling includes adding irrigation water to the plant on scientific basis that takes into account several factors affecting plant water requirement. It is recognized that the adoption of appropriate irrigation scheduling practices can lead to increased yield and greater profit for farmers, significant water savings, reduced environmental impacts of irrigation and improved sustainability of irrigated agriculture [24].

TABLE 10.1   Examples of Available Options for Improving Irrigation Efficiency at the Farm Level [19, 26]

Water Use Efficiency	
1. Improved crop husbandry.	4. Introduction of higher yielding varieties.
2. Adoption of cropping strategies that maximize cropped area during periods of low potential evaporation and periods of high rainfall.	5. Laser leveling of flood irrigation schemes to improve irrigation uniformity.
3. Introduction of more efficient irrigation methods, such as drip/sprinkler irrigation.	

Water Conveyance Efficiency	
1. Improve uniformity and reduce drainage.	4. Adoption of demand-based irrigation scheduling systems.
2. Improved maintenance of equipment.	5. Use of deficit scheduling.
3. Better use and management of saline and waste water.	6. Introduction of efficient irrigation methods: Drip and sprinkler irrigation methods.

Water Application Efficiency	
1. Improve the physical properties of soil, supplement soil by natural conditioners, industrial conditioners.	3. Improved agricultural practices i.e. choosing appropriate sowing date, selection of appropriate agricultural density, good and proper fertilization, weed and pest etc.
2. Expansion in the use of greenhouses.	

The use of different WDM tools depends on soil factors, the nature of the plant and most important are the climatic factor. Alderfasi, et al. [2, 3] indicated that there was no significant difference in the yield per hectare of wheat in the Central Region of Saudi Arabia, when the crop was irrigated one or two times per week. This will result in a significant reduction in the amount of water needed to irrigate wheat.

Experiments also showed that tomatoes and cucumbers can be produced in greenhouses with 30% and 40% less water requirements, respectively without significantly reducing the productivity of these two crops [1].

Much research has focused on improving wheat productivity under drought conditions either by selection of drought-resistance varieties or by improvement of irrigation water management. The use of treated municipal waste water (TMWW) has been encouraged in Saudi Arabia during the last decade to increase the efficient use of irrigation water in crop production. For this reason, field experiments were conducted during the 1994–1995 and 1995–1996 seasons at the Agricultural Experimental Station in Dirab, Riyadh on a calcareous soil. The study concluded that the use of treated municipal waste water (TMWW) as a source of irrigation water led to a significant increase in crop productivity and stability especially in water-limited environments.

Irrigation methods such as drip irrigation can be used effectively to increase the yield, crop quality and WUE of many crops. Further, genetic engineering techniques and selection of the drought resistant genotypes can be used to reduce the amount of irrigation water. It is also possible to increase WUE by 25–40% through modifying practices that involve tillage and by 15 to 25% through soil nutrient management [14].

10.2 METHODOLOGY

Data about the irrigation methods used, irrigated areas and total production of different crops in the different regions in 2009 was obtained from the agricultural statistical year book which is published by Ministry of Agriculture [16]. Productivity of the different crop groups in the different regions was calculated. The least significant difference test (LSD) was used to examine whether the difference in the productivity of the different crop groups in the different regions is statistically significant.

10.3 RESULTS AND DISCUSSION

Since the scarcest agricultural resource in Saudi Arabia is water, the cropping pattern should be based on maximizing return per unit of water. This will enhance both food and water security. Table 10.2 reveals that the maximum return per unit of water is obtained from vegetables, followed by dates and lastly wheat. Thus to enhance water security through water demand management in the long-run, agricultural development plans should emphasize expansion of vegetable and fruit production and make use of 'Virtual Water' [6] by importing cereals for human and animals uses.

TABLE 10.2   Estimated Returns for Selected Crops Per Unit of Water, KSA [16]

Crop	Rate of water consumption	Average productivity	Producer price	Returns on water
	m³/ha/year	ton/ha	SAR/ton	SAR per 1000 m³ of water use
Dates	26000	6.25	6100	1466
Vegetables (tomatoes, onions, potatoes)	15000	20.5	1500	2050
Wheat	8000	4.88	1000	610

SAR = Saudi Arabian Riyal; 1.00 SAR = 0.2664 US$ on November 4, 2014.

On the other hand, in the short-run and under the existing cropping patterns, water security can be enhanced by benefiting from the comparative advantage of the different areas in producing the crops that give the highest yield. For example, the cereals crops give very high productivity in Jouf, Tabouk and Hail regions amounting to 6.81, 6.40 and 6.26 tons/ha, respectively compared to the other regions (Table 10.3). However, it was 4.86 tons/ha in Riyadh region, which ranks third in comparative advantage in cereal production. Table 10.3 indicates that there is no significant difference in cereals productivity in Jouf and Tabuk regions and Tabuk and Hail regions; and the difference between Jouf and Hail is modest at 0.05 level of significance. On the other hand, cereal productivity in these three regions is significantly higher at $P = 0.05$ than in any of the other regions (Table 10.3). Ironically under the current cropping patterns, Jazan region, which has the lowest productivity of cereals (2.2 tons/ha), has the largest area cultivated with cereals (i.e., 27.03% of the cereals total area) as shown in Table 10.8. On the other hand, Tabuk region, which ranks second in term of cereal productivity, ranks seventh in terms of cereal area with only a share of 5.53% of the cereal total area. These figures clearly indicate the potential of enhancing water and/or food security by encouraging cereal production in Jouf, Tabuk and Hail regions.

It worth mentioning that the Eastern region which ranks 1st in terms of vegetable productivity and Asseer region which ranks 3rd, come in the 6th and 11th rank in terms of cereals productivity respectively, (Tables 10.3 and 10.4). Under the current cropping system, the

TABLE 10.3 Average Productivity of Various Crop Groups in Different Regions of KSA (ton/ha)

Cereals		Vegetables		Fodders		Fruits	
Region	**ton/ha**	**Region**	**ton/ha**	**Region**	**ton/ha**	**Region**	**ton/ha**
Jouf	6.81	Eastern	39.79	Jouf	18.71	Eastern	12.24
Tabuk	6.40	Tabuk	28.08	Tabuk	18.55	Tabuk	9.41
Hail	6.26	Aseer	27.72	Hail	18.46	Najran	9.19
Riyadh	4.86	Hail	26.81	Qaseem	17.09	Baha	8.28
Qaseem	4.76	Qaseem	25.12	Riyadh	16.94	Makkah	8.20
Eastern	4.33	Jouf	24.47	Eastern	16.34	Aseer	8.19
Northern	4.30	Najran	24.21	Baha	16.12	Hail	6.99
Madenah	3.91	Baha	24.07	Jazan	15.78	Jazan	6.59
Baha	3.27	Riyadh	21.69	Aseer	15.60	Madenah	6.33
Najran	3.25	Madenah	18.02	Madenah	15.40	Riyadh	6.15
Aseer	2.80	Makkah	16.78	Najran	15.03	Northern	6.10
Makkah	2.27	Jazan	13.90	Northern	12.63	Jouf	5.44
Jazan	2.20	Northern	13.48	Makkah	11.23	Qaseem	5.15

Eastern region which gives the highest productivity of vegetables, it ranks 5th in term of vegetables cultivated area with only 6.24% of the vegetables total area and Asser which ranks 3rd has only 2.57% of the vegetables total area (Table 10.8). On the other hand, Riyadh region which ranks 9th in terms of vegetable productivity (Table 10.5) ranks 1st in terms of vegetables cultivated area (45.62%) of the vegetables total area (Table 10.8).

As depicted in Table 10.3, the top five regions that give the highest productivity of fodder crops are Jouf, Tabuk, Hail, Quasseem and Riyadh. The least significant difference test (LSD) has indicated no statistically significant difference in fodder crops productivity between any of these five regions (Table 10.6). Apart from Riyadh region, the other four regions rank top in productivity of cereal and vegetable crops. Thus, it is intuitive that Riyadh is the region where fodder production should be concentrated. Since fodder crops have high water requirements and are not perishable, its domestic production should be limited and instead be imported to make use of virtual water. Unlike other crops, fodder production under the current cropping pattern seems to be proper where Riyadh region which appears to

TABLE 10.4 Differences Between Mean Productivity of Cereals in Different Regions (ton/hectare)

Region	Jouf	Tabuk	Hail	Riyadh	Qaseem	Eastern	Northern	Madenah	Baha	Najran	Aseer	Makkah
Jouf												
Tabuk	0.411											
Hail	0.547*	0.136										
Riyadh	1.988***	1.577***	1.44***									
Qaseem	2.041***	1.63***	1.494***	0.052								
Eastern	2.478***	2.067***	1.931***	0.49**	0.437							
Northern	2.507***	2.095***	1.96***	0.518**	0.465	0.028						
Madenah	2.902***	2.491***	2.355***	0.914***	0.861**	0.424	0.395					
Baha	3.537***	3.125***	2.99***	1.548***	1.495***	1.058***	1.03***	0.634**				
Najran	3.565***	3.154***	3.018***	1.577***	1.524***	1.087***	1.058***	0.662**	0.028			
Aseer	4.005***	3.594***	3.4558***	2.017***	1.964***	1.527***	1.498***	1.102***	0.468	0.44		
Makkah	4.538***	4.127***	3.991***	2.55***	2.497***	2.06***	2.031***	1.635***	1.001***	0.972***	0.532**	
Jazan	4.605***	4.194***	4.058***	2.617***	2.564***	2.127***	2.098***	1.702***	1.068***	1.04***	0.6*	0.067

*sig. at 0.05; **sig. at 0.01; ***sig. at 0.001.

TABLE 10.5 Differences Between Mean Productivity of Vegetables in Different Regions (ton/hectare)

Region	Eastern	Tabuk	Aseer	Hail	Qaseem	Jouf	Najran	Baha	Riyadh	Madenah	Makkah	Jazan
Eastern												
Tabuk	11.711***											
Aseer	12.072***	0.361										
Hail	12.977***	1.265	0.904									
Qaseem	14.668***	2.957	2.595	1.691								
Jouf	15.317***	3.605*	3.244	2.340	0.648							
Najran	15.581***	3.87*	3.508	2.604	0.912	0.264						
Baha	15.722***	4.011*	3.65*	2.745	1.054	0.405	0.141					
Riyadh	18.094***	6.382***	6.021***	5.117**	3.425	2.777	2.512	2.371				
Madenah	21.770***	10.058***	9.697***	8.792***	7.101***	6.452***	6.188***	6.047***	3.675*			
Makkah	23.004***	11.292***	10.931***	10.027***	8.335***	7.687***	7.422***	7.281***	4.91**	1.234		
Jazan	25.887***	14.175***	13.814***	12.91***	11.218***	10.570***	10.305***	10.164***	7.792***	4.117*	2.882	
Northern	26.310***	14.598***	14.237***	13.332***	11.641***	10.992***	10.728***	10.587***	8.215***	4.540*	3.305	0.422

sig. at 0.05; **sig. at 0.01; ***sig. at 0.001.

TABLE 10.6 Differences Between Mean Productivity of Fodders in Different Regions (ton/hectare)

Region	Jouf	Tabuk	Hail	Qaseem	Riyadh	Eastern	Baha	Jazan	Aseer	Madenah	Najran	Northern
Jouf												
Tabuk	0.16											
Hail	0.247	0.087										
Qaseem	1.621	1.461	1.374									
Riyadh	1.768	1.608	1.521	0.147								
Eastern	2.372*	2.212*	2.125*	0.751	0.604							
Baha	2.595*	2.435*	2.348*	0.974	0.827	0.222						
Jazan	2.932**	2.772**	2.685*	1.311	1.164	0.560	0.337					
Aseer	2.962**	2.802**	2.715*	1.341	1.194	0.590	0.367	0.03				
Madenah	3.307***	3.147**	3.06**	1.685	1.538	0.934	0.711	0.374	0.344			
Najran	3.685***	3.527***	3.438**	2.064	1.917	1.312	1.09	0.752	0.722	0.378		
Northern	6.057***	5.915***	5.828***	4.454***	4.307***	3.702***	3.48***	3.142**	3.112**	2.768**	2.39*	
Makkah	7.480***	7.320***	7.232***	5.858***	5.711***	5.107***	4.884***	4.547***	4.517***	4.172***	3.794***	1.404

sig. at 0.05; **sig. at 0.0;1 ***sig. at 0.001.

have the best relative advantage in fodders productions ranks 1st in terms of fodder cultivated area with a share of 48.14% as shown in Table 10.8.

The six regions that give the highest productivity of vegetables in a descending order are the Eastern, Tabuk, Najran, Baha, Makkah and Asser (Table 10.3). Apart from the eastern region there is no statistically significant difference in the mean productivity of fruits in the other five regions (Table 10.7). Most of these regions, except Baha and Makkah regions, have clear relative advantage (high productivity) in production of other crops; cereals and vegetables. Thus, to enhance food and water security in the country fruit production should be concentrated in Baha and Makkah regions, which have less relative advantage in production of other crops. Ironically under the current cropping pattern Baha and Makkah regions which rank 4th and 5th in terms of fruits' productivity, rank 12th and 7th in terms of percentage of the fruit's total area with shares of 1.4% and 6.9%, respectively (Table 10.8).

10.4   CONCLUSIONS

These results indicate that there is a great potential for enhancing food and water security in Saudi Arabia through altering the existing cropping pattern by encouraging production of different crops in regions that have clear relative advantage in their production (i.e., give the highest productivity). For example, production of cereals could be concentrated in Jouf and Tabuk areas, while vegetables production in the eastern region and Asser area. This is particularly possible at the present time since different regions and parts of the country are well connected with a network of highways, which makes it possible to transport agricultural products from its area of production to any local market in a very short time. Nevertheless, this study did not take into consideration the total area suitable for agricultural production and the country's total demand for different crops, which are factors that deserve close attention when deciding on the final optimum cropping pattern. Thus, the results of this study should be looked at as providing signals for potential of enhancing food and water security in the country by reconsidering the existing cropping pattern, but a detailed socioeconomic and farming system study is needed to work out the details of optimum farming system.

TABLE 10.7 Differences Between Mean Productivity of Fruits in Different Regions (ton/hectare)

Region	Eastern	Tabuk	Najran	Baha	Aseer	Makkah	Hail	Jazan	Madenah	Riyadh	Northern	Jouf	Qaseem
Eastern	2.830***												
Tabuk													
Najran	3.34***	0.51											
Baha	3.955***	1.125*	0.615										
Aseer	4.048***	1.218*	0.708	0.092									
Makkah	4.034***	1.204*	0.694	0.078	0.014								
Hail	5.245***	2.415***	1.905***	1.29*	1.197*	1.211*							
Jazan	5.642***	2.812***	2.302***	1.687**	1.594***	1.608**	0.397						
Madenah	5.91***	3.08***	2.57***	1.954***	1.861***	1.875***	0.664	0.267					
Riyadh	6.087***	3.257***	2.747***	2.131***	2.038***	2.052***	0.841	0.444	0.177				
Northern	6.138***	3.308***	2.798***	2.182***	2.09***	2.104***	0.892	0.495	0.228	0.051			
Jouf	6.798***	3.968***	3.458***	2.842***	2.75***	2.764***	1.552**	1.155	0.888	0.711	0.66		
Qaseem	7.084***	4.254***	3.744***	3.128***	3.035***	3.05***	1.838***	1.441*	1.174*	0.997	0.945	0.285	

*sig. at 0.05; **sig. at 0.01; ***sig. at 0.001.

TABLE 10.8 Area Cultivated by Main Crop Groups in Different Region as Percentage of the Total Area Cultivated by Respective Group Fruits

	Cereal		Vegetables		Fodders		Fruits	
Region	Region	Area as % of total area	Region	Area as % of total area	Region	Area as % of total area	Region	Area as % of total area
Jazan	Riyad	27.04	Riyad	45.62	Riyadh	48.14	Riyadh	21.98
Jouf	Qaseem	20.74	Qaseem	10.69	Qaseem	11.38	Qaseem	18.45
Hail	Makah	12.44	Makah	10.58	Madena	1.78	Jouf	11.29
Riyadh	Hail	10.61	Hail	9.64	Jouf	9.44	Hail	10.04
Eastern	Eastern	9.56	Eastern	6.24	Hail	7.45	Madena	9.95
Qaseem	Jouf	8.79	Jouf	3.87	Jazan	7.16	Eastern	7.49
Tabuk	Tabuk	5.53	Tabuk	3.57	Tabuk	6.46	Makkah	6.89
Makkah	Jazan	2.72	Jazan	3.47	Makkah	2.77	Tabuk	4.14
Aseer	Aseer	2.06	Aseer	2.58	Eastern	2.67	Aseer	3.61
Najran	Madena	0.24	Madena	1.91	Najran	1.47	Najran	2.49
Baha	Najran	0.20	Najran	1.40	Aseer	1.16	Jazan	2.27
Madenah	Baha	0.06	Baha	0.34	Baha	0.13	Baha	1.38
Northern	Northern	0.004	Northern	0.07	Norther	0.004	Northern	0.02
Total	Total	100	Total	100	Total	100	Total	100

10.5 SUMMARY

Fresh water is one of the most important and scarcest natural resources around the globe. In Saudi Arabia fresh water shortage for different uses represents one of the main challenges hindering development efforts in the country. As such there is a need to pursue a water demand management policy, which should include any measure or initiative that will result in reduction of water usage and/or water demand. This study is intended to explore the potential of proper cropping pattern as a water demand management tool to enhance water and food security in Saudi Arabia. Secondary data were obtained from the ministry of Agriculture and Ministry of Electricity and Water. Analyzes revealed that existing cropping pattern did not benefit from the relative advantage of different regions in producing different crops. Thus, there is a great potential for enhancing food and water security in Saudi Arabia through altering existing cropping pattern by encouraging production of different crops in the regions that have clear relative advantage in their production, that is, give the highest productivity. For example, production of cereals could be concentrated in Jouf and Tabuk areas, while vegetables production in the eastern region and Asser area.

KEYWORDS

- arid region
- cereal
- cropping pattern
- cucumber
- FAO
- farming system
- food security
- fresh water
- irrigation water
- King Saud University, KSU
- Kingdom of Saudi Arabia, KSA

- **Ministry of Agriculture, MOA**
- **Ministry of Electricity and Water**
- **Saudi Arabia**
- **tomato**
- **vegetable production**
- **water demand management**
- **water scarcity**

ACKNOWLEDGEMENTS

The authors are grateful to the King Saud University, Deanship of Scientific Research, College of Food and Agricultural Sciences, Research Center, for extending all possible assistance, cooperation and needed support towards the completion of the project.

REFERENCES

1. Abdulgader, A. M., A. A. Ali, A. Alwathig, and K. Ownginiko, 2001. The effect of different quantities of irrigation water on productivity of tomato and cucumber produced in green houses in Saudi Arabia: A technical and economic study. Proceedings of the Arabian Gulf Water 5th Conference held in Doha (Qatar): March 24–28, pp. 339–350.
2. Alderfasi, A. A., M. S. AL-Sewailem, F. A. AL-Yahya, K. A. Kamel, and A. Aleter, 2002. Effect of irrigation with treated municipal waste water on wheat production under drought stress conditions. *J. King Saud Univ., Agric. Sci.,* 14(1):57–73.
3. Alderfasi, A. A., M. O. Ghandorah, and K. A. Moustafa, 1999. Evaluation of some wheat genotypes under drought stress in arid region of Saudi Arabia. *Alex. J. Agric. Res.,* 44(3):209–217.
4. Allan, J. A., 1999. Productive efficiency and allocative efficiency: why better water management may not solve the problem. *Agricultural Water Management,* 40:71–75.
5. Allan, J. A., 1998. Watersheds and problem sheds: explaining the absence of armed conflict over water in the Middle East. *Middle East Review of International Affairs* 2(1).
6. Allan, J. A., 1996. Policy responses to the closure of water resources: regional and global issues. In: *Howsam, P., and Carter, R.C. (Eds.), Water policy: allocation and management in practice.* E&FN Spon, London, pp. 3–12.

7. Allan, J. A., 1997. Virtual water: a long-term solution for water short Middle Eastern economies? In: *Proceedings of the Paper Presentation at the 1997 British Association Festival of Sciences*, University of Leeds, Water and Development Session, 9 September. Available at http://www2.soas.ac.uk/geography/waterissues/.

8. Azhar, A. H., M. N. Bhutta, and M. Latif, 2010. Reclamation of irrigated agriculture through tile at fourth drainage project, Faisalabad. *J. Anim. Plant Sci.*, 20(3):211–216.

9. Azhar, A. H., 2011. *Irrigation analytical report: Field visit of Diamer Basha dam (DBD) project area – Chilas*. Land Acquisition & Resettlement (LA&R), WAPDA, Lahore – Pakistan, pp. 45.

10. Azhar, A. H. and B. J. C. Perera, 2011. *Planning and operation of irrigation systems – judicial use of limited water resources for enhanced water productivity*. VDM Verlag Dr. Muller GmbH & Co. KG, Germany.

11. Azhar, A. H., M. M. Alam, and M. Latif. 2011a. Subsurface drainage impacts on cropping intensity in Pakistan. *The J. Anim. Plant Sci.*, 21(1):97–103.

12. Azhar, A. H., C. M. Ashraf, and M. Ahmad, 2011b. *Modern irrigation techniques and technologies – efficient utilization of scarce water resources*. VDM Verlag Dr. Muller GmbH & Co. KG, Germany, pp. 153.

13. Bouwer, H., 2000. Integrated water management: emerging issues and challenges. *Agric. Water Manage.*, 45:217–228.

14. Hatfield, J. L., T. J. Sauer, and J. H. Prueger, 2001. Managing soils to achieve greater water use efficiency. *Agron. J.* 93:271–280.

15. Molden, D., D. Binb, R. Loevec, R. Barkera, and T. P. Tuongd, 2007. Agricultural water productivity and savings: policy lessons from two diverse sites in China. *Water Policy*, 9(S1):29–44.

16. Ministry of Agriculture, 2009. *Agricultural Statistical Yearbook*. Riyadh (Saudi Arabia). Ministry of Agriculture, KSA.

17. Ministry of Economy and Planning, 2005. *The 8th Development Plan*. Riyadh (Saudi Arabia), Ministry of Economy and Planning, KSA.

18. Ministry of Economy and Planning, 2009. *Achievements of the 8th development plan*. Riyadh (Saudi Arabia), Ministry of Economy and Planning, KSA.

19. Omran, A. A. M., 2008. *Water requirements for irrigation and rationalization*. King Saud University Press.

20. Pretty, J. N., 1995. *Regenerating agriculture: policies and practice for sustainability and self-reliance*. Earthscan, London.

21. Qunaibet, M. H., 2002. Agriculture and Water in Saudi Arabia. A paper presented to: The future vision of the Saudi Economy Symposium. Riyadh (Saudi Arabia). Ministry of Economy and Planning, KSA.

22. Jensen, M. E., W. R. Rangeley, and P. J. Dieleman, 1990. Irrigation trends in world agriculture. In: *Stewart, B. A., Nielsen, D.R. (Eds.), Irrigation of Agricultural Crops*. Agronomy Monograph 30, ASA, Wisconsin, USA, pp. 31–67.

23. Qadir, M., T. M. Boers, S. Schubert, A. Ghafoor, and G. Murtaza, 2003. Review: Agricultural water management in water-starved countries: challenges and opportunities. *Agricultural Water Management*, 62:165–185.

24. Smith, M., L. S. Pereira, J. Beregena, B. Itier, J. Goussard, R. Ragab, R. L. Tollefson, and P. V. Hoffwegen (Eds.), 1996. *Irrigation Scheduling: From Theory to Practice*. FAO Water Report 8, ICID and FAO, Rome.

25. The Global Development Research Centre, 2010. *Urban Water Resources Management.* http://gdrc.org/uem/water.
26. Wallace, J. S., and C. H. Bachelor, 1997. Managing water resources for crop production. *Phil. Trans, R. Soc. London. B.*, 352:937–947.
27. Zahrani, K. H., E. M. Alshfei, S. Algahtani and A. Alhamid, 2005. *Changing the household water consumption pattern as an approach for water demand management.* Riyadh (Saudi Arabia): King Saud University, Scientific Research Deanship.

APPENDIX

Use of sprinkler irrigation system for fodder crops production in Saudi Arabia.

Wheat production in Saudi Arabia.

CHAPTER 11

IRRIGATION SCHEDULING METHODS FOR DRIP IRRIGATED TOMATO: KSA

MOHAMED SAID ABDALLA EL MARAZKY

CONTENTS

In this chapter: One U.S. dollar = 3.75 riyals (KSA currency).

11.1   INTRODUCTION

Water scarcity and drought are the major factors constraining agricultural crop production in arid and semi-arid zones of the world. Innovations for saving water in irrigated agriculture and thereby improving water use efficiency (WUE) are of paramount importance in water-scarce regions. Therefore, use of new irrigation technologies in agriculture have aimed at increasing crop production, and new developments in irrigation technologies are of great importance. Efficient use of irrigation is becoming increasingly important, and drip irrigation (DI) may contribute substantially to the best use of water for agriculture, improving irrigation efficiency [28]. Therefore, adoption of modern irrigation techniques must be emphasized to increase WUE. These modern techniques must result in efficient water delivery and high productivity while minimizing water use [1].

In the arid and semi-arid areas, DI is frequently used to reach the maximum WUE [15]. DI has been used in horticultural operations since the middle of the 20th century [19] and conventional drip irrigation is considered one of the most efficient irrigation systems. DI has the potential to use scarce water resources most efficiently to produce vegetables [22]. However, DI is an irrigation system whereby water is supplied under low pressure directly to the plant roots [26]. DI is the most effective way to save water by using water more efficiently to increase crop yield and improve the irrigation uniformity [5, 27, 34].

DI can distribute water uniformly, precisely control water amount, increase plant yield, reduce evapotranspiration (ET) and deep percolation, and decrease harmful effects of soil degradation and salinity [7, 8, 21]. The trend in recent years has been towards conversion of surface irrigation to DI to improve plant quality and yield. While, at present, some farmers in Saudi Arabia are not sure when and how much water they should apply under drip irrigation, and they tend to confirm irrigation timing and amount according to field experience, and then, induce new water loss under new technology. So, an easy-operation irrigation scheduling method is very stringent for tomato under drip irrigation.

Earlier studies have shown that DI is the most suitable method for vegetable crops and it is possible to increase WUE by modern irrigation scheduling methods, such as cucumber [33], eggplant [6], potato [13], and tomato [9, 30]. Many studies comparing sprinkler or furrow irrigation with

DI in tomatoes and in other crops have shown that DI generally resulted in higher WUE and crop yields [16, 18, 30].

Irrigation events may be scheduled based on: measured soil moisture, climatic parameters and estimated evapotranspiration (ET) coupled with crop coefficient specific to the region. Many methods of irrigation scheduling have been proposed in order to measure the amount of water use by a crop. There are three methods for matching irrigation with crop water requirements: the weather-based methods using ETr [3, 25], the soil water-based methods using soil moisture sensors [14], and the soil–water-balance calculations and plant stress-sensing techniques [20].

There are a variety of techniques can be used to reduce water use [24, 26]. These techniques include ET control devices and soil moisture controllers. Mayer et al. [23] found that ET controllers reduced irrigation by 6.1%; and it was found that 56.7% of the sites were responsible for a significant decrease in irrigation application, while 41.8% were responsible for a significant increase. Davis et al. [12] demonstrated that the ET controllers applied only half of the irrigation calculated for the theoretical requirement for each irrigation event, and irrigation adequacy was decreased when the ET controllers were allowed to irrigate any day of the week. Davis and Dukes [11] found that ET controllers can match irrigation application with seasonal demand and in particular reduce irrigation in the winter when plant demands are dramatically reduced. In addition, they indicate that when ET controllers are applied to sites irrigating at levels less than plant demand, those controllers will likely increase irrigation.

Automation of DI systems based on evapotranspiration controllers or soil moisture sensors may further improve WUE. Development of automated site-specific drip irrigation systems allow producers to maximize irrigation efficiency, while minimizing negative effects on their productivity [29]. Adoption of modern water-saving technology is often cited as a key to increasing WUE while maintaining current levels of production [10, 17]. Though, this technology has not been tested with field crop in a hyper arid region such as Saudi Arabia, yet such systems technique can be used to determine crop yield and performance to irrigation criteria.

This chapter discusses effects of different irrigation scheduling management strategies on fruit yield and quality of drip-irrigated fresh market tomatoes, water use efficiency and irrigation application efficiency.

11.2   MATERIALS AND METHODS

During the two seasons of 2011 and 2012, the experiments were conducted at the Experimental Farm of the College of Food and Agriculture
Sciences of King Saud University, Riyadh (24°43′ N latitude, 46°43′ E
longitude and 635 m altitude). The experimental site was irrigated by a
subsurface drip irrigation system. Before the start of the experiment, intact
soil cores were collected from different locations in the field to determine
soil physical properties including soil mechanical analysis. Locations were
selected to represent the dominant soil conditions in the field. Three soil
samples were taken from the field at three different depths (0–20, 20–30
and 30–60 cm) to determine soil texture. The soil was loamy sand (85.9%
sand, 6% silt and 8.1% clay).

Surface drip irrigation systems were installed in the field. Buffer zone
of approximately 3 m separated each plot to reduce interactions between
the treatments. The DI system consisted of 16 mm inside diameter (I.D.)
thin-wall lateral drip lines with welded-on emitters (GR, 50 cm dripper
spacing) with a nominal emitter discharge of 4 lph at a design pressure
of 200 kPa. Drip lines were buried 25 cm deep directly under the soil
beds in plots 1, 2 and 3 (Figure 11.1a). After the ID installation, the
soil surface was leveled and firmed. Irrigation amounts were metered
separately in each plot using commercial municipal-grade flow accumu-
lators. The irrigation duration varied among treatments because of the
three different methods of irrigation scheduling. The hydraulic aspects
of the design for each system were aimed to give uniform application of
irrigation water.

The uniformity of water application for each scheduling method below
the soil surface through the soil profile was determined by measuring
gravimetric moisture contents from soil samples taken 24 and 48 hours
after irrigation. The samples were collected parallel and perpendicular to
the lateral line at distances of 0, 5, 10, 15, 20 and 25 cm from the emitter
location as shown in (Figure 11.1b). The gravimetric soil samples at each
depth (0, 10, 20, 30, 40, 50 and 60 cm) were repeated three times after
irrigation (24 and 48 hr). These measurements were taken from each plot
three times during mid-season of tomato crop.

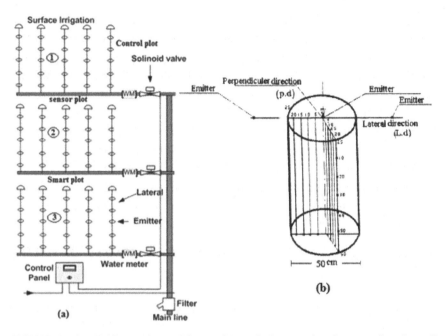

FIGURE 11.1   Field experimental layout (a); and diagram showing sampling for soil moisture contents parallel and perpendicular for the direction of the irrigation line (b).

Three methods of irrigation scheduling were used to determine the duration and amount of water to be applied to a tomato crop by subsurface drip irrigation system. The irrigation scheduling in plot 3 was controlled by evapotranspiration controller (ET controller). The ET-based controllers consider weather based parameters when determining irrigation events. Depending on the manufacturer, each controller functions differently but typically can be programed with various conditions specific to the field. These conditions can include soil type, plant type, root depth, sun and shade, etc. The ET controller has the ability to add water to the crop when it was needed based on controlled evapotranspiration and weather data. The controller Hunter Pro-c was purchased locally and was programed according to site specific conditions.

Plot 2 was controlled by automatic watermark soil moisture sensors. The initiation and termination of irrigation in the scheduling technique was based on soil moisture measured by watermark sensors installed at 5 cm above the drip line. Automatic scheduling for plot 2 was set at 10% soil moisture content as the lower limit and 15% as the upper limit (on – off).

The scheduling treatment in plot 1 (control treatment) was manually irrigated based on weather data collected from an automatic weather station installed at the experimental site. Penman–Monteith equation was used to calculate evapotranspiration (ET).

Each plot was approximately 4.5 m wide and 7 m long and had 5 rows of drip lines spaced 0.9 m apart running from west to east. Tomato plants were spaced 0.50 m apart in each row. The 5 drip lines in each plot were connected to a common sub-main irrigation line at the inlet side of the plot; and a common flush line and flush valve at the distal end of the plot (Figure 11.1a).

11.2.1   WATER USE EFFICIENCY AND UNIFORMITY OF WATER DISTRIBUTION

Irrigation water used efficiency (IWUE) is the ratio between the total fresh yield (FY) and the seasonal applied irrigation water $(Dg)_t$. While, water use efficiency (WUE) is the relationship between the yield and the ETc.

$$WUE = \left(\frac{Y}{ETc} \right) \qquad (1)$$

$$IWUE = \left(\frac{Y}{(Dg)_t} \right) \qquad (2)$$

In these equations, Y is the economical yield (kg.m^{-3}), ETc is evapotranspiration (mm), and $(Dg)_t$ is the total amount irrigation water (mm) during the crop season.

To calculate the ETc and the irrigation water requirement of tomato, daily ET_o values were first determined using the meteorological data and then multiplied by the crop coefficient. The irrigation system in each plot was operated based on the scheduling method used; turned on and off manually in the control treatment and automatically in ET controller and watermark sensor treatments. The depths of irrigation water (Dg) applied in each irrigation event for all plots were calculated separately from the differences of flow meter reading before and after irrigation.

11.2.1.1   Assessment of the Uniformity of Water Redistribution in the Soil Profile

The coefficient of uniformity by Christiansen [3] was calculated by using soil gravimetric moisture contents measured at seven soil depths (0, 10, 20, 30, 40, 50, and 60 cm), and at different distances from emitter (10, 15, 20 and 25 cm in parallel and perpendicular directions to the drip line), as shown in Figure 11.1b). The soil water contents were measured 24 and 48 hours after irrigation was ceased. The evaluation tests were carried out four times starting from the beginning until the end of season. The following equation was used to evaluate the uniformity (Cu_s) of water redistribution below the soil:

$$Cu_s = 100 \left(1 - \frac{\sum |\theta_i - \bar{\theta}|}{N\bar{\theta}} \right)$$ (3)

$$\bar{\theta} = \left| \sum_{i=1}^{N} \theta_i \right| / N$$

where: Cu_s = Christiansen's coefficient of uniformity of soil water content below soil surface, θ_i = the measured gravimetric soil water content at depth i, $\bar{\theta}$ = the mean gravimetric soil water content, and N = number of measured points.

11.3   RESULTS AND DISCUSSION

11.3.1   CROPS EVAPOTRANSPIRATION (ETc)

The daily and weekly averages of the ETc rates for tomato crop in control treatment (plot 1) were calculated using the daily climatic records during the two growing seasons (Table 11.1). The values of ETc were estimated by the product of the reference evapotranspiration (ETo) and the crop coefficient (Kc) for different stages of tomato crop. From this table, it can be concluded that ETc values were small in early 2 weeks and then increased with the development of plants.

TABLE 11.1   Average Weekly ETc for Tomato Under Subsurface Drip System for Control Treatment During the Two Seasons

Growth period	ETo	Kc	ETc
Week	mm/day	–	mm/day
1	4.22	0.70	2.95
2	4.65	0.70	3.25
3	4.98	0.93	4.54
4	5.56	1.15	6.39
5	5.61	1.15	6.46
6	5.78	1.15	6.64
7	5.28	1.15	6.08
8	5.92	1.03	6.30
9	6.71	1.03	6.84
10	6.67	0.90	6.00
11	6.54	0.90	5.89
12	6.87	0.90	6.18
13	6.56	0.83	5.53
14	6.64	0.83	5.53
15	7.49	0.90	6.74
16	6.96	0.75	5.22
17	7.17	0.75	5.38
Avg.	Average ETc, (mm/day)		**5.64**
Sum	Total ETc/season, (mm)		**671.57**

11.3.2   IRRIGATION MANAGEMENT

The averages of weekly and seasonal total water applied (m^3), irrigation depth (Dg) and the accumulated $(Dg)_t$ to the tomato crop by three scheduling technique (ET controller, automatic watermark and control treatments) are presented in Table 11.2. It can be observed that the average total amount of water added during crop season were 10.83 m^3, 11.12 m^3 and 13.03 m^3 in ET controller, automatic watermark and control treatments, respectively. There was water saving of 4.66% and 18.64% in ET controller treatment compared to other two treatments, respectively. Also, watermark sensor technique used less water by 14.67% compared to the

TABLE 11.2 Averages of Irrigation Water Depths Applied to Tomato Crop During Two Growing Seasons by the Different Irrigation Scheduling Methods

Growth period	ET controllers plot 3			Watermark sensor plot 2			Control treatment plot 1		
	Water added	Irrigation depth Dg	Acc. depth (Dg)t	Water added	Irrigation depth Dg	Acc. depth (Dg)t	Water added	Irrigation depth Dg	Acc. depth (Dg)t
week	m³	mm	mm	m³	mm	mm	m³	mm	mm
1	0.47	37.23	37.23	0.60	47.90	47.90	1.33	105.60	105.60
2	0.43	34.40	71.63	0.44	34.56	82.46	0.79	62.76	168.36
3	0.42	33.44	105.07	0.54	42.66	125.12	0.64	50.96	219.32
4	0.45	35.60	140.67	0.59	46.83	171.94	0.82	65.46	284.78
5	0.61	48.28	188.95	0.53	42.10	214.04	0.60	47.61	332.38
6	0.30	24.06	213.01	0.22	17.67	231.71	0.27	21.13	353.52
7	0.30	24.13	237.13	0.15	11.70	243.40	0.66	52.06	405.58
8	0.56	44.83	281.97	0.57	45.07	288.48	0.80	63.49	469.07
9	0.92	73.10	355.06	0.47	37.63	326.10	0.74	58.98	528.05
10	0.29	22.63	377.69	0.88	69.91	396.02	0.77	61.10	589.14
11	0.67	53.17	430.86	0.59	46.93	442.94	0.73	58.02	647.16
12	0.89	70.63	501.49	0.90	71.48	514.43	0.73	57.56	704.72
13	0.96	76.58	578.07	0.53	42.22	556.65	0.79	63.00	767.72
14	0.67	53.21	631.29	0.85	67.60	624.25	0.77	61.26	828.98
15	0.89	70.29	701.57	1.00	79.13	703.38	0.71	56.10	885.09
16	0.86	68.03	769.61	0.82	64.69	768.07	0.80	63.43	948.52
17	0.91	71.90	841.51	1.44	114.53	882.60	1.08	85.81	1034.33
Sum	10.60	841.51		11.12	882.60		13.03	1034.33	

control treatment. Consequently, the use of ET controller or watermark methods conserves water and this superiority in saving water may be due to the fact that the two methods have the feature of increasing or reducing irrigation water automatically according to the plant needs compared to the control treatment.

11.3.3   AGRONOMICAL CHARACTERISTICS

This study revealed that both irrigation-scheduling techniques had a clear impact on the agronomical characteristics of the plants as shown in Table 11.3. In the same context, it was found that the average yields for the two seasons were 39.22, 35.35 and 30.23 ton/ha in the ET controller, automatic watermark and control methods, respectively. This shows that the variation between the yields in the ET controller between automatic watermark and control treatments was 10 to 23%, respectively. Meanwhile, the agronomical data (Table 11.3) for the ET controller treatment reveled a significant difference in plant height (cm), number of branches, fruit length (cm), average fruit weight (g), total yield (Kg.m^{-2}), total yield (ton/ha) and WUE/IWUE (Kg.m^{-3}) compared to the automatic watermark and control treatments.

TABLE 11.3   Average Response of Tomato Growth to Irrigation Treatments During the Two Seasons

Character	Units	Treatment		
		Smart ET	**Sensor**	**Control**
Plant height	cm	49.6	48.4	53.2
Number of branches	—	5.31	5.5.24	5.12
Fruit length	cm	5.62	5.7	6.44
Fruit diameter	cm	5.18	5.07	5.2
Fruit shape index	—	1.28	1.25	1.23
Avg. fruit weight	g	92.3	91.8	88.9
Early yield	ton/ha	22.23	20.15	23.04
Total yield	ton/ha	39.22	35.35	30.23
WUE	kg m^{-3}	7.26	6.08	4.50
IWUE	kg m^{-3}	4.66	4.01	2.92

11.3.4 WATER USE EFFICIENCY

Table 11.4 demonstrates the effects of the three scheduling techniques (ET controller, automatic watermark and control treatments) on tomato WUE during the growing seasons. The data in Table 11.5 revealed that the values of WUE and IWUE were higher in the ET controller treatment. The tomato yield, in the case of ET controller treatment, was higher (39.22 ton/ha) compared to the yield in the two other scheduling methods (Table 11.3). Similar trend was observed for WUE and IWUE. The maximum and minimum values of WUE and IWUE for ET controller methods were 7.26 and 4.66 kg.m^{-3}, while WUE and IWUE for watermark and control methods were 6.08, 4.01 and 4.50,

TABLE 11.4   Average WUE and IWUE Under Different Scheduling Methods During the Two Seasons

Irrigation treatments	Drip irrigation				WUE	IWUE
	ETc		AIW			
	mm	m³/ha	mm	m³/ha	kg m^{-3}	kg m^{-3}
Smart ET	540.42	5404.20	841.51	8415.1	7.26	4.66
Sensor	581.23	5812.30	882.6	8826.0	6.08	4.01
Control	671.57	6715.70	1034.33	10343.3	4.50	2.92

TABLE 11.5   Cu$_s$ as Function of Soil Depth After 24 and 48 Hours of Irrigation, for Three Irrigation Scheduling Methods

Soil depth cm	Values of Cu$_s$ for three scheduling methods					
	ET controller		Watermark sensor		Control	
	48 hr	24 hr	48 hr	24 hr	48 hr	24 hr
0	89.70	70.48	86.25	83.61	86.37	84.44
10	86.92	70.35	86.86	81.79	92.12	89.24
20	92.51	71.67	89.92	84.27	97.62	94.69
30	89.45	72.23	91.68	83.22	94.87	88.72
40	86.85	71.07	92.03	81.19	94.72	91.29
50	86.99	79.98	93.95	82.26	94.07	91.90
60	88.29	86.15	92.01	81.32	93.40	92.15
Average	**88.67**	**74.56**	**90.38**	**82.52**	**93.31**	**90.69**
	81.62		**86.45**		**92.00**	

2.92 kg.m^{-3}, respectively (Table 11.4). However, the results indicated that irrigation water was used more effectively through ET controller treatment.

The Table 11.4 shows that the highest and lowest values of IWUE for tomato crop were 4.66 and 2.92 kg.m^{-3} in ET controller and control treatments, respectively. The increase of IWUE value was 37.34% for ET controller compared with the control treatment. In contrast, the smallest amount of irrigation water was 540.42 mm during the entire season in ET controller treatment, while the largest amount was 671.57 mm in control treatment. Data in Tables 11.3 and 11.4 indicated that scheduling method of ET controller gave the highest values of total yield, WUE and IWUE and applied less irrigation water compared to watermark sensor method and control method, respectively.

The lower amounts of water used with ET controller method correspond inversely to higher WUE. This agrees with the results by Faberio et al. [15], Topak et al. [31] and Almarshadi and Ismail [4]. Similar findings were also obtained by Wan and Kang [32], who found a low irrigation frequency. The decreased values of WUE and IWUE under the watermark sensor and control methods can be attributed to the increasing level of applied irrigation water. Under conditions of the three irrigation treatments in the both growing season, ET controller resulted in the highest WUE and IWUE, followed by watermark sensor and then control treatment. It was apparent that the WUE and IWUE of tomato decreased with more water applied in irrigation.

11.3.5 UNIFORMITY OF WATER DISTRIBUTION

The water redistribution patterns under drip irrigation systems for the three scheduling methods were determined at different depths below the soil surface, as shown in Figure 11.2. The Table 11.5 and Figure 11.2 show the average of uniformity coefficient (Cu_s) and patterns for Cu_s in parallel and perpendicular locations to the drip line at different depths for each scheduling method after 24 and 48 hour from irrigation was ceased. The highest uniformity was obtained in the control treatment and the lowest with ET controller treatment for 24 and 48 hours after irrigation. Generally, the average values of Cu_s of control scheduling technique were higher than those of both ET controller and automatic watermark systems at any depth of soil profile and time of measurements (24 and 48 hour).

However, in general, the water within the soil profile at any depth was uniformly distributed through soil profile. This can be explained by the

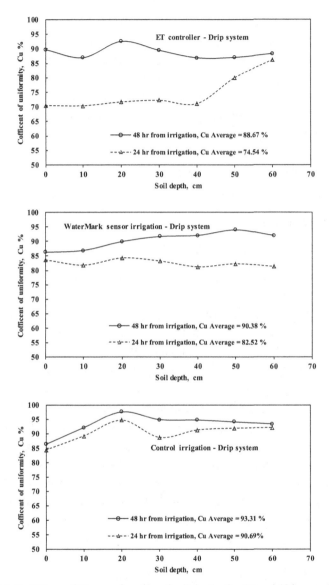

FIGURE 11.2   Values of *Cus* as a function of soil depth after 24 and 48 hours of irrigation, for the three irrigation-scheduling methods.

hydraulic gradients within the irregularly wetted soil, which causes water movement within the soil profile parallel and perpendicular to the irrigation lines, resulting in the water movement within the soil to be more uniformly distributed. Also, the results showed that the average of Cu_s values were 81.62% for ET controller, 86.45% for watermark sensor and 92% for control treatment.

Also, the values of Cu_s were decreased slightly with the increase in soil depth (Figure 11.2) due to the soil diffusivity, but increased with the time of measurements due to the accomplishment of equilibrium within the soil [2].

In general, the Cu values were low in depths near the soil surface and increased with depth for all scheduling methods. However, this increase with depth was higher in control treatment compared to the increase in ET controllers and automatic watermark scheduling techniques.

11.4 CONCLUSIONS

This study concludes that the ET controller methods offered a significant advantage in managing the irrigation of tomato crops in both seasons under severely arid conditions. In compression with the other treatments, the ET controller significantly reduced water use by 18%. Consequently, water was used most commendably with the ET controller treatment. It was also found that the values of yield, WUE and IWUE were superior with ET controller compared to corresponding values in automatic watermark and control treatments.

The coefficient of uniformity for control treatments was 10.4% higher than for ET controller irrigation scheduling method, while variations in *Cus* values were not significant among the three scheduling techniques. ET controller technique gave the best crop yield, WUE and IWUE.

It can be concluded that there was an economic advantage when applying advance scheduling irrigation techniques using drip irrigation system with ET controller under arid conditions, such as Saudi Arabia.

11.5 SUMMARY

Irrigation is necessary in order to produce tomato in arid region such as Saudi Arabia, but water supplies are becoming limited. Drip irrigation (DI) is being adopted in areas to conserve water while maintaining economical crop production. Field experiment was carried out to study effects of different irrigation scheduling management strategies on fruit yield and quality of drip-irrigated fresh market tomatoes, water use efficiency and irrigation application efficiency. The experiment included three scheduling methods, which were used to irrigate tomato field: Evapotranspiration (ET) controller, soil moisture sensor (watermark sensor) and control treatment based on weather data. All irrigation-scheduling methods were effective, prescribing different amount of water for a given season. Irrigation amount increased from 841.5 mm (ET controller) to 882.60 mm (watermark sensor) and 1034.33 mm (control treatment), during two seasons. Both WUE and IWUE increased with decrease in irrigation water depth. The maximum average WUE (7.26 kg-m^{-3}) and IWUE (4.66 kg-m^{-3}) values were for ET controller as average, during the two seasons. In addition, ET controller method conserved up to 7 % and 18% water compared to the watermark sensor and control treatments, respectively. Based on results in this chapter, it is recommended that if the tomatoes are well irrigated based on ET controller, the farmer can get higher tomato yield in areas experiencing severe drought, such as Saudi Arabia.

ACKNOWLEDGMENTS

This project was supported by King Saudi University, Deanship of Scientific Research, College of Food AND Agriculture Sciences, and Research Center. The use of the trade name does not imply promotion of the product; it is mentioned for research purposes only.

KEYWORDS

- arid region
- automatic irrigation
- drip irrigation
- evapotranspiration
- irrigation controller
- irrigation scheduling
- irrigation scheduling

- irrigation water use efficiency
- King Saudi University
- Saudi Arabia
- tomato
- water application efficiency
- water use efficiency
- watermark sensor

REFERENCES

1. Acar, B., Topak, R. and Direk, M., 2010. Impacts of pressurized irrigation technologies on efficient water resources uses in semi-arid climate of Konya Basin of Turkey. *Int. J. of Sustainable Water & Environmental Systems*, 1(1):1–4.
2. Al-Ghobari, H. A., 2003. A study on selecting the proper applied water under different irrigation systems. *Misr Journal of Agricultural Engineering*, Egypt, 20:320–327.
3. Allen, G. R., Pereira, S. L., Dirk, R., and Martin, S., 1998. *Crop evapotranspiration—guidelines for computing crop water requirements.* FAO irrigation and drainage paper 56, Rome, pp. 7–28.
4. Almarshadi, M. H., and Ismail, S. M., 2011. Effects of precision irrigation on productivity and water use efficiency of alfalfa under different irrigation methods in Arid climates. *J. Appl. Sci. Res.,* 7(3):299–308.
5. Al-Omran, A. M., Sheta, A. S., Falatah, A. M., and Al-Harbi, A. R., 2005. Effect of drip irrigation on squash yield and water use efficiency in sandy calcareous soils amended with caly deposits. *Agric. Water Manag.,* 73(1):43–55.
6. Aujla, M. S., Thind, H. S., and Buttar, G. S., 2007. Fruit yield and water use efficiency of eggplant as influenced by different quantities of nitrogen and water applied through drip and furrow irrigation. *Sci. Hortic.,* 112:142–148.
7. Ayars, J. E., Phene, C. J., Hutmacher, R. B., Davis, K. R., Schoneman, R. A., Vail, S. S., and Mead, R. M., 1999. Subsurface drip irrigation of row crops: a review of 15 years of research at the Water Management Research Laboratory. *Agric. Water Manage.,* 42:1–27.
8. Batchelor, C. H., Lovell, C. J., and Murata, M., 1996. Simple microirrigation techniques for improving irrigation efficiency on vegetable gardens. *Agric. Water Manage.,* 32:37–48.
9. Cetin, O., and Uygan, D., 2008. The effect of drip line spacing, irrigation regimes and planting geometries of tomato on yield, irrigation water use efficiency and net return. *Agric. Water Manage.,* 95:949–958.
10. Comprehensive Assessment of Water Management in Agriculture, 2007. *Water for Food, Water for Life: A Comprehensive Assessment of Water Management in Agriculture.* Earthscan/International Water Management Institute, London/Colombo.

11. Davis, S. L., and Dukes, M. D., 2012. Landscape irrigation with evapotranspiration controller in a humid climate. *Trans. ASABE*, 55(2):571–580.
12. Davis, S. L., Dukes, M. D., and Miller, G. L., 2010. Irrigation scheduling performance by evapotranspiration-based controllers. *Agr. Water Manage.*, 98(1):19–28.
13. Erdem, T., Erdem, Y., Orta, H., and Okursoy, H., 2006. Water–yield relationships of potato under different irrigation methods and regimens. *Sci. Agric.*, 3:226–231.
14. Evett, S. R., 2008. Neutron moisture meter (in field estimation of soil water content): A practical guide to methods, instrumentation, and sensor technology. *International Atomic Energy Agency:* Vienna, Austria, Chapter 3, pp. 39–54.
15. Faberio, C., and Martin de Juan, J. A., 2002. Production of muskmelon under controlled deficit irrigation in a semi-arid climate, *Agricultural Water Management*, 54:93–105.
16. Flowers, T. J., Ragab, R., Malash, N., Abdel Gawad, G., Cuartero, J., and Arslan, A., 2005. Sustainable strategies in salt-prone Mediterranean: SALMED. *Agr. Water Manage.*, 78:3–14.
17. Green, G., Sunding, D., Zilberman, D., and Parker, D., 1996. Explaining irrigation technology choices: a micro parameter approach. *Am. J. Agric. Econ.*, 78:1064–1072.
18. Hebbar, S. S., Ramachandrappa, B. K., Nanjappa, H. V., and Prabhakar, M., 2004. Studies on NPK drip fertigation in field grown tomato (*Lycopersicon esculentum* Mill.). *Eur. J. Agron.*, 21:117–127.
19. Hillel, D., 2008. 40 Years of Drip Irrigation. *CSA News*, 53(9):3–7.
20. Jones, H., 2004. Irrigation scheduling: advantages and pitfalls of plant based methods. *J. Exp. Bot.*, 55(407):2427–2436.
21. Karlberg, L., and Frits, W. T. P. V., 2004. Exploring potentials and constraints of low-cost drip irrigation with saline water in sub-Saharan Africa. *Phys. Chem. Earth*, 29:1035–1042.
22. Locascio, J. S., 2005. Management of irrigation for vegetables: past, present, future. *Hort. Technology*, 15(3):482–485.
23. Mayer, P., DeOreo, W., Hayden, M., Davis, R., Caldwell, E., Miller, T. and P. J. Bickel, 2009. Evaluation of California weather-based "smart" irrigation controller programs. Final report for the California Department of Water Resources. Sacramento, CA: California Urban Water Conservation Council.
24. McCready, M. S., Dukes, M. D., and Miller, G. L., 2009. Water conservation potential of smart irrigation controllers on St. Augustine grass. *Agr. Water Manage.*, 96(11):1623–1632.
25. Monteith J. L., 1985. Evaporation from land surfaces: progress in analysis and prediction since 1948. In: *Advances in evapotranspiration*, proceedings of the national conference on advances in evapotranspiration, Chicago, IL, ASAE, St. Joseph, Michigan, pp. 4–12.
26. Nautiyal, M., Grabow, G., Miller, G., and Huffman, R. L., 2010. Evaluation of two smart irrigation technologies in Cary, North Carolina. ASABE Annual International Meeting, June 20–June 23.
27. Schwankl, L. J., and Hanson, B. R., 2007. Surface drip irrigation. In: *Microirrigation for crop production*, Eds. Lamm, F. R., Ayars, J. E., and Nakayama. Elsevier, pp. 431–472.
28. Sezen, S. M., Yazar, A., and Eker, S., 2006. Effect of drip irrigation regimes on yield and quality of field grown bell pepper. *Agric. Water Manage.*, 81:115–131.

29. Shock, C. C., Feibert, E. B. G., Saunders, L. D., and Eldredge, E. P., 2002. Automation of Subsurface Drip Irrigation for crop research. *ASAE. Publication Number 701P0301*, pp. 809–816, St. Joseph, MI.

30. Sigandhupe, R. B., Rao, G. G. S. N., Patil, N. G., Brahmanand, P. S., 2003. Fertigation studies and irrigation scheduling in drip irrigation system in tomato crop (*Lycopersicon esculentum* Mill.). *Eur. J. Agron.*, 19:327–340.

31. Topak, R., Suheri, S., and Acar, B., 2011. Effect of different drip irrigation regimes on sugar beet (*Beta vulgaris L.*) yield, quality and water use efficiency in Middle Anatolian, Turkey. *Irrig. Sci.*, 29:79–89.

32. Wan, S., and Kang, Y., 2006. Effect of drip irrigation frequency on radish (*Raphanus sativus L.*) growth and water use. *Irrigation Science*, 24:161–174.

33. Yuan, B. Z., Sun, J., Kang, Y., and Nishiyama, S., 2006. Response of cucumber to drip irrigation water under a rain shelter. *Agric. Water Manage.*, 81:145–158.

34. Zotarelli, L., Scholberg, J. M., Dukes, M. D., Munoz-Carpena, R., and Icerman, J., 2009. Tomato yield, biomass accumulation, root distribution and irrigation water use efficiency on a sandy soil, as affected by nitrogen rate and irrigation scheduling. *Agricultural Water Management*, 96:23–34.

APPENDIX A

CONVERSION SI AND NON-SI UNITS

To convert the Column 1 in the Column 2	Column 1 Unit	Column 2 Unit	To convert the Column 2 in the Column 1
Multiply by	SI	Non-SI	Multiply by

LINEAR

0.621 —— kilometer, km (10^3 m)		miles, mi ——	1.609
1.094 —— meter, m		yard, yd ——	0.914
3.28 —— meter, m		feet, ft ——	0.304
3.94×10^{-2} — millimeter, mm (10^{-3})		inch, in ——	25.4

SQUARES

2.47 —— hectare, he		acre ——	0.405
2.47 —— square kilometer, km^2		acre ——	4.05×10^{-3}
0.386 —— square kilometer, km^2		square mile, mi^2 —	2.590
2.47×10^{-4} — square meter, m^2		acre ——	4.05×10^{-3}
10.76 —— square meter, m^2		square feet, ft^2 ——	9.29×10^{-2}
1.55×10^{-3} —— mm^2		square inch, in^2 ——	645

CUBICS

9.73×10^{-3} — cubic meter, m^3		inch-acre ——	102.8
35.3 —— cubic meter, m^3		cubic-feet, ft^3 ——	2.83×10^{-2}
6.10×10^4 —— cubic meter, m^3		cubic inch, in^3 ——	1.64×10^{-5}
2.84×10^{-2} —— liter, L (10^{-3} m^3)		bushel, bu ——	35.24
1.057 —— liter, L		liquid quarts, qt ——	0.946
3.53×10^{-2} —— liter, L		cubic feet, ft^3 ——	28.3

(Modified and reprinted with permission from: Megh R. Goyal, 2012. Appendices. Pages 317–332. In: *Management of Drip/Trickle or Micro Irrigation* edited by Megh R. Goyal. New Jersey, USA: Apple Academic Press Inc.)

0.265 ——— liter, L	gallon ——————— 3.78
33.78 ——— liter, L	fluid ounce, oz — 2.96×10^{-2}
2.11 ——— liter, L	fluid dot, dt ——— 0.473

WEIGHT

2.20×10^{-3} — gram, g (10^{-3} kg)	pound, ——————— 454
3.52×10^{-2} — gram, g (10^{-3} kg)	ounce, oz ——————— 28.4
2.205 ——— kilogram, kg	pound, lb ——————— 0.454
10^{-2} ——— kilogram, kg	quintal (metric), q — 100
1.10×10^{-3} — kilogram, kg	ton (2000 lbs), ton — 907
1.102 ——— mega gram, mg	ton (US), ton ——————— 0.907
1.102 ——— metric ton, t	ton (US), ton ——————— 0.907

YIELD AND RATE

0.893 ——— kilogram per hectare	pound per acre ——— 1.12
7.77×10^{-2} — kilogram per cubic meter	pound per fanega ——— 12.87
1.49×10^{-2} — kilogram per hectare	pound per acre, 60 lb – 67.19
1.59×10^{-2} — kilogram per hectare	pound per acre, 56 lb – 62.71
1.86×10^{-2} — kilogram per hectare	pound per acre, 48 lb – 53.75
0.107 ——— liter per hectare	galloon per acre ——— 9.35
893 ——— ton per hectare	pound per acre ——— 1.12×10^{-3}
893 ——— mega gram per hectare	pound per acre ——— 1.12×10^{-3}
0.446—— ton per hectare	ton (2000 lb) per acre – 2.24
2.24 ——— meter per second	mile per hour ——— 0.447

SPECIFIC SURFACE

10 ——— square meter per kilogram	square centimeter per gram ——— 0.1
10^3 ——— square meter per kilogram	square millimeter per gram ——— 10^{-3}

PRESSURE

9.90 ——— megapascal, MPa	atmosphere ——— 0.101
10 ——— megapascal	bar ——————— 0.1
1.0 ——— megagram per cubic meter	gram per cubic ——— 1.00 centimeter

2.09×10^{-2} — pascal, Pa	pound per square feet— 47.9
1.45×10^{-4} — pascal, Pa	pound per square inch— 6.90×10^3

TEMPERATURE

1.00 (K-273)—Kelvin, K	centigrade, °C — 1.00 (C+273)
(1.8 C + 32)—centigrade, °C	Fahrenheit, °F — (F–32)/1.8

ENERGY

9.52×10^{-4} — Joule J	BTU ——————— 1.05×10^3
0.239 ——— Joule, J	calories, cal ——————— 4.19
0.735 ——— Joule, J	feet-pound ——————— 1.36
2.387×10^5 —— Joule per square meter	calories per square — 4.19×10^4 centimeter
10^5 ——— Newton, N	dynes ——————— 10^{-5}

WATER REQUIREMENTS

9.73×10^{-3} — cubic meter	inch acre ——————— 102.8
9.81×10^{-3} — cubic meter per hour	cubic feet per second – 101.9
4.40 ——— cubic meter per hour	galloon (US) per — 0.227 minute
8.11 ——— hectare-meter	acre-feet ——————— 0.123
97.28 —— hectare-meter	acre-inch ——————— 1.03×10^{-2}
8.1×10^{-2} — hectare centimeter	acre-feet ——————— 12.33

CONCENTRATION

1 ——— centimol per kilogram	milliequivalents ——— 1 per 100 grams
0.1 ——— gram per kilogram	percents ——————— 10
1 ——— milligram per kilogram	parts per million — 1

NUTRIENTS FOR PLANTS

2.29 —— P	P_2O_5 ——————— 0.437
1.20 —— K	K_2O ——————— 0.830
1.39 —— Ca	CaO ——————— 0.715
1.66 —— Mg	MgO ——————— 0.602

NUTRIENT EQUIVALENTS

Column A	Column B	Conversion A to B	Equivalent B to A
N	NH_3	1.216	0.822
	NO_3	4.429	0.226
	KNO_3	7.221	0.1385
	$Ca(NO_3)_2$	5.861	0.171
	$(NH_4)_2SO_4$	4.721	0.212
	NH_4NO_3	5.718	0.175
	$(NH_4)_2HPO_4$	4.718	0.212
P	P_2O_5	2.292	0.436
	PO_4	3.066	0.326
	KH_2PO_4	4.394	0.228
	$(NH_4)_2HPO_4$	4.255	0.235
	H_3PO_4	3.164	0.316
K	K_2O	1.205	0.83
	KNO_3	2.586	0.387
	KH_2PO_4	3.481	0.287
	Kcl	1.907	0.524
	K_2SO_4	2.229	0.449
Ca	CaO	1.399	0.715
	$Ca(NO_3)_2$	4.094	0.244
	$CaCl_2 \times 6H_2O$	5.467	0.183
	$CaSO_4 \times 2H_2O$	4.296	0.233
Mg	MgO	1.658	0.603
	$MgSO_4 \times 7H_2O$	1.014	0.0986
S	H_2SO_4	3.059	0.327
	$(NH_4)_2SO_4$	4.124	0.2425
	K_2SO_4	5.437	0.184
	$MgSO_4 \times 7H_2O$	7.689	0.13
	$CaSO_4 \times 2H_2O$	5.371	0.186

APPENDIX B

PIPE AND CONDUIT FLOW

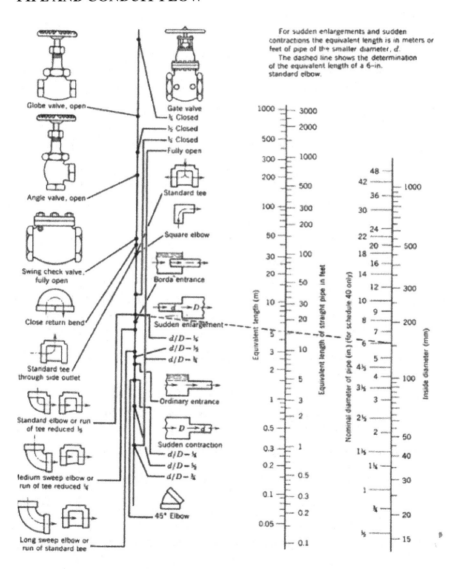

For sudden enlargements and sudden contractions the equivalent length is in meters or feet of pipe of the smaller diameter, *d*.

The dashed line shows the determination of the equivalent length of a 6–in. standard elbow.

APPENDIX C

PERCENTAGE OF DAILY SUNSHINE HOURS: FOR NORTH AND SOUTH HEMISPHERES

Latitude	Jan	Feb	Mar	Apr	May	Jun	Jul	Aug	Sep	Oct	Nov	Dec
					NORTH							
0	8.50	7.66	8.49	8.21	8.50	8.22	8.50	8.49	8.21	8.50	8.22	8.50
5	8.32	7.57	8.47	3.29	8.65	8.41	8.67	8.60	8.23	8.42	8.07	8.30
10	8.13	7.47	8.45	8.37	8.81	8.60	8.86	8.71	8.25	8.34	7.91	8.10
15	7.94	7.36	8.43	8.44	8.98	8.80	9.05	8.83	8.28	8.20	7.75	7.88
20	7.74	7.25	8.41	8.52	9.15	9.00	9.25	8.96	8.30	8.18	7.58	7.66
25	7.53	7.14	8.39	8.61	9.33	9.23	9.45	9.09	8.32	8.09	7.40	7.52
30	7.30	7.03	8.38	8.71	9.53	9.49	9.67	9.22	8.33	7.99	7.19	7.15
32	7.20	6.97	8.37	8.76	9.62	9.59	9.77	9.27	8.34	7.95	7.11	7.05
34	7.10	6.91	8.36	8.80	9.72	9.70	9.88	9.33	8.36	7.90	7.02	6.92
36	6.99	6.85	8.35	8.85	9.82	9.82	9.99	9.40	8.37	7.85	6.92	6.79
38	6.87	6.79	8.34	8.90	9.92	9.95	10.1	9.47	3.38	7.80	6.82	6.66
40	6.76	6.72	8.33	8.95	10.0	10.1	10.2	9.54	8.39	7.75	6.72	7.52
42	6.63	6.65	8.31	9.00	10.1	10.2	10.4	9.62	8.40	7.69	6.62	6.37
44	6.49	6.58	8.30	9.06	10.3	10.4	10.5	9.70	8.41	7.63	6.49	6.21
46	6.34	6.50	8.29	9.12	10.4	10.5	10.6	9.79	8.42	7.57	6.36	6.04
48	6.17	6.41	8.27	9.18	10.5	10.7	10.8	9.89	8.44	7.51	6.23	5.86
50	5.98	6.30	8.24	9.24	10.7	10.9	11.0	10.0	8.35	7.45	6.10	5.64
52	5.77	6.19	8.21	9.29	10.9	11.1	11.2	10.1	8.49	7.39	5.93	5.43
54	5.55	6.08	8.18	9.36	11.0	11.4	11.4	10.3	8.51	7.20	5.74	5.18
56	5.30	5.95	8.15	9.45	11.2	11.7	11.6	10.4	8.53	7.21	5.54	4.89
58	5.01	5.81	8.12	9.55	11.5	12.0	12.0	10.6	8.55	7.10	4.31	4.56
60	4.67	5.65	8.08	9.65	11.7	12.4	12.3	10.7	8.57	6.98	5.04	4.22
					SOUTH							
0	8.50	7.66	8.49	8.21	8.50	8.22	8.50	8.49	8.21	8.50	8.22	8.50
5	8.68	7.76	8.51	8.15	8.34	8.05	8.33	8.38	8.19	8.56	8.37	8.68
10	8.86	7.87	8.53	8.09	8.18	7.86	8.14	8.27	8.17	8.62	8.53	8.88
15	9.05	7.98	8.55	8.02	8.02	7.65	7.95	8.15	8.15	8.68	8.70	9.10
20	9.24	8.09	8.57	7.94	7.85	7.43	7.76	8.03	8.13	8.76	8.87	9.33

Latitude	Jan	Feb	Mar	Apr	May	Jun	Jul	Aug	Sep	Oct	Nov	Dec
25	9.46	8.21	8.60	7.74	7.66	7.20	7.54	7.90	8.11	8.86	9.04	9.58
30	9.70	8.33	8.62	7.73	7.45	6.96	7.31	7.76	8.07	8.97	9.24	9.85
32	9.81	8.39	8.63	7.69	7.36	6.85	7.21	7.70	8.06	9.01	9.33	9.96
34	9.92	8.45	8.64	7.64	7.27	6.74	7.10	7.63	8.05	9.06	9.42	10.1
36	10.0	8.51	8.65	7.59	7.18	6.62	6.99	7.56	8.04	9.11	9.35	10.2
38	10.2	8.57	8.66	7.54	7.08	6.50	6.87	7.49	8.03	9.16	9.61	10.3
40	10.3	8.63	8.67	7.49	6.97	6.37	6.76	7.41	8.02	9.21	9.71	10.5
42	10.4	8.70	8.68	7.44	6.85	6.23	6.64	7.33	8.01	9.26	9.8	10.6
44	10.5	8.78	8.69	7.38	6.73	6.08	6.51	7.25	7.99	9.31	9.94	10.8
46	10.7	8.86	8.90	7.32	6.61	5.92	6.37	7.16	7.96	9.37	10.1	11.0

APPENDIX D

PSYCHOMETRIC CONSTANT (γ) FOR DIFFERENT ALTITUDES (Z)

$$\gamma = 10^{-3}\,[(C_p.P) \div (\varepsilon.\lambda)] = (0.00163) \times [P \div \lambda]$$

γ, psychrometric constant [kPa C^{-1}]
c_p, specific heat of moist air = 1.013 [kJ kg^{-1} °C^{-1}]
P, atmospheric pressure [kPa].

ε, ratio molecular weight of water vapor/dry air = 0.622
λ, latent heat of vaporization [MJ kg^{-1}]
= 2.45 MJ kg^{-1} at 20 °C.

Z (m)	γ kPa/°C	z (m)	γ kPa/°C	z (m)	γ kPa/°C	z (m)	γ kPa/°C
0	0.067	1000	0.060	2000	0.053	3000	0.047
100	0.067	1100	0.059	2100	0.052	3100	0.046
200	0.066	1200	0.058	2200	0.052	3200	0.046
300	0.065	1300	0.058	2300	0.051	3300	0.045
400	0.064	1400	0.057	2400	0.051	3400	0.045
500	0.064	1500	0.056	2500	0.050	3500	0.044
600	0.063	1600	0.056	2600	0.049	3600	0.043
700	0.062	1700	0.055	2700	0.049	3700	0.043
800	0.061	1800	0.054	2800	0.048	3800	0.042
900	0.061	1900	0.054	2900	0.047	3900	0.042
1000	0.060	2000	0.053	3000	0.047	4000	0.041

APPENDIX E

SATURATION VAPOR PRESSURE [e$_s$] FOR DIFFERENT TEMPERATURES (T)

Vapor pressure function = e$_s$ = [0.6108] × exp{[17.27T]/[T + 237.3]}

T °C	e$_s$ kPa	T °C	e$_s$ kPa	T °C	e$_s$ kPa	T °C	e$_s$ kPa
1.0	0.657	13.0	1.498	25.0	3.168	37.0	6.275
1.5	0.681	13.5	1.547	25.5	3.263	37.5	6.448
2.0	0.706	14.0	1.599	26.0	3.361	38.0	6.625
2.5	0.731	14.5	1.651	26.5	3.462	38.5	6.806
3.0	0.758	15.0	1.705	27.0	3.565	39.0	6.991
3.5	0.785	15.5	1.761	27.5	3.671	39.5	7.181
4.0	0.813	16.0	1.818	28.0	3.780	40.0	7.376
4.5	0.842	16.5	1.877	28.5	3.891	40.5	7.574
5.0	0.872	17.0	1.938	29.0	4.006	41.0	7.778
5.5	0.903	17.5	2.000	29.5	4.123	41.5	7.986
6.0	0.935	18.0	2.064	30.0	4.243	42.0	8.199
6.5	0.968	18.5	2.130	30.5	4.366	42.5	8.417
7.0	1.002	19.0	2.197	31.0	4.493	43.0	8.640
7.5	1.037	19.5	2.267	31.5	4.622	43.5	8.867
8.0	1.073	20.0	2.338	32.0	4.755	44.0	9.101
8.5	1.110	20.5	2.412	32.5	4.891	44.5	9.339
9.0	1.148	21.0	2.487	33.0	5.030	45.0	9.582
9.5	1.187	21.5	2.564	33.5	5.173	45.5	9.832
10.0	1.228	22.0	2.644	34.0	5.319	46.0	10.086
10.5	1.270	22.5	2.726	34.5	5.469	46.5	10.347
11.0	1.313	23.0	2.809	35.0	5.623	47.0	10.613
11.5	1.357	23.5	2.896	35.5	5.780	47.5	10.885
12.0	1.403	24.0	2.984	36.0	5.941	48.0	11.163
12.5	1.449	24.5	3.075	36.5	6.106	48.5	11.447

APPENDIX F

SLOPE OF VAPOR PRESSURE CURVE (Δ) FOR DIFFERENT TEMPERATURES (T)

$$\Delta = [4098.\ e^\circ(T)] \div [T + 237.3]^2$$
$$= 2504\{\exp[(17.27T) \div (T + 237.2)]\} \div [T + 237.3]^2$$

T °C	Δ kPa/°C	T °C	Δ kPa/°C	T °C	Δ kPa/°C	T °C	Δ kPa/°C
1.0	0.047	13.0	0.098	25.0	0.189	37.0	0.342
1.5	0.049	13.5	0.101	25.5	0.194	37.5	0.350
2.0	0.050	14.0	0.104	26.0	0.199	38.0	0.358
2.5	0.052	14.5	0.107	26.5	0.204	38.5	0.367
3.0	0.054	15.0	0.110	27.0	0.209	39.0	0.375
3.5	0.055	15.5	0.113	27.5	0.215	39.5	0.384
4.0	0.057	16.0	0.116	28.0	0.220	40.0	0.393
4.5	0.059	16.5	0.119	28.5	0.226	40.5	0.402
5.0	0.061	17.0	0.123	29.0	0.231	41.0	0.412
5.5	0.063	17.5	0.126	29.5	0.237	41.5	0.421
6.0	0.065	18.0	0.130	30.0	0.243	42.0	0.431
6.5	0.067	18.5	0.133	30.5	0.249	42.5	0.441
7.0	0.069	19.0	0.137	31.0	0.256	43.0	0.451
7.5	0.071	19.5	0.141	31.5	0.262	43.5	0.461
8.0	0.073	20.0	0.145	32.0	0.269	44.0	0.471
8.5	0.075	20.5	0.149	32.5	0.275	44.5	0.482
9.0	0.078	21.0	0.153	33.0	0.282	45.0	0.493
9.5	0.080	21.5	0.157	33.5	0.289	45.5	0.504
10.0	0.082	22.0	0.161	34.0	0.296	46.0	0.515
10.5	0.085	22.5	0.165	34.5	0.303	46.5	0.526
11.0	0.087	23.0	0.170	35.0	0.311	47.0	0.538
11.5	0.090	23.5	0.174	35.5	0.318	47.5	0.550
12.0	0.092	24.0	0.179	36.0	0.326	48.0	0.562
12.5	0.095	24.5	0.184	36.5	0.334	48.5	0.574

APPENDIX G

NUMBER OF THE DAY IN THE YEAR (JULIAN DAY)

Day	Jan	Feb	Mar	Apr	May	Jun	Jul	Aug	Sep	Oct	Nov	Dec
1	1	32	60	91	121	152	182	213	244	274	305	335
2	2	33	61	92	122	153	183	214	245	275	306	336
3	3	34	62	93	123	154	184	215	246	276	307	337
4	4	35	63	94	124	155	185	216	247	277	308	338
5	5	36	64	95	125	156	186	217	248	278	309	339
6	6	37	65	96	126	157	187	218	249	279	310	340
7	7	38	66	97	127	158	188	219	250	280	311	341
8	8	39	67	98	128	159	189	220	251	281	312	342
9	9	40	68	99	129	160	190	221	252	282	313	343
10	10	41	69	100	130	161	191	222	253	283	314	344
11	11	42	70	101	131	162	192	223	254	284	315	345
12	12	43	71	102	132	163	193	224	255	285	316	346
13	13	44	72	103	133	164	194	225	256	286	317	347
14	14	45	73	104	134	165	195	226	257	287	318	348
15	15	46	74	105	135	166	196	227	258	288	319	349
16	16	47	75	106	136	167	197	228	259	289	320	350
17	17	48	76	107	137	168	198	229	260	290	321	351
18	18	49	77	108	138	169	199	230	261	291	322	352
19	19	50	78	109	139	170	200	231	262	292	323	353
20	20	51	79	110	140	171	201	232	263	293	324	354
21	21	52	80	111	141	172	202	233	264	294	325	355
22	22	53	81	112	142	173	203	234	265	295	326	356
23	23	54	82	113	143	174	204	235	266	296	327	357
24	24	55	83	114	144	175	205	236	267	297	328	358
25	25	56	84	115	145	176	206	237	268	298	329	359
26	26	57	85	116	146	177	207	238	269	299	330	360
27	27	58	86	117	147	178	208	239	270	300	331	361
28	28	59	87	118	148	179	209	240	271	301	332	362
29	29	(60)	88	119	149	180	210	241	272	302	333	363
30	30	—	89	120	150	181	211	242	273	303	334	364
31	31	—	90	—	151	—	212	243	—	304	—	365

APPENDIX H

STEFAN-BOLTZMANN LAW AT DIFFERENT TEMPERATURES (T)

$$[\sigma(T_K)^4] = [4.903 \times 10^{-9}], \text{ MJ K}^{-4} \text{ m}^{-2} \text{ day}^{-1}$$

where: $T_K = \{T[°C] + 273.16\}$

T	$\sigma*(T_K)^4$	T	$\sigma*(T_K)^4$	T	$\sigma*(T_K)^4$
			Units		
°C	MJ m^{-2} d^{-1}	°C	MJ m^{-2} d^{-1}	°C	MJ m^{-2} d^{-1}
1.0	27.70	17.0	34.75	33.0	43.08
1.5	27.90	17.5	34.99	33.5	43.36
2.0	28.11	18.0	35.24	34.0	43.64
2.5	28.31	18.5	35.48	34.5	43.93
3.0	28.52	19.0	35.72	35.0	44.21
3.5	28.72	19.5	35.97	35.5	44.50
4.0	28.93	20.0	36.21	36.0	44.79
4.5	29.14	20.5	36.46	36.5	45.08
5.0	29.35	21.0	36.71	37.0	45.37
5.5	29.56	21.5	36.96	37.5	45.67
6.0	29.78	22.0	37.21	38.0	45.96
6.5	29.99	22.5	37.47	38.5	46.26
7.0	30.21	23.0	37.72	39.0	46.56
7.5	30.42	23.5	37.98	39.5	46.85
8.0	30.64	24.0	38.23	40.0	47.15
8.5	30.86	24.5	38.49	40.5	47.46
9.0	31.08	25.0	38.75	41.0	47.76
9.5	31.30	25.5	39.01	41.5	48.06
10.0	31.52	26.0	39.27	42.0	48.37
10.5	31.74	26.5	39.53	42.5	48.68
11.0	31.97	27.0	39.80	43.0	48.99

T	$\sigma*(T_K)^4$	T	$\sigma*(T_K)^4$	T	$\sigma*(T_K)^4$
		Units			
°C	MJ m^{-2} d^{-1}	°C	MJ m^{-2} d^{-1}	°C	MJ m^{-2} d^{-1}
11.5	32.19	27.5	40.06	43.5	49.30
12.0	32.42	28.0	40.33	44.0	49.61
12.5	32.65	28.5	40.60	44.5	49.92
13.0	32.88	29.0	40.87	45.0	50.24
13.5	33.11	29.5	41.14	45.5	50.56
14.0	33.34	30.0	41.41	46.0	50.87
14.5	33.57	30.5	41.69	46.5	51.19
15.0	33.81	31.0	41.96	47.0	51.51
15.5	34.04	31.5	42.24	47.5	51.84
16.0	34.28	32.0	42.52	48.0	52.16
16.5	34,52	32.5	42.80	48.5	52.49

APPENDIX I

THERMODYNAMIC PROPERTIES OF AIR AND WATER

1. Latent Heat of Vaporization (λ)

$$\lambda = [2.501 - (2.361 \times 10^{-3})\,T]$$

where: λ = latent heat of vaporization [MJ kg^{-1}]; and T = air temperature [°C].

The value of the latent heat varies only slightly over normal temperature ranges. A single value may be taken (for ambient temperature = 20 °C): λ = 2.45 MJ kg^{-1}.

2. Atmospheric Pressure (P)

$$P = P_o\,[\{T_{Ko} - \alpha(Z - Z_o)\} \div \{T_{Ko}\}]^{(g/(\alpha.R))}$$

where: P, atmospheric pressure at elevation z [kPa]
$\quad$ P$_o$, atmospheric pressure at sea level = 101.3 [kPa]
$\quad$ z, elevation [m]
$\quad$ z$_o$, elevation at reference level [m]
$\quad$ g, gravitational acceleration = 9.807 [m s^{-2}]
$\quad$ R, specific gas constant == 287 [J kg^{-1} K^{-1}]
$\quad$ α, constant lapse rate for moist air = 0.0065 [K m^{-1}]
$\quad$ T$_{Ko}$, reference temperature [K] at elevation z$_o$ = 273.16 + T
$\quad$ T, means air temperature for the time period of calculation [°C]
When assuming P$_o$ = 101.3 [kPa] at z$_o$ = 0, and T$_{Ko}$ = 293 [K] for T = 20 [°C], above equation reduces to:

$$P = 101.3[(293 - 0.0065Z)\,(293)]^{5.26}$$

3. Atmospheric Density (ρ)

$$\rho = [1000P] \div [T_{Kv}\,R] = [3.486P] \div [T_{Kv}], \text{ and } T_{Kv} = T_K[1 - 0.378(e_a)/P]^{-1}$$

where: ρ, atmospheric density [kg m^{-3}]

R, specific gas constant = 287 [J kg$^{-1\,K-1}$]

$T_{Kv,}$ virtual temperature [K]

$T_{K,}$ absolute temperature [K]: $T_K = 273.16 + T$ [°C]

$e_{a,}$ actual vapor pressure [kPa]

T, mean daily temperature for 24-hour calculation time steps.

For average conditions (e_a in the range 1–5 kPa and P between 80–100 kPa), T_{Kv} can be substituted by: $T_{Kv} \approx 1.01\,(T + 273)$

4. Saturation Vapor Pressure function (e_s)

$$e_s = [0.6108] \times \exp\{[17.27 \times T]/[T + 237.3]\}$$

where: e_s, saturation vapor pressure function [kPa],

T, air temperature [°C]

5. Slope Vapor Pressure Curve (Δ)

$$\Delta = [4098.\ e°(T)] \div [T + 237.3]^2$$
$$= 2504\{\exp[(17.27T) \div (T + 237.2)]\} \div [T + 237.3]^2$$

where: Δ, slope vapor pressure curve [kPa C^{-1}]

T, air temperature [°C]

e°(T), saturation vapor pressure at temperature T [kPa]

In 24-hour calculations, Δ is calculated using mean daily air temperature. In hourly calculations T refers to the hourly mean, T_{hr}.

6. Psychrometric Constant (γ)

$$\gamma = 10^{-3}\ [(C_p.P) \div (\varepsilon.\lambda)] = (0.00163) \times [P \div \lambda]$$

where: γ, psychrometric constant [kPa C^{-1}]

c_p, specific heat of moist air = 1.013 [kJ kg$^{-1\,°C-1}$]

P, atmospheric pressure [kPa]: equations 2 or 4

ε, ratio molecular weight of water vapor/dry air = 0.622

λ, latent heat of vaporization [MJ kg^{-1}]

7. Dew Point Temperature (T_{dew})

When data is not available, T_{dew} can be computed from e_a by:

$$T_{dew} = [\{116.91 + 237.3\text{Log}_e(e_a)\} \div \{16.78 - \text{Log}_e(e_a)\}]$$

where: T_{dew}, dew point temperature [°C]

   e_a, actual vapor pressure [kPa]

For the case of measurements with the Assmann psychrometer, T_{dew} can be calculated from:

$$T_{dew} = (112 + 0.9T_{wet})[e_a \div (e^\circ\ T_{wet})]^{0.125} - [112 - 0.1T_{wet}]$$

8. Short Wave Radiation on a Clear-Sky Day (R_{so})

The calculation of R_{so} is required for computing net long wave radiation and for checking calibration of pyranometers and integrity of R_{so} data. A good approximation for R_{so} for daily and hourly periods is:

$$R_{so} = (0.75 + 2 \times 10^{-5}\ z)R_a$$

where: z, station elevation [m]

   R_a extraterrestrial radiation [MJ m^{-2} d^{-1}]

Equation is valid for station elevations less than 6000 m having low air turbidity. The equation was developed by linearizing Beer's radiation extinction law as a function of station elevation and assuming that the average angle of the sun above the horizon is about 50°.

For areas of high turbidity caused by pollution or airborne dust or for regions where the sun angle is significantly less than 50° so that the path length of radiation through the atmosphere is increased, an adoption of Beer's law can be employed where P is used to represent atmospheric mass:

$$R_{so} = (R_a)\ \exp[(-0.0018P) \div (K_t \sin(\Phi))]$$

where: K_t, turbidity coefficient, $0 < K_t < 1.0$ where $K_t = 1.0$ for clean air and

   $K_t = 1.0$ for extremely turbid, dusty or polluted air.

   P, atmospheric pressure [kPa]

   Φ, angle of the sun above the horizon [rad]

   R_a, extraterrestrial radiation [MJ m^{-2} d^{-1}]

For hourly or shorter periods, Φ is calculated as:

$$\sin \Phi = \sin \varphi \sin \delta + \cos \varphi \cos \delta \cos \omega$$

where: φ, latitude [rad]

       δ, solar declination [rad] (Eq. (24) in Chapter 3)

       ω, solar time angle at midpoint of hourly or shorter period [rad]

For 24-hour periods, the mean daily sun angle, weighted according to R_a, can be approximated as:

$$\sin(\Phi_{24}) = \sin[0.85 + 0.3\ \varphi \sin\{(2\pi J/365)-1.39\}-0.42\ \varphi^2]$$

where: Φ_{24}, average Φ during the daylight period, weighted according to R_a [rad]

       φ, latitude [rad]

       J, day in the year

The Φ_{24} variable is used to represent the average sun angle during daylight hours and has been weighted to represent integrated 24-hour transmission effects on 24-hour R_{so} by the atmosphere. Φ_{24} should be limited to >0. In some situations, the estimation for R_{so} can be improved by modifying to consider the effects of water vapor on short wave absorption, so that: $R_{so} = (K_B + K_D)\ R_a$ where:

$$K_B = 0.98\exp[\{(-0.00146P) \div (K_t \sin \Phi)\}-0.091\{w/\sin \Phi\}^{0.25}]$$

where: K_B, the clearness index for direct beam radiation

       K_D, the corresponding index for diffuse beam radiation

       $K_D = 0.35-0.33\ K_B$ for $K_B > 0.15$

       $K_D = 0.18 + 0.82\ K_B$ for $K_B < 0.15$

       R_a, extraterrestrial radiation [MJ m^{-2} d^{-1}]

       K_t, turbidity coefficient, $0 < K_t < 1.0$, where $K_t = 1.0$ for clean air and $K_t = 1.0$ for extremely turbid, dusty or polluted air.

       P, atmospheric pressure [kPa]

       Φ, angle of the sun above the horizon [rad]

       W, perceptible water in the atmosphere [mm] $= 0.14\ e_a\ P + 2.1$

       e_a, actual vapor pressure [kPa]

       P, atmospheric pressure [kPa]

APPENDIX J

PSYCHROMETRIC CHART AT SEA LEVEL

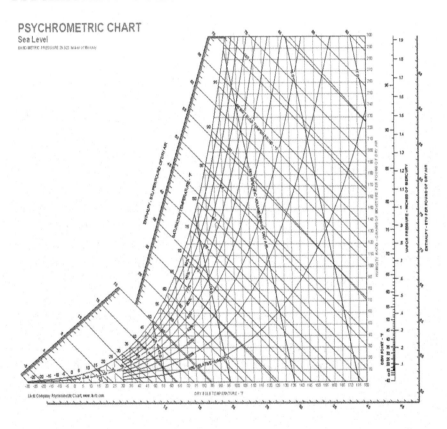

APPENDIX K

[<http://www.fao.org/docrep/T0551E/t0551e07.htm#5.5%20field%20
management%20practices%20in%20wastewater%20irrigation>]

1. Relationship between applied water salinity and soil water salinity
    at different leaching fractions (FAO 1985)

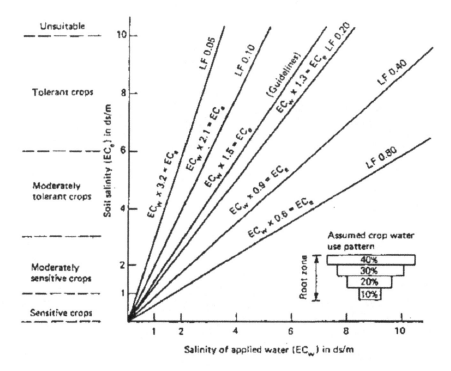

2. Schematic representations of salt accumulation, planting positions, ridge shapes and watering patterns.

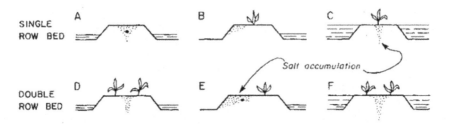

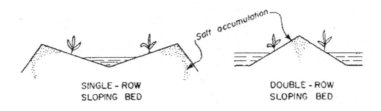

SINGLE - ROW
SLOPING BED

DOUBLE - ROW
SLOPING BED

3. Main components of general planning guidelines for wastewater reuse (Cobham and Johnson 1988).

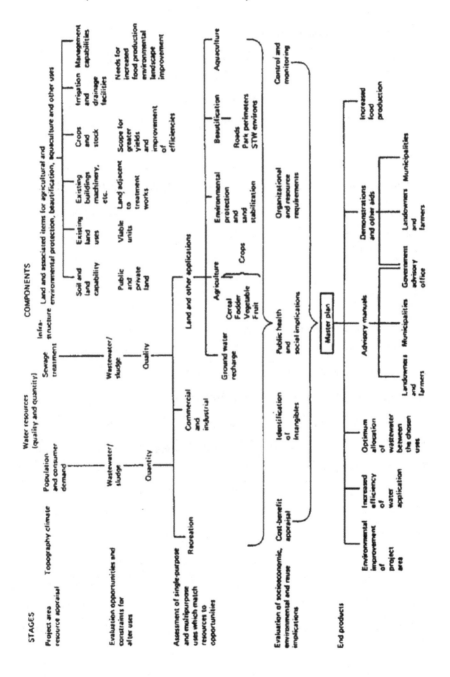

APPENDIX L

1. Uniformity classification.

Classification	Statistical Uniformity	Emission Uniformity
Excellent	For U = 100–95%	100–94%
Good	For U = 90–85%	87–81%
Fair	For U = 80–75%	75–68%
Poor	For U = 70–65%	62–56%
Not Acceptable	For U < 60%	<50%

2. Acceptable intervals of uniformity in a drip irrigation system.

Type of dripper	Slope	Uniformity interval, %
Point Source: located in planting distance > 3.9 m.	Level*	90–95
	Inclined**	85–90
Point Source: located in planting distance < 3.9 m.	Level*	85–90
	Inclined**	80–90
Drippers inserted in the lines for annual row crops.	Level*	80–90
	Inclined**	75–85

* Level = Slope less that 2%. ** Inclined = Slope greater than 2%.

3. Confidence limits for field uniformity (U).

Field uniformity	18 drippers		36 drippers		72 drippers	
	Confidence limit		Confidence limit		Confidence limit	
	N Sum*	%	N Sum	%	N Sum	%
100%	3	U ± 0.0	6	U ± 0.6%	12	U ± 0.0%
90%	3	U ± 2.9	6	U ± 2.0%	12	U ± 1.4%
80%	3	U ± 5.8	6	U ± 4.0%	12	U ± 2.8%
70%	3	U ± 9.4	6	U ± 6.5%	12	U ± 4.5%
60%	3	U ± 13.3	6	U ± 9.2%	12	U ± 6.5%

*N Sum = 1/6 part of the total measured drippers. This is a number of samples that will be added to calculate T_{max} and T_{min}.

(From: *Vincent F. Bralts*, 2015. Chapter 3: Evaluation of the uniformity coefficients. In: *Sustainable Micro Irrigation Management for Trees and Vines, Volume 3*, edited by Megh R. Goyal, Apple Academic Press Inc.).

4. Nomograph for statistical uniformity.

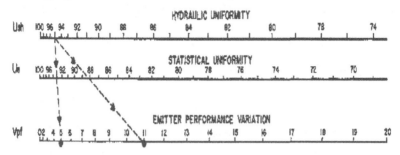

5. The field uniformity of an irrigation system based on the dripper times and the dripper flow rate.

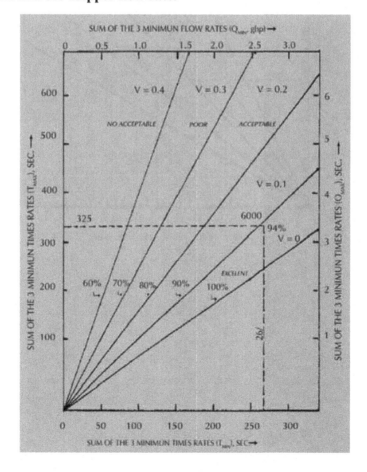

6. **The field uniformity of a drip irrigation system based on the time to collect a known quantity of water or based on pressure for hydraulic uniformity.**

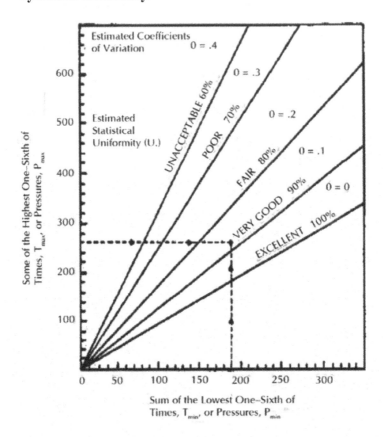

INDEX

Drip micro sprinkler irrigation, 41
Drip system, 7, 8, 24, 38–42, 72, 79,
    80–83, 86, 88, 184, 250, 275, 277,
    282, 301, 307, 308
    analysis, 232
    fertigation, 22
    technology, 17–19, 24, 32, 38, 71,
        73, 90
Dripper discharge, 80, 81
Drought stress, 192
Drought-resistance varieties, 320
Dynamic calibration, 107

E

Economic
    analysis, 267–269
    crops production, 177
    decision, 114
    efficiency, 48, 318
    solution, 117
    viability, 19
Economics, 19, 34, 38, 43, 48–50, 269,
    299, 310
    banana cultivation, 34
    scale, 58, 61
Effective pressure, 125, 126, 132, 133,
    135–137, 140–146, 150, 152–155,
    162, 163
Effective utilization of fertilizers, 37
Effects of fertigation, 192
    fertilizer use efficiency, 192
    water use efficiency, 192
Efficient water management practices,
    15
Egypt, 101, 103–105, 122, 177, 185,
    196–198, 243, 252, 270, 275
Egyptian conditions, 104, 161
Electric power, 97, 318
Electrical conductivity, 201, 202
Electronic digital caliper, 122
Emitter
    clogging, 216–221, 225
    first order interaction, 221
    flow rate, 203
    second order interactions, 221

spacing, 66, 198, 253, 298, 299
Empirical data, 117
Employment opportunities, 29, 40
Energy conservation, 109
Equipment performance, 177
Equivalent length method, 116
Eutrophic soil, 193
Evapotranspiration, 33, 196, 205, 251,
    276, 279, 281–283, 287, 289, 293,
    300, 301, 336, 337, 339–341, 349
Expansion requirements, 181
Experiment layout, 206
    bubbler heights, 159, 160
    design, 254
    evaluation of emitter clogging, 207
    farm, 276, 338
    field, 201
    layout, 206, 339
    nitrogen use efficiency, 208
    site, 252–254

F

Factorial experimental design, 206
Factors affecting fertilizer selection,
    179–180
FAO (Food and Agricultural
    Organization), 274, 281, 293, 330
Farm
    categories, 49, 50, 60, 65
    households general characteristics, 78
    level constraints, 66
    production, 197
    size, 49, 55–57, 60, 65, 66, 77–79
Farm of Agriculture Faculty, 122
Feddan, 101, 222, 233, 243, 249, 254
Ferrous sulfate, 33
Fertigation, 25–28, 32–34, 37, 64, 72,
    75, 78, 82, 83, 87, 177–180, 184,
    187–190, 192–195, 210, 232–234,
    241, 244
    advantages, 178
    area, 210
    disadvantages, 178
    improved management, 178
    nutrients, 178

analysis, 27
application efficiency, 281, 293, 319, 350
application rates, 8
changes, 187
conflicts, 318
consumption, 103, 275, 317
demand management, 2, 44, 48, 67, 316, 319, 321, 330, 331
distribution efficiency, 177, 241
distribution uniformity, 340, 346
logging, 14, 90
management, 3, 15, 24, 29, 30, 36, 44, 48, 67, 251, 317, 321
poverty, 103, 196
pricing, 90
resource degradation, 90
resource development, 48
resources of India and Tamil Nadu, 15
saving, 8, 10, 16, 17, 22, 24, 28, 30, 32, 41, 48, 65, 269, 284, 342
scarcity, 3, 11–15, 38, 42, 44, 70, 90, 93, 104, 285, 316, 317, 331
soluble fertilizers, 87
Water Technology Center, 6, 41
Water use efficiency (WUE), 14, 48, 70, 177, 192, 207, 225, 239, 241, 243, 251, 288–290, 291, 318, 336, 340, 345–346

first order interactions, 227
improved WUE at farm level, 319
irrigation levels and nitrogen treatments, 227
irrigation scheduling and methods, 319
second order interactions, 227–229
uses, 7, 14, 15, 48, 70, 90, 190, 192, 207, 225, 244, 251, 269, 283, 288, 291, 292, 298, 316–319, 336, 340
consuming sector, 48
scarce, 48, 336
stressed regions, 48
mark sensor, 340, 342, 346, 348–350
mark soil moisture sensors, 339
mark systems, 346
shed development, 48, 70, 90
shed, 48, 67, 70, 90
Wind-breakers, 33
Wooden stakes, 125
World War II, 3, 44
WTC, 8, 11, 44, 72

Z

Zero land slope, 104
Zinc sulfate, 33

Printed in the United States
by Baker & Taylor Publisher Services